LINEAR INTEGRATED CIRCUITS: OPERATION AND APPLICATIONS

J. Michael McMenamin

Prentice-Hall, Inc., Englewood Cliffs, New Jersey 07632

Library of Congress Cataloging in Publication Data

McMenamin, J. Michael.
    Linear integrated circuits.

    Includes index.
    1. Linear integrated circuits. I. Title.
TK7874.M36 1984   621.381'73   84-17720
ISBN 0-13-537333-6

Editorial/production supervision and
   interior design: Barbara Palumbo
Cover design: Edsal Enterprises
Manufacturing buyer: Anthony Caruso

Printed in the United States of America

10 9 8 7 6 5 4 3 2 1

ISBN 0-13-537333-6

PRENTICE-HALL INTERNATIONAL, INC., *London*
PRENTICE-HALL OF AUSTRALIA PTY. LIMITED, *Sydney*
EDITORA PRENTICE-HALL DO BRASIL, LTDA., *Rio de Janeiro*
PRENTICE-HALL CANADA INC., *Toronto*
PRENTICE-HALL OF INDIA PRIVATE LIMITED, *New Delhi*
PRENTICE-HALL OF JAPAN, INC., *Tokyo*
PRENTICE-HALL OF SOUTHEAST ASIA PTE. LTD., *Singapore*
WHITEHALL BOOKS LIMITED, *Wellington, New Zealand*

Contents

CHAPTER 2   Operational Amplifier Applications 36

CHAPTER 3   Linear Voltage Regulators 82

CHAPTER 4   Switching Regulators 112

Contents   v

CHAPTER 5   Voltage Reference and Current Sources *132*

CHAPTER 6   Voltage Comparators *148*

Preface

In recent years there has been a gradual shift from discrete transistor circuits to integrated packages, which in themselves perform a complete circuit function. This is a prepackaged design concept that has allowed for more complex equipment design. The design engineer is no longer concerned with the biasing and interconnection of transistors to form a complete circuit, but rather with the integration of integrated circuit packages to generate a complete equipment design.

Integrated circuit devices in use today typically fall into two groups—digital and linear (analog). Digital devices perform two state logic functions while the linear devices process signals that can assume any voltage level consistent with the power supply used.

This text will describe the operation and applications of a select number of linear integrated circuits. The actual devices selected are the result of surveying industry and determining which devices have the greatest usage potential or are devices that illustrate ''state of the art'' concepts.

It is assumed that the reader has had exposure to transistor circuits, simple logic circuits, and also has a math background through college algebra. Some exposure to microcomputer operations will also be helpful.

The material presented in this text is at a level which should enable the reader to understand and, if required, design basic circuits using the devices described. This book will be useful to students, technicians, and practicing engineers.

At the end of each chapter is a suggested laboratory experiment to reinforce the concepts introduced in the chapter. A ''cookbook'' procedure is not given for these experiments, but rather they require that the reader use initiative in determining how to set up the circuit with recourse to the text and manufacturers' specifications. This prepares the student for real-world situations. Although initially somewhat apprehensive about using this technique, I have had good results with this method over the last few years. This has been reinforced by students indicating that, although the labs were tougher, they learned more about circuit operation. It is recommended, however, that the Guidelines for Laboratory Experiments in Appendix C be read before attempting the experiments.

The majority of the material used in this book was derived from manufacturers' specification sheets and device descriptions. I have taken the role of an interpretor who has attempted to translate the manufacturers' data into a form that can be more easily understood. In addition, I have tried to show practical applications for the various devices.

Data sheets and device descriptions have been used with the kind permission of:

National Semiconductor Corporation
Intersil Incorporated
RCA Solid State Division
Signetics Corporation

I take full responsibility for any errors in the interpretation of this data.

Finally, I want to thank Richard Petrytyl for his encouragement to start this book and for his fine illustrations, and James Sketoe (engineer at Ford Motor Company), along with those college instructors (unnamed) who provided constructive criticisms of this text. Also, I want to thank Lori Pogoda and Susan Shannahan for spending hot summer nights typing from the manuscript.

J. Michael McMenamin

1

Operational Amplifiers

OBJECTIVES

Upon completion of this chapter you should know the following operational amplifier characteristics:
1. Open loop gain
2. Closed loop gain for:
   a) Inverting amplifiers,
   b) Non-inverting amplifiers, and
   c) Followers
3. Input and output resistance
4. Offset voltage and current
5. Frequency response
6. Slew rate
7. Power supply change rejection
8. Amplifier noise
9. Current mode (Norton) op-amps.

1-1 INTRODUCTION

The operational amplifier (op-amp) was originally designed to perform various mathematical *operations* in the analog computer. The first op-amp circuit built consisted of a series of vacuum tubes connected together to give a very high voltage gain. It was discovered that the gain of the op-amp could be precisely controlled and reduced through a negative feedback loop (closed loop operation) designed into the circuit. This precise control of circuit gain allowed the op-amp to perform with great accuracy such mathematical operations as addition, subtraction, integration, and differentiation.

Also discovered was the fact that the overall circuit gain was independent of the individual vacuum tube gain. Thus, defective vacuum tubes could be replaced without adversely affecting circuit gain. The next configuration of op-amps was constructed from transistors which decreased the size and increased the reliability.

Today the op-amp has evolved from discrete transistor circuits to its present form—the integrated circuit (IC), which is normally housed in a dual in-line package (DIP). This has resulted in the use of millions of these devices in both military and industrial equipment. The basic op-amp is designed into many other linear devices that are also extensively used.

Op-amps are represented on schematic diagrams by the symbol shown in Fig. 1-1.

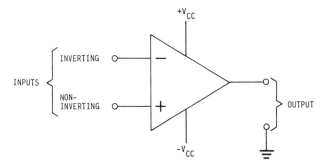

Figure 1-1 Operational Amplifier Schematic Diagram Symbol

A triangle with one input and one output terminal is the standard representation for an amplifier. The op-amp shown in Fig. 1-1 utilizes the triangle except that it has two input terminals, designated positive and negative, and one output terminal. A negative sign (−) on the input indicates that the signal applied to this terminal is inverted at the output (180° phase shift) and a positive sign (+) on the input indicates that the signal applied to this terminal is not inverted at the output (non-inverting). The use of two input terminals on an op-amp is referred to as a differential input. A differential input means that the output signal of the op-amp is the amplified difference of the two input signals. If the same signal is applied to both inputs (zero difference), the output signal is zero.

Power supply voltages normally used for op-amps are both positive and negative. This allows the op-amp to have both positive and negative output voltage swings with reference to circuit ground.

1-2 OPEN LOOP GAIN

We call the circuit voltage gain with no feedback (no connection from output to input) the open loop gain. The most common op-amp is the type 741 (see specification sheets in Appendix D). Its open loop gain (A_{VOL}) is typically 200,000. This means that a 10-μV signal difference at the inputs would be amplified to 2 V (200,000 $\times$ 10 μV) at the output. This is indicated in Fig. 1-2.

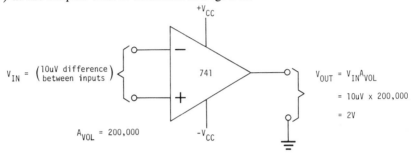

Figure 1-2   Voltage Gain of a Typical Open Loop Amplifier

The problem with operating an op-amp with open loop gain is that we cannot effectively control it. The gain is usually too high for most applications; we are also subject to gain variations between op-amps. Thus, open loop operation is normally reserved for non-linear (output not proportional to input) applications such as for comparator and shaping circuits (to be described later).

1-3 CLOSED LOOP GAIN

We will find that closed loop operation makes the circuit gain independent of the op-amp and reduces waveform distortion. Most applications for op-amps are with closed loop configurations. Closed loop means that we are applying negative (180° out of phase) feedback from the output to the input. We obtain negative feedback by feeding a portion of the output voltage (V_F) back to the inverting input. Figure 1-3 shows a basic closed loop circuit configuration.

V_S is the input signal to be amplified, while resistor R_1 and feedback resistor R_F form a voltage divider that feeds the output signal back to the inverting output. Before we proceed, we need to make two assumptions: (1) The op-amp gain is infinite, and (2) the resistance seen looking into the op-amp input terminals is infinite. The last assumption implies that no current is flowing into the op-amp terminals. With no current flow into the inverting input, the feedback voltage V_F applied to the inverting input terminal from the output is:

$$V_F = V_{out}\left(\frac{R_1}{R_1 + R_F}\right) \tag{1-1}$$

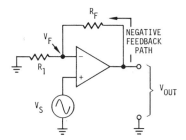

Figure 1-3 Negative Feedback—Non-Inverting Amplifier

Using the assumption of infinite gain, the voltage difference between the two op-amp inputs needs to be slightly greater than zero in order to have an output signal. Thus we can assume $V_F = V_S$. Substituting in Eq. 1-1 and solving for the non-inverting circuit gain (A_V):

$$A_V = \frac{V_{out}}{V_S}$$

$$A_V = \frac{(R_F + R_1)}{R_1} = \frac{R_F}{R_1} + 1 \tag{1-2}$$

The significance of Eq. 1-2 is that the circuit gain is independent of the op-amp gain and can be as accurate as the resistors used. In the practical situation, this gain equation is typically within $\pm 1\%$ of the predicted value if the open loop gain is at least a hundred times greater than the closed loop gain.

Internal distortion is reduced in a closed loop configuration because distortion will change the op-amp output signal, but this change will be fed back to the op-amp input. However, since the two input leads of the op-amp cannot have a voltage difference between them, the distortion is cancelled from the op-amp output.

Figure 1-4 illustrates the op-amp inverting amplifier circuit.

Again, using the assumption that the gain is infinite, we can say that the signal at the inverting input is the same as that on the non-inverting input. In Fig. 1-4, the non-inverting input is $\emptyset$ volts or ground potential and the inverting input is considered to be

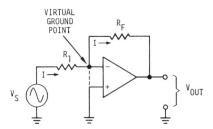

Figure 1-4 Inverting Amplifier

virtually grounded. Virtually grounded implies that the signal at the inverting input is cancelled by the negative feedback signal developed from the output. Thus, the voltage level at the inverting input is a very small value that is considered to be practically $\emptyset$ volts (ground potential), hence the term *virtual ground.*

Since the inverting input is virtually grounded, $V_S = IR$, and since we have infinite input impedance, the current through R_1 is the same as the current through R_F. Therefore:

$$A_V = -\frac{V_{\text{out}}}{V_S} = -\frac{IR_F}{IR_1} = -\frac{R_F}{R_1} \tag{1-3}$$

(Note: The minus sign indicates the output signal is inverted.)

The follower circuit shown in Fig. 1-5 is a special case of the non-inverting amplifier with a zero value for R_F and an infinite value for R_1. Using these values in Eq. 1-2:

$$A_V = \frac{R_F}{R_1} + 1 = \frac{0}{\infty} + 1 = 1$$

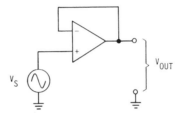

Figure 1-5   Follower Circuit

Thus, as expected, the follower has unity gain ($A_V = 1$). In all the circuits described thus far, the maximum output voltage swing is about 1 V less than the supply voltage; therefore, with $\pm$ 12-V supplies, the maximum output voltage is +11 V or −11 V.

1-4 INPUT AND OUTPUT RESISTANCE

The input resistance R_i of the inverting amplifier is simply R_1 since the inverting input is virtually grounded. For the non-inverting amplifier and follower, the input resistance is the resistance seen between the two input terminals (typically 2 MΩ for a 741) times the loop gain. Loop gain A_{LOOP} is defined as the open loop gain A_{OL} divided by the closed loop gain A_V or:

$$A_{\text{LOOP}} = \frac{A_{OL}}{A_V} \tag{1-4}$$

The reason the input resistance R_i is increased by the loop gain is that the voltage feedback to the junction of R_1 and R_F is almost the same as the input voltage. Therefore, little current flows through R_i, and its resistance is effectively raised.

The output resistance R_{OUT} for all three op-amp circuit configurations is the output resistance of the op-amp (typically 75 Ω for a 741) divided by the loop gain A_{LOOP}. This is because as we reduce the load resistance, the output voltage tends to reduce; but this reduces the amount of negative feedback signal, so the output voltage raises. The net effect is to give an apparent lower internal resistance. Figure 1-6 shows the resistance relationships. *Note:* At high frequencies the capacitance loading should be taken into account.

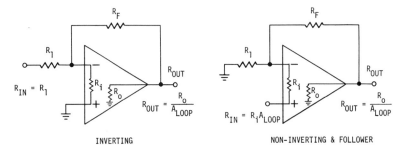

INVERTING   NON-INVERTING & FOLLOWER

Figure 1-6   Input and Output Resistance

1-5 OFFSET CURRENT AND VOLTAGE

The voltages applied to an op-amp input normally are fed directly to an internal differential amplifier consisting of two transistors (bipolar or FET) as illustrated in Fig. 1-7.

The output voltage V_{OUT} is the amplified signal difference between the two inputs. In order for the bipolar transistors Q_1 and Q_2 to turn on, bias current must flow through resistors R_1 and R_2. This current is in the nanoamp range (600 nA maximum for a 741) and causes a voltage drop across the resistors. If the voltage drop across the resistors is not equal, then an error voltage appears at the output.

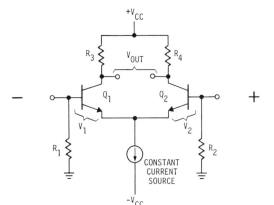

Figure 1-7   Bi-Polar Op-Amp Input Circuit

Consider a situation where R_1 and R_2 are equal, but the bias current through R_1 is larger than the bias current through R_2. There will be more voltage drop across R_1 and therefore the inverting input terminal will be at a lower potential than the non-inverting input terminal. This difference in DC voltage will be amplified by the op-amp and will cause a voltage offset or DC shift (positive) at the output. The shift in the output voltage can cause an error signal if we are using the op-amp as a DC amplifier, or it can shift the output center point, causing clipping, if we are amplifying sine wave signals. We shall now look at the equations for determining the amount of voltage offset for a given amount of bias current.

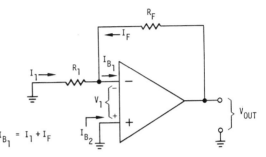

Figure 1-8   Current Offset $I_{B_1} = I_1 + I_F$

Notice in Fig. 1-8 that the two bias currents are represented by I_{B_1} and I_{B_2}. Since there is no resistor on the non-inverting side, no voltage drop will occur. The resistor R_1 includes the internal resistance of the driving circuit, which causes a voltage drop $V_1 = I_{B_1} \times R_1$ to appear as an input voltage to the op-amp. This voltage will be negative with respect to ground and will be amplified by the open loop gain to cause a positive output voltage V_{OUT}, determined as follows:

$$V_{OUT} = V_1 A_{OL}$$

but
$$V_{OUT} = I_F R_F - V_1$$

(assuming V_{OUT} goes to ground) and substituting for V_1:

$$V_{OUT} = I_F R_F - V_{OUT}/A_{OL}$$

rearranging
$$I_F R_F = V_{OUT}(1 + 1/A_{OL})$$

if we assume infinite gain: $I_F R_F = V_{OUT}$

therefore
$$V_1 = \emptyset$$

and we conclude
$$I_1 = \emptyset \quad \text{and} \quad I_{B_1} = I_F$$

We are finally left with the relationship that:

$$V_{OUT} = I_{B_1} R_F \tag{1-5}$$

Equation 1-5 indicates that R_F should be as low as possible to reduce offset. If R_F is too low, the op-amp output could be loaded down. Usually R_F should have a resistance value in the 10-kΩ to 100-kΩ range.

We can reduce the amount of output offset by causing a voltage drop on the non-inverting input side. This is accomplished by adding a "compensating" resistor R_2 in series with the non-inverting input as shown in Fig. 1-9.

The voltage drop V_2 caused by the flow of I_{B_2} through R_2 is amplified by a non-inverting gain factor of $(R_F/R_1 + 1)$, resulting in an output voltage of:

$$V_{OUT}^+ = V_2 \left(\frac{R_F}{R_1} + 1 \right) = I_{B_2} R_2 \left(\frac{R_F}{R_1} + 1 \right) \tag{1-6}$$

This reduces the output offset caused by the inverting side, since the output voltage polarity is different.

For zero output voltage:

$$V_{OUT}^+ + V_{OUT}^- = 0V$$

and

$$I_{B_2} R_2 \left(\frac{R_F}{R_1} + 1 \right) - I_{B_1} R_F = 0V$$

If we assume $I_{B_1} = I_{B_2}$ and solve for R_2:

$$R_2 = \frac{R_F R_1}{R_F + R_1} \tag{1-7}$$

If we use this value for R_2 when I_{B_1} is not equal to I_{B_2}, the output voltage is:

$$V_{OUT} = |I_{B_1} - I_{B_2}| R_F \tag{1-8}$$

The difference between I_{B_1} and I_{B_2} is called the *current offset* on op-amp specification sheets.

Referring to Fig. 1-7, we have base-to-emitter voltage drops V_1 and V_2. These are the forward-biased voltage drops of the base-emitter junctions of the transistors and are approximately 0.6 V each. Normally, there is a slight difference between these two voltages (6 mV maximum for a 741) which is amplified and causes an

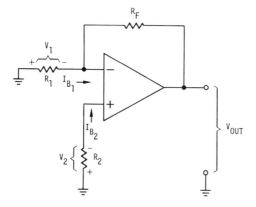

Figure 1-9 Offset Compensation

output offset voltage. The slight difference between the two voltages is referred to as the input offset voltage (V_i) and can be represented by a small battery placed in series with either the inverting or non-inverting input as shown in Fig. 1-10.

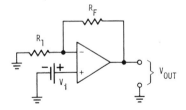

Figure 1-10   Input Voltage Offset

This input offset voltage is amplified by the non-inverting gain:

$$V_{OUT} = V_i \left(\frac{R_F}{R_1} + 1 \right) \qquad (1\text{-}9)$$

The total output offset voltage is the combined effect of the input current and voltage offsets. Depending on the polarity of the two types of off-sets, there can be an overall reduction or increase in the output voltage.

The combined output offset can be nulled out by injecting a voltage into the amplifier which will cancel out the offset. Methods vary with different op-amps, but a recommended technique using a 10-kΩ potentiometer is shown for the 741 op-amp in Fig. 1-11.

The disadvantage of using a potentiometer for offset nulling is the cost, mounting of the added component, and drift of the bias current caused by temperature changes.

Equipment that does experience wide temperature variations uses FET input op-amps. These devices have extremely low bias currents (picoamps), and therefore the effect is insignificant.

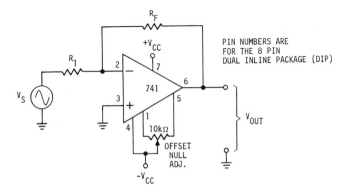

Figure 1-11   Offset Nulling Circuit (741)

1-6 OP-AMP FREQUENCY RESPONSE

At high frequencies a point is reached where the op-amp's output voltage is reduced because of capacitance to ground within the amplifier. Inside the op-amp, a combination of resistance and capacitance to ground acts as low pass filters. These filters cause a phase shift as well as a reduction in amplitude, with the phase shift being a maximum of 90° for each RC filter section. The result of this internal phase shift is an output signal from the op-amp that is in phase with the inverting input rather than 180° out of phase. This means that negative feedback can be converted to positive feedback, which will cause the op-amp to become unstable and break into oscillation if the overall gain is greater than unity.

We can determine whether an op-amp will be stable by examining a "Bode plot," which is a straight-line approximation of a frequency response curve. Figure 1-12 shows a frequency response curve with respect to a Bode plot.

Note that the "corner" frequencies for the Bode plot are at the 3 dB down points on the conventional plot. The -20 dB/decade slope corresponds to a single RC filter, which means that as the frequency is increased by a factor of 10, the output amplitude is reduced by a factor of 10. A maximum 90° phase shift can occur with this slope. The steeper slope of -40 dB/decade is for two RC sections, and as frequency is increased by a factor of 10 (decade), the amplitude is reduced to one hundredth. Maximum phase shift for this slope is 180°, which can cause instability in an op-amp.

We can represent the open loop frequency response for an op-amp with a Bode diagram as shown in Fig. 1-13.

This particular plot indicates that the op-amp has the equivalent of three RC sections with corner frequencies of f_1, f_2, and f_3. In order to find the closed loop response of the op-amp, we determine the gain (dB) and draw a horizontal line at that value. A closed loop gain of 78 dB was chosen in Fig. 1-12, and the dotted line indicates the frequency response. Note that the dotted closed loop plot intersects the open loop curve on a slope of 40 dB per decade. This means that the phase shift can be 180°, negative feedback can become positive, and oscillation could result.

To guarantee stability, the closed loop plot should intercept the open loop on a slope no greater than -20 dB/decade. From the graph, the minimum stable gain is about 80 dB for this op-amp. In order to operate with lower gains, the op-amp is compensated with a capacitor that rolls off the frequency response at a lower frequency, thus ensuring that the slope is never greater than -20 dB/decade, as illustrated in Fig. 1-13. The use of compensation reduces the frequency response in the graph from f_2 to $f_\emptyset$. Some op-amps like the 741 have a built-in capacitor which ensures that the op-amp will be stable at any gain setting. Others, like the 748, have provisions for connecting an external capacitor to allow selection of the frequency response.

The product of gain (ratio) and frequency response on the sloping part of the plot is called the *gain bandwidth product*. An open loop curve for the internally compensated 741 op-amp is shown in Fig. 1-14. Note that the open loop frequency response is only 10 Hz.

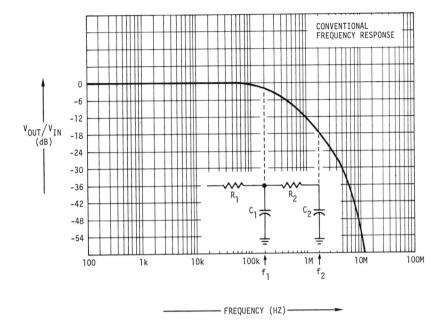

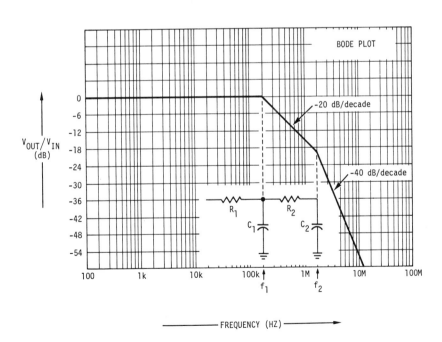

Figure 1-12   Frequency Response of RC Sections

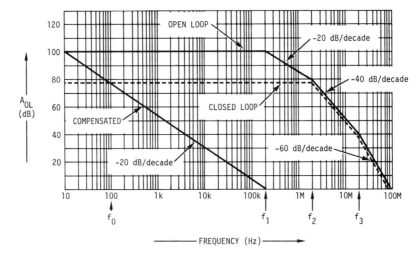

Figure 1-13 Bode Plot for an Operational Amplifier

We again draw horizontal lines corresponding to the required closed loop gain; if we select 60 dB (1000), we get a bandwidth of 1 kHz. If we choose a closed loop gain of 40 dB (100), we get a frequency response of 10 kHz. The gain bandwidth product in both cases is 1 MHz (1000 × 1 kHz and 100 × 10 kHz). Thus, higher closed loop gains mean less frequency response. The follower circuit with a gain of Ø dB (1) has a bandwidth of 1 MHz.

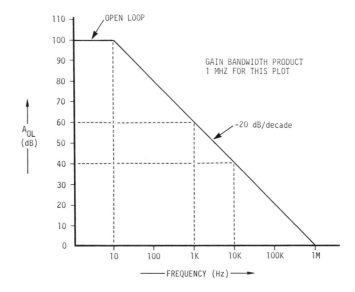

Figure 1-14 741 Open Loop Response

1.7 GAIN EQUATIONS FOR THE NON-IDEAL AMPLIFIER

In the derivation of the op-amp closed loop equations, it was assumed that the open loop gain was infinite. But from Fig. 1-14, we can see that the open loop gain for the 741 is only 100 (40 dB) at a frequency of 10 kHz.

We need to reexamine the equations for both the inverting and non-inverting amplifiers when the open loop gain is not infinite. Let us first derive the gain equation for the inverting amplifier with reference to Fig. 1-15.

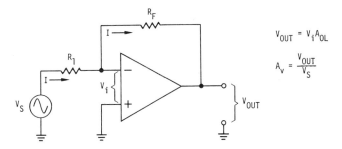

Figure 1-15   Non-ideal Inverting Amplifier

The current I is:

$$I = \frac{V_S - V_i}{R_1} = \frac{V_O - V_i}{R_F}$$

From these equations: $R_F(V_S - V_i) = R_1(V_i - V_O)$

but:

$$V_i = -\frac{V_O}{A_{OL}}$$

then:

$$R_F\left(V_S + \frac{V_O}{A_{OL}}\right) = R_1\left(-\frac{V_O}{A_{OL}} - V_O\right)$$

$$R_F V_S = -\left(\frac{R_F V_O}{A_{OL}} + \frac{R_1 V_O}{A_{OL}} + R_1 V_O\right)$$

The closed loop gain A_V is:

$$A_V = -\frac{V_O}{V_S}$$

Solving for this:

$$A_V = \frac{-R_F}{R_1 + \dfrac{R_1}{A_{OL}} + \dfrac{R_F}{A_{OL}}} \qquad (1\text{-}10)$$

Comparing this with the gain of the ideal op-amp of R_F/R_1, we can see that the terms R_1/A_{OL} and R_F/A_{OL} reduce the gain.

For example, for the 741 op-amp, $A_{OL} = 100$ at 10 kHz. If $R_F = 100$ kΩ and $R_1 = 1$ kΩ, then putting these values into Eq. 1-10 yields:

$$A_V = \frac{100 \text{ K}\Omega}{1\,\text{k}\Omega + \dfrac{1\,\text{k}\Omega}{100} + \dfrac{1\,\text{k}\Omega}{100}} \sim 50$$

This shows the actual gain is half of the 100 we would have expected with the ratio of R_F over R_1.

Using a similar derivation for the non-ideal gain of non-inverting amplifier, we find:

$$A_V = \frac{R_F + R_1}{R_1 + \dfrac{R_F}{A_{OL}} + \dfrac{R_1}{A_{OL}}} \tag{1-11}$$

We see the same reduction in closed loop gain due to the lower open loop gain. The only exception is the follower, where the closed loop gain remains at one and is reduced only when the open loop gain is less than one.

In a similar fashion, the output resistance is raised and the input resistance is lowered because of the reduction in loop gain as the open loop gain decreases.

The Bode diagram in Fig. 1-14 can be redrawn to provide a curve of closed loop gain, which is always a factor of 100 less than the open loop gain. This new plot is shown in Fig. 1-16.

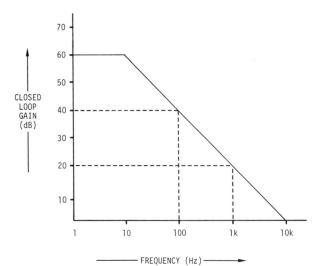

Figure 1-16 741 Closed Loop Gain Response

We can use this curve without having a gain error of greater than $\pm 1\%$ for given R_1 and R_F resistor values. Note that we are restricted to a gain bandwidth product (GBP) of 10 kHz rather than the previously specified 1 MHz for the open loop (or follower) response. Actually, the 3 dB down point of the closed loop response occurs where the open loop gain is roughly three times the computed closed loop gain. This means the gain bandwidth product is reduced by a factor of 3—if we are using the usual definition of a bandwidth, to the 3 dB down point.

To summarize: If we want to determine the GBP at which the closed loop gain equations are valid (within $\pm 1\%$ error), we divide the open loop GBP by 100. For frequency response (3 dB reduction in closed loop gain), we divide the open loop GBP by 3.

With the 741 we get:

Full closed loop gain GBP = 1 MHz/100 = 10 kHz

Frequency response GBP = 1 MHz/3 = 333 kHz

Thus, if the required closed loop gain is 10, this gain holds to a frequency of 1 kHz, while the frequency response is to 33.3 kHz.

1-8 SLEW RATE

If we put a sharp step-type waveform into the input of an op-amp, the output will not rise as quickly as the input. This is because internal capacitors (including the compensation capacitor) have to be charged to the new output voltage from limited current sources. The rate of rise of the op-amp output is called the slew rate; it is measured in V/μs (typically 0.5 V/μs for a 741). Slew rate can cause square wave inputs to become triangular at the output and limit the maximum output amplitude of sine wave signals since it creates crossover distortion. The maximum rate of change for a sine wave occurs at the zero crossing point. It can be shown that the maximum undistorted slew rate limited for a sine wave is:

$$f_{SR} = \frac{\text{Slew Rate}}{2 \pi E_p} \tag{1-12}$$

Equation 1-12 indicates that the higher the frequency, the lower the possible undistorted output waveform. This is due to the rate of rise at the zero crossing point, which is greater at higher frequencies.

Figure 1-17 shows the slew rate response for a square wave and sine wave.

When is the output of the op-amp limited by slew rate and when is it limited by rise time? Let us work an example to point out the difference.

Assume we have a 741 op-amp set up for a gain of 10, as demonstrated earlier. The closed loop gain bandwidth product of the 741 is 333 kHz, so with a gain of 10, the bandwidth is:

$$f_{3\text{dB}} = \frac{333 \text{ kHz}}{10}$$

$$= 33 \text{ kHz}$$

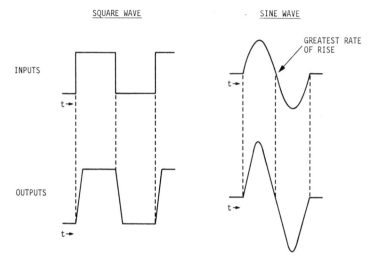

SQUARE WAVE   SINE WAVE

Figure 1-17   Effect of Slew Rate on Output

We can convert from amplifier bandwidth to rise time with the following equation:

$$t_r = \frac{0.35}{f_{3\,\mathrm{dB}}}$$

$$= 10.5 \ \mu s \ (10\% \ \text{to} \ 90\% \ \text{points})$$

If we assume a 10-mV peak-to-peak square wave input to the circuit, the output will, with a gain of 10, be 100 mV. The steepest (greatest slope) part of an exponential rising waveform is at the start with a slope (S_m) of:

$$S_m = \frac{E}{RC}$$

But $t_r = 2.2 \ RC$ and $E_{\text{PEAK-TO-PEAK}} = 100$ mV, thus:

$$RC = \frac{t_r}{2.2}$$

and $$S_m = \frac{2.2E}{t_r} \qquad\qquad (1\text{-}13)$$

$$= \frac{2.2 \times 100\,\text{mV}}{10.5\ \mu\text{s}}$$

$$= 0.021\ V/\mu\text{s}$$

Since the slew rate for the 741 op-amp is 0.5 V/μs, the output is rise time limited. What if we now apply a 1 V peak-to-peak square wave to the amplifier? The output will now be 10 V peak-to-peak and the maximum slope:

$$S_m = \frac{2.2 \times 10\ V}{10.5\,\mu\text{s}}$$

$$= 2.1\ V/\mu\text{s}$$

The op-amp slew rate cannot handle this fast a change; thus the output is slew rate limited. We conclude, then, that low-level output signals are rise time limited, while large output signals are slew rate limited.

1-9 POWER SUPPLY CHANGE REJECTION

When describing the effect of voltage offset on the output of an op-amp, it was indicated that a shift in the DC level occurred. A similar voltage error can occur if the power supply voltages are changed. This is because the power supply change causes an internal offset in the op-amp. We can reference this change to the input and handle it like an input voltage offset. The output offset can be calculated by multiplying this input offset by the non-inverting gain of the amplifier.

Manufacturers specify the power supply sensitivity in various ways. It can be specified as *power supply sensitivity* (μV/V) or as *power supply rejection ratio* (in dB or ratio). The 741 power supply rejection expressed as a ratio has a minimum value of 7000 (77 dB).

The circuit shown in Fig. 1-18 is operated from ±12-V supplies. A drop in the power line voltage causes a lowering of the supply voltages to ±11 V, which is an overall change of 2 V.

We need to know the change in the output voltage of the op-amp with this power supply change. The effective input offset to the op-amp is calculated by dividing the 2-V change by the power supply rejection ratio of 7000 or:

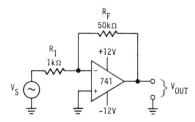

Figure 1-18   Power Supply Change Example

$$V_{id} = \frac{2 \text{ V}}{7000}$$

$$\sim 0.3 \text{ mV}$$

This offset is multiplied by the non-inverting gain because the offset appears internally between the inverting and non-inverting terminal. The output change V_{OO} is as follows:

$$V_{OO} = 0.3 \text{ mV} \left(\frac{50 \text{ k}\Omega}{1 \text{ k}\Omega} + 1 \right)$$

$$= 15.3 \text{ mV}$$

We can conclude from the result that op-amps are not terribly sensitive to power supply voltage change, particularly if the circuit gain is low.

Another type of change that can occur in the power supply output voltages is *ripple*. If we had 2 V of 120-Hz total ripple riding on the supplies used in the circuit in Fig. 1-16, the output ripple would be 15.3 mV. The calculation is the same as was previously conducted with the DC voltage change.

As the frequency of ripple (or noise) on the power supply increases, however, capacitance coupling within the op-amp causes the power supply rejection to decrease. Thus, more ripple voltage appears at the op-amp output.

1-10 AMPLIFIER NOISE CONSIDERATIONS

Without noise, we could amplify the weakest signal and bring it up to a usable level. But noise exists in all electronic components, being higher in active devices that amplify, like transistors. Thus noise, as well as signal, is amplified, and even though the signal level is high at the output of an amplifier, it can still be "washed out" by a higher level of noise.

The most basic form of noise is *thermal* or *Johnson noise*. This noise exists in any component that has resistance and is caused by the random movement of electrons within the material. Increasing temperature causes this type of noise to increase, thus the term *thermal noise*. In transistor circuits, additional noise sources are *flicker noise*, or 1/F noise, which has highest amplitude at low frequencies; and *shot noise*, which increases with the DC current and increases with frequency.

We don't need to know the actual noise mechanisms in order to determine the noise output from an op-amp. Manufacturers provide graphs like those shown in Fig. 1-19, which provide an equivalent input noise voltage and noise current at a given frequency. These values can be used in Eq. 1-14 below to find the op-amp output noise voltage (V_N).

$$V_N = A_V \sqrt{(e_N)^2 B + (i_N)^2 (R_{EQ})^2 B} \qquad (1\text{-}14)$$

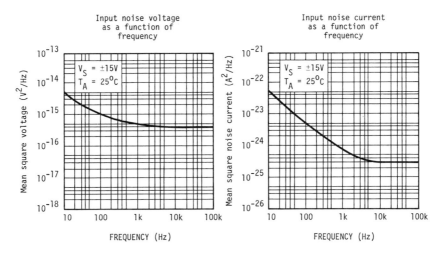

Figure 1-19 Noise Graphs for 741 Op-Amp   REV A

where e_N and i_N are taken from the graph, and B is the circuit bandwidth in hertz (Hz) and R_{EQ} is the equivalent resistance at the op-amp input terminals (for a simple amplifier—R_1 in parallel with R_F).

In calculating the *total* noise output from an amplifier, we must also consider the thermal noise from the input circuit resistors. This noise is expressed as:

$$V_N = \sqrt{4KTBR} \qquad (1\text{-}15)$$

where K is the Boltzman constant (1.38×10^{-23} J/°K), T is the absolute temperature (degrees centigrade plus 273°), the bandwidth B in Hz and R is the equivalent resistance which generates the thermal noise—again, R_1 in parallel with R_F for a simple amplifier. The noise equivalent circuit for an amplifier is shown in Fig. 1-20.

The current and voltage noise generators are shown external to the amplifier for clarity. Since the input resistance of the amplifier is very high, the noise current only flows through the parallel combination of R_1 and R_F.

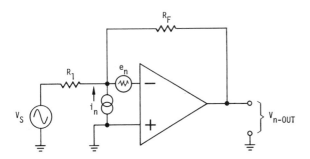

Figure 1-20 Equivalent Noise Circuit

Equation 1-14 can be modified to include the thermal noise of R_1 and R_F (R_{EQ}). From Eq. 1-15 the total output noise V_{NT} is:

$$V_{NT} = \sqrt{(e_N{}^2 + i_N{}^2\,R_{EQ}{}^2 + 4KTR_{EQ})\,B} \qquad (1\text{-}16)$$

But what bandwidth B do we use for this equation? We have to know something about the desired signal in order to answer this question. The bandwidth must be only wide enough to accommodate the required signal frequencies. Anything wider than this will allow more noise to pass. If we desire to pass only a single sine wave frequency, the bandwidth must be wide enough to allow for any drift of this frequency from its original value.

Let us use Eq. 1-16 and Fig. 1-20 to determine the noise output for an inverting amplifier with the following assumptions:

1. Circuit has a center frequency of 1 kHz and a bandwidth of 100 Hz.
2. R_F is 100 kΩ and R_1 is 1 kΩ (R_{EQ} is approximately 1 kΩ).
3. The amplifier is a 741 and is operating at 25°C (298°K).

From the graph at 1 kHz:

$$e_N{}^2 = 5 \times 10^{-16}\ V^2/\text{Hz}$$

$$i_N{}^2 = 8 \times 10^{-25}\ A^2/\text{Hz}$$

$$V_{NT} = 100\,\sqrt{[5 \times 10^{-16}\,V^2/\text{Hz} + 8 \times 10^{-25}\,A^2/\text{Hz} \times (1\text{k}\Omega)^2 + (4 \times 1.38 \times 10^{-23} \times 298°\text{K} \times 1\text{k}\Omega)]\,100\ \text{Hz}}$$

$$= 100\,\sqrt{[5 \times 10^{-16}\,V^2/\text{Hz} + 8 \times 10^{-19}\,V^2/\text{Hz} + 1.65 \times 10^{-17}\,V^2/\text{Hz}]\,100\ \text{Hz}}$$

$$= 22\text{uV}$$

Note that in the example, the maximum noise contribution is from the amplifier noise voltage (e_N). The reason for this is that the low equivalent resistance (R_{EQ}) of the external resistors used in this example keeps the voltage generated by the noise current and the thermal noise voltage to low values. A rule with low-noise amplifier circuits, then, is to keep the circuit resistance as low as possible.

Carbon resistors have additional internal noise sources caused by intergranular contact. Thus, those resistors should not be used in low-noise amplifier circuits. A better choice is the low-noise metal film resistors.

Returning to the results of the computation, we find that we have 22 μV of noise at the amplifier's output. If we arbitrarily assume that we want an output signal-to-noise ratio of 10 to 1, we would require an output signal of 220 μV. Referring this signal back to the input by dividing by the gain gives a minimum input signal requirement of 2.2 μV.

We assumed a bandwidth of 100 Hz in this example, but how can we achieve this? One solution would be to make the amplifier into a band pass filter by adding resistors and capacitors, or we could follow the amplifier with another op-amp connected as a band pass filter circuit.

1-11 OPERATION WITH A SINGLE POWER SUPPLY

Most op-amps are designed to operate with both a positive and negative power supply, which allows the output under no signal conditions to be at ground potential. However, if we are just concerned with AC signals, we could use a single positive (or negative) supply and bias the output up to half the supply voltage. Then, taking the output signal through a capacitor will center the output signal (symmetrically) about ground. A circuit for operating with a single power supply is shown in Fig. 1-21.

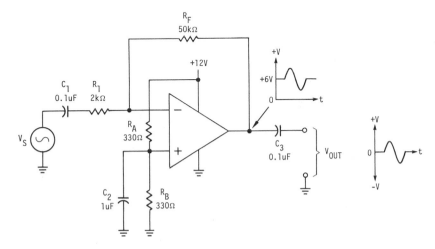

Figure 1-21   Single Power Supply Operation

Referring to Fig. 1-21, the 330-Ω divider provides 6 V (half the supply voltage) at the non-inverting input. The 0.1 μF coupling capacitor C_C is connected in series with the inverting input, thus making the circuit look like a follower to the +6-V input. The +6 V, then, appears at the output.

Assuming that the reactance of C_C is negligible, the input signal V_S is amplified by the 50 kΩ/2 kΩ factor, or 25. The 50-kΩ resistor in series with the non-inverting terminal is to minimize bias offset. Connecting the 1-μF capacitor across the 330-Ω resistor maintains the divider point at AC ground for the frequencies of interest.

1-12 COMPARISON OF OP-AMP TYPES

Thus far in this chapter we have referred only to the 741 op-amp specifications when we had described the various op-amp characteristics. Considering only the 741 op-amp, however, gives us a rather narrow view of op-amp capabilities. Op-amps available today are considerably improved over the 741 op-amp in terms of basic specifications. We should understand that specification improvement usually means that the device will cost more, so higher-performance op-amps are only used where circuit requirements dictate such usage.

Table 1-1 lists some other op-amp types along with their key specifications at a temperature of 25°C. Let us use the 741 op-amp as a yardstick when studying this table. Notice that each op-amp listed has a feature that can be desirable for some application. For example, if a high slew rate is required, the HA-2539 would be the best choice. The cost factor indicates how much more expensive the op-amp is relative to the 741 type. The LF351 with a factor of 2 is twice as expensive as the 741.

If bias current is a concern, the AD515L would be used. An improved version of the 741 is the LM356, with better specifications in all areas except voltage offset.

The LM363 instrumentation amplifier has high common mode and power supply rejection and also has low internal noise. This amplifier would be used for amplification of very small input signals.

The LM10C device provides a reference output voltage along with the amplifier, giving the device greater versatility. The reference output voltage can be used to establish a precise DC component at the amplifier's output upon which AC waveforms can be superimposed.

1-13 CURRENT MODE (NORTON) OP-AMPS

Conventional op-amps are considered voltage-operated devices. We put in a voltage and an amplified voltage appears at the output. There are a class of op-amps, however, that operate in the current mode—i.e., a current is driven into the input to cause a voltage change at the output.

The most common Norton amplifier is the LM3900, which has four amplifiers in one package. It is shown in schematic form in Fig. 1-22.

A current mirror circuit is formed by transistors Q_6 and Q_7. Notice that Q_7 is connected like a diode, with the base and collector shorted. Since the bases of both transistors are at the same potential, the collector currents must be the same, so $I^+ = I^-$.

If the current at the inverting input increases, Q_5 collector current increases, which in turn causes more current to flow in the emitter circuit of Q_3. This causes an increased voltage drop across Q_2, which is used as a high-value load resistor. The decreasing emitter voltage of Q_3 is coupled via the base of Q_1 to the output. So an increase in current at the inverting terminal causes a decrease in the output voltage. A similar analysis reveals that increasing the current at the non-inverting input will increase the voltage at the output. Q_8 serves as a high-value resistor like Q_2, while Q_4 provides a low resistance path to ground (sink circuit) when there are large voltage swings at the output. Voltage V_1 and V_2 are fixed internal voltages which make Q_8 and Q_2 constant current sources.

Table 1-1 Selected Op-Amp Specifications

Op-Amp type	Min. open loop gain	Typ. gain bandwidth product	Typ. slew rate (V/μS)	Max. bias current	Max. current offset	Max. voltage offset (mV)	Min. common mode reject (dB)	Min. power supply reject (dB)	Noise voltage @ 1 KHz (N²/Hz) 10^{-16}	Cost factor	Key feature
741C	20,000	1 MHz	0.5	500nA	200nA	6.0	70	77	2.5	1	Low cost.
101H											Low bias current
LF351	25,000	4 MHz	13	100pA	25pA	10.0	70	70	2.5	2	High slew rate.
LM318	25,000	15 MHz	70	500nA	200nA	10.0	70	65	2.5	5	High CMRR & P.S. rejection.
LF357	50,000	20 MHz	50	200pA	50pA	10.0	80	80	2.5	4	Wide bandwidth.
HA2539	10,000	600 MHz	600	20uA	6uA	15.0	60	60	2.5	34	High Slew Rate.
AD515L	40,000	0.35 MHz	0.3	0.075pA	—	1.0	70	70	—	—	Ultra low bias current.
											Instrumentation amp.
LM363	1,000,000	2 MHz	—	10nA	3nA	0.1	94	100	0.5	45	Low noise. High rejection.
LM356	25,000	5 MHz	10	200pA	50pA	10.0	80	80	2.5	3	Improved 741.
LM10C	25,000	100 KHz	—	40nA	3nA	5.0	90	87	2.5	11	Has on chip Volt. Ref.
LM108A	80,000	1 MHz	0.3	2nA	0.2nA	1.0	96	96	2.5	—	Precision low drift.

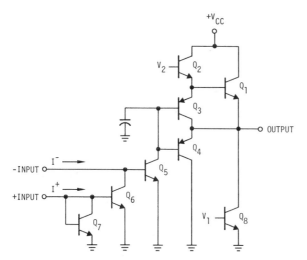

Figure 1-22   Non-Inverting Amplifier

We can set the current I^+ by connecting a resistor $(2R_F)$ from the positive supply to the non-inverting input as shown in Fig. 1-23.

$$I^+ = (V_{CC} - 0.7 \text{ V})/2R_F \tag{1-17}$$

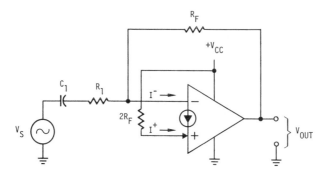

Figure 1-23   Inverting Amplifier

Now if we want to have maximum output voltage swing without clipping of the waveform, the DC voltage at the output should be half the supply voltage or $V_{CC}/2$. This means we have half the voltage drop across the feedback resistor to the inverting input that we have across the resistor connected to the non-inverting input. Since the current through these two resistors is the same (because of the current mirror circuit), we can now see why one resistor has the value R_F and the other $2R_F$.

Again with reference to Fig. 1-23, the input signal is coupled via the DC blocking capacitor C_1 to R_1. If the input signal is positive, current will flow through R_1 into the amplifier inverting input. The current into the inputs are constant so current through R_1 causes a reduction in the current through R_F. But the voltage at the output must go lower in order to reduce the current through R_F. With R_F typically greater than R_1, the voltage change at the output is greater than the voltage change at the input—so voltage amplification occurs and the inverting gain is simply:

$$A_V = - R_F/R_1 \qquad (1\text{-}18)$$

Moving the capacitor C_1 and resistor R_1 to the non-inverting side, as shown in Fig. 1-24, creates a non-inverting amplifier.

Again, a positive input signal will cause current flow through R_1, but now we have current flow into the non-inverting terminal. Current through the dynamic resistance (r_d) of the diode connected to transistor (Q_7) will cause a small voltage drop. The increase in input current will be mirrored by a similar increase in the output voltage. With R_F greater than R_1, the voltage gain will be:

$$\text{Non-inverting Gain } A_V = R_F/(R_1 + r_d) \qquad (1\text{-}19)$$

$$= R_F/R_1 \text{ if } R_1 \gg r_d$$

This differs from the conventional op-amp equation by not having the plus one term.

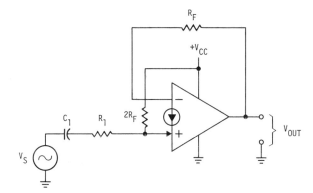

Figure 1-24 Non-Inverting Amplifier

The main advantages of the Norton amplifier are as follows:

1. High input voltages can be accommodated—just increase R_1 to maintain the required current.
2. Single power supply operation with output swings from ground to 0.7 V from the supply voltage.
3. Wider bandwidth—the LM3900 has a gain bandwidth product of 2.5 MHz as compared with 1 MHz for a 741 op-amp.

It should be clear that because of the different mode of operation, conventional op-amp circuits cannot be used with a Norton type amplifier. See the National Semiconductor Application note AN72-2 for more detailed information and circuits for the LM3900.

1-14 OP-AMP MECHANICAL CONFIGURATIONS

Most op-amps are contained within two basic packages—the dual in-line package (DIP) and the transistor-type TO-5 metal can.

The DIP configuration is the most common type used and is contained in a black plastic case with eight or fourteen connecting leads. The eight-pin variety has a single op-amp inside, while the fourteen-pin type can have up to four op-amps in the package. Leads are numbered around the package from a reference point in a horseshoe or U-shaped pattern as shown in Fig. 1-25.

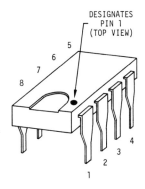

DUAL IN-LINE PACKAGE

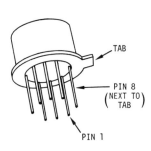

TO-5 CONFIGURATION

Figure 1-25   Common Op-Amp Package

Better sealing and electrical shielding are afforded by the TO-5 package, but not all op-amps are available in this form. The TO-5 type also has a higher maximum operating temperature (150°C versus 125°C for the DIP) and a lower thermal resistance, which allows easier flow of heat out of the op-amp device.

1-15 PRACTICAL EXAMPLES

To reinforce some of the concepts presented in this chapter, we shall work through some example problems.

EXAMPLE 1-1
Design an inverting amplifier circuit using a 741 op-amp with a gain of 20 using an R_F value of 100 kΩ and assuming an input sine wave of 0.5 V peak. Find the following:

   a) The value of R_1.
   b) The value of R_2 for offset compensation.
   c) The peak output voltage.
   d) The maximum output offset.
   e) The maximum sine wave input frequency.
   f) Input and output resistance.

Solution:

a) The value of R_1 can be determined from Eq. 1-3:

$$R_1 = \frac{R_F}{-A_V} = \frac{100 \text{ k}\Omega}{20} = 5 \text{ k}\Omega$$

b) For the value of the compensating resistor R_2, we use Eq. 1-7:

$$R_2 = \frac{R_F R_1}{R_F + R_1} \sim 5 \text{ k}\Omega$$

c) The peak output voltage is the input (0.5 V) times the gain (20)—or 10 V.

d) Total output offset is the sum of the effects of both current and voltage offset. The maximum difference between I_{B_1} and I_{B_2} (called the offset current on spec sheets) is 200 mA for the 741. Using Eq. 1-8:

$$V_{\text{INPUT OFFSET}} = |I_{B_1} - I_{B_2}| R_F$$

$$= \pm 200 \text{ nA} \times 100 \text{ k}\Omega$$

$$= \pm 20 \text{ mV}$$

maximum input offset spec is 6 mV. Using this value in Eq. 1-8:

$$V_{\text{OFFSET}} = V_i \left(\frac{R_F}{R_1} + 1 \right)$$

$$= \pm 6 \text{ mV} \left(\frac{100 \text{ k}\Omega}{5 \text{ k}\Omega} + 1 \right)$$

$$= \pm 126 \text{ mV}$$

The plus and minus terms are used because we don't know the sense of the input offset. Total offset is:

$$V_{\text{OFFSET TOTAL}} = \pm(20 + 126)\text{mV}$$

$$= \pm 146 \text{ mV}$$

e) In considering frequency response, we need to look at both closed loop gain bandwidth product and slew rate. For the 741, the gain bandwidth product specified is 333 kHz. We want a gain of 20, so the 3 dB down frequency is:

$$f_{3\text{dB}} = \frac{333 \text{ kHz}}{20}$$

$$= 16.67 \text{ kHz}$$

With slew rate, we must consider the peak output voltage, which is 10 V in this example (part c) and the slew rate of 0.5 V/μs for the 741. Rearranging Eq. 1-9:

$$f = \frac{\text{Slew Rate}}{2 \pi E_{\text{PEAK}}}$$

$$= \frac{0.5 \text{ V}}{2\pi \times 10^{-6}\text{s} \times 10 \text{ V}}$$

$$= 7957 \text{ Hz}$$

We can see that to avoid distortion we must limit the input frequency to 7.957 kHz rather than 16.67 kHz. If the output amplitude had been less, we would have used the maximum frequency response of 16.67 kHz.

f) Since the inverting input terminal is at "virtual ground," the input resistance is simply R_1 or 5 kΩ for this example. Output resistance is determined by dividing the open loop output resistance (75 Ω for a 741) by the loop gain as determined by Eq. 1-4, as follows:

$$A_{\text{LOOP}} = \frac{100,000}{20}$$

$$= 5000$$

and

$$R_{\text{OUT}} = \frac{75\Omega}{5000}$$

$$= 15 \text{ m}\Omega$$

EXAMPLE 1-2

For the circuit shown in Fig. 1-26, find the following:

a) The circuit gain.

b) The maximum expected DC shift at the output of the op-amp if the +12-V supply drops 1 V.

c) The minimum 100 Hz input signal level if we desire a 10-to-1 signal-to-noise ratio at the amplifier output (V_{OUT}) with a bandwidth of 10 Hz and a temperature at 25°C.

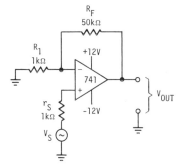

Figure 1-26 Non-Inverting Amplifier

Solution:

a) The circuit gain is simply the ratio of R_F divided by R_1 plus one:

$$A_V = \frac{R_F}{R_1} + 1$$

$$= \frac{50 \text{ k}\Omega}{1 \text{ k}\Omega} + 1$$

$$= 51$$

b) Dropping the positive power supply by 1 V (from +12 V to +11 V) will cause the DC level at the output of the op-amp to change. We can find the change in the output voltage by multiplying the 1 V by the non-inverting circuit gain and then dividing by the power supply rejection ratio (about 7000 for the 741 op-amp) or:

$$V_{\text{OUT}(\Delta \text{DC})} = \frac{1 \text{ V} \times 51}{7000}$$

$$= 7.3 \text{ mV}$$

c) In order to find the minimum input signal level, we need to find the noise level for the amplifier at the center frequency of 100 Hz with a bandwidth of 10 Hz. From the graphs of Fig. 1-18, we find at 100 Hz:

$$e_N{}^2 = 10^{-15} \ V^2/\text{Hz}$$

and

$$i_N{}^2 = 5 \times 10^{-24} \ A^2/\text{Hz}$$

The equivalent input resistance is R_F in parallel with R_1 plus R_S (noise current also flows through this resistor).

$$R_{\text{EQ}} = \frac{1 \text{ k}\Omega \times 50 \text{ k}\Omega}{1 \text{ k}\Omega + 50 \text{ k}\Omega} + 1 \text{ k}\Omega$$

$$\sim 2 \text{ k}\Omega$$

The total noise at the op-amp output is from Eq. 1-16.

$$V_{NT} = 51 \sqrt{[10^{-15} \ V^2/\text{Hz} + 5 \times 10^{-24} \ A^2/\text{Hz} \times (2 \text{ k}\Omega)^2 + (4 \times 1.38 \times 10^{-23} \times 298°\text{K} \times 2 \text{ k}\Omega)}$$

$$= 5.23 \ \mu\text{V}$$

For a 10-to-1 signal-to-noise ratio at the op-amp output, the signal should be 52.3 μV. The required input signal is the voltage divided by the gain of 51, or 1.03 μV.

EXAMPLE 1-3

A particular op-amp circuit has a closed loop gain of 50. If the open loop gain bandwidth product of the op-amp is 100 MHz and the slew rate is 50 Volts/μs microsecond, find:

    a) The frequency response of the circuit.

    b) The frequency at which an output 10-V peak sine wave just starts to distort due to slew rate limiting.

Solution:

    a) We determined earlier that we can find the closed loop gain bandwidth product from the open loop GBP by dividing by 3. Therefore:

$$GBP_{CL} = \frac{100 \text{ MHz}}{3}$$

$$= 33.3 \text{ MHz}$$

    b) With a circuit gain of 50, the frequency response is:

$$f_{3 \text{ dB}} = \frac{33.3 \text{ MHz}}{50}$$

$$= 666 \text{ kHz}$$

The frequency at which slew rate limiting starts is:

$$f = \frac{\text{SLEW RATE}}{2 \pi E_{\text{PEAK}}}$$

$$= \frac{50 \text{ V}}{10^{-6}\text{s} \times 2 \pi \times 10 \text{ V}}$$

$$= 796 \text{ kHz}$$

In this example the frequency response is rolling off before slew rate is a problem.

1-16 CHAPTER 1—SUMMARY

- An operational amplifier approaches an ideal amplifier because it has high gain, high input resistance, and low output resistance.
- By using negative feedback (closed loop) we can make the gain only a function of R_F and R_1, i.e., R_F/R_1 for the inverting amplifier and $R_F/R_1 + 1$ for the non-inverting amplifier.
- Current and voltage offset causes a shift in the op-amp DC output voltage.

- An op-amp will oscillate if the closed loop gain intersects the open loop gain on a slope greater than 20 dB per decade.
- Slew rate indicates how fast the op-amp output voltage can change and, if too low, can cause distortion in the output voltage waveform.
- The theoretical closed loop gain will not be achieved if the open loop gain is too small, because of roll-off at higher frequencies.
- Changes in the power supply voltages create a voltage offset within the amplifier.
- The limitation on how small a signal can be amplified by an op-amp is determined by the op-amp internal noise and the value and type of input resistors.

1-17 CHAPTER 1—EXERCISE PROBLEMS

1.1. A 741 op-amp is connected in an open loop configuration. If the input signal is 7 mV, determine the amplified signal at the output of the op-amp.

1.2. The circuit of Figure 1-26 shows a closed loop amplifier. Determine the input signal voltage (V_S) if the output signal voltage is 10 V peak to peak.

1.3. For the circuit in Figure 1-27, determine the output signal voltage if the input voltage is a sine wave with a peak voltage of 5 mv.

1.4. State the two assumptions which were made in order to derive the gain equations for the op-amp.

1.5. With reference to Figure 1-27, what is the actual voltage between the two input leads of the op-amp?

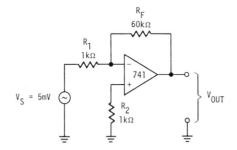

Figure 1-27   Inverting Amplifier

1.6. Determine the effective input resistance of the circuit in Figure 1-26, and also Figure 1-27.

1.7. Determine the effective output resistance of the circuit in Figure 1-27.

1.8. Find the maximum output voltage offset due to input voltage offset for the circuit in Figure 1-27.

1.9. Replace R_F in Figure 1-27 with a 1 MΩ resistor and replace R_2 with a short circuit. Now find the maximum output offset voltage caused by input biasing current.

1.10. If we remove the short from R_2 in Problem 9 (R_F still 1 MΩ) find the new total output offset voltage caused by input biasing currents and input offset voltage.

1.11. An op-amp has a maximum open loop gain of 100 dB(100,000). If the first corner frequency is at 100 Hz and the second at 10 KHz, find the minimum stable closed loop gain.

1.12. If the gain bandwidth product for an amplifier is 20 MHz, find the bandwidth if the closed loop gain is 50.

1.13. Determine the actual closed loop gain for an inverting amplifier with a R_F of 50 kΩ and R_1 of 2 kΩ if the open loop gain is 500.

1.14. The input signal to a 741 op-amp is a sine wave with a frequency of 5 kHz and a peak to peak amplitude of 100 mV. Find the maximum gain of the amplifier before slew rate distortion occurs.

1.15. Determine whether the output square wave from a LF351 op-amp with a circuit gain of 100 is rise time or slew rate limited. Assume an input signal of 150 mV peak to peak and ± 12 V power supplies.

1.16. If a LF357 op-amp circuit has a gain of 100, find the amount of change at the op-amp output with a total (plus to minus) power supply voltage change of 3 V.

1.17. Find the minimum input signal level at 500 Hz in order to have an output signal to noise ratio of ten for the circuit in Figure 1-28. The required bandwidth is 40 Hz. [Hint: $R_{eq} = R_1R_F/(R_1 + R_F) + R_2$]

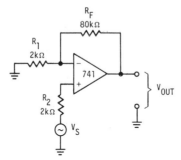

Figure 1-28   Non-Inverting Amplifier

1.18. For the circuit shown in Figure 1-29, compute:

a) The AC gain at 10 kHz

b) The DC voltage at the output

1.19. Select an Op-amp from Table 1-1 to be used as a low level compensated DC inverting amplifier with a gain of 100 and a R_F of 100 KΩ.

1.20. Find the lowest cost Op-amp in Table 1-1 which has a common mode rejection and a power supply rejection of at least 80 db.

1.21. Determine the value of the mirror resistor for a Norton Amplifier if we desire the output voltage to be one third of the supply voltage. Assume R_F is 1.2 MΩ.

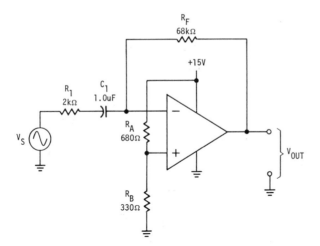

Figure 1-29   AC Amplifier

1.22. What value of R_1 should be used if we desire a gain of 10 from a Norton Non-inverting Amplifier and R_F is 3.3 MΩ?

1.23. Determine the value of the coupling capacitor for a Norton Amplifier with $R_1 = 100$ KΩ and the minimum signal frequency of 100 Hz. Assume a maximum gain deviation caused by the capacitor of 1%.

1-18 CHAPTER ONE—LABORATORY EXPERIMENTS

Experiment 1A

Purpose: To demonstrate the operation of a 741 op-amp amplifier circuit.

Requirements: This amplifier is to be connected as a non-inverting configuration with $R_1 = 1$K ohm and $R_F = 100$K ohm. Power supplies to be ±12 volts.

Measure:

1. Voltage gain and compare with computed value (don't forget plus one).
2. Frequency response and compare with predicted curve (see 741 Bode plot).
3. Slew rate and compare with spec value given in the 741 data sheets (use a square wave input signal).

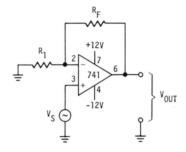

Figure 1-30   Non-inverting Amplifier

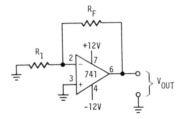

Figure 1-31   Inverting Amplifier

Experiment 1B

Purpose: To show how voltage and current offset can change the DC level at the op-amp output.

Requirements: Connect up a 741 inverting amplifier with a gain of 1000 using ±12-V power supplies. (Reference Figure 1-31.)

Measure:

1. Voltage offset.   Use a low value for R_F (10kΩ) to make current offset effect at output negligible. Determine the input offset by dividing the DC voltage at the output by the gain. Compare with the spec value given on the 741 data sheets.
2. Measure I_{B_1}.   Increase R_F to 10 MΩ and increase R_1 to maintain same gain. Determine the voltage change from the previous step. Compute I_{B_1} by dividing the voltage change by R_F. Compare with the specified value.
3. Measure I_{B_0}.   Add R_2 resistor of 10 kΩ to non-inverting input and record the voltage change from step 2. Divide the voltage change by R_F to find I_{B_0}. Compare with specified value.
4. Connect up an offset potentiometer and adjust to provide zero offset at the op-amp output.

Experiment 1C

Purpose: To show operation of an op-amp with a single power supply.

Requirements: Circuit to operate off a single $+12$-V supply as shown in the Fig. 1-32. Select values of R_a and R_b to provide $+6$ V at the output and R_1 and R_F to give a gain of 47. (Keep reactance of C_1 less than one hundredth of R_1 for the lowest frequency of interest.)

Measure:

1. DC voltage at output of op-amp. Compare with predicted value.
2. AC gain of the amplifier at 1 kHz and compare with predicted value.
3. Drop the power supply voltage to 11 V and measure the change in DC voltage at the output of op-amp. Compare with the computed change.

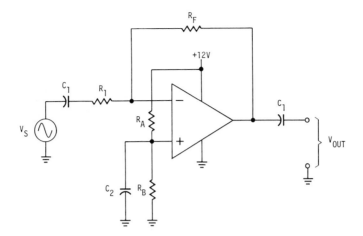

Figure 1-32   Single Supply Operated Op-Amp

2

Operational Amplifier Applications

OBJECTIVES

Upon completion of this chapter you will be able to recognize and explain the operation of the following op-amp circuits:
1. Summing, averaging, and subtracting circuits
2. Instrumentation amplifiers
3. Current boosters
4. Integrators and differentiators
5. Basic filter circuits
6. Sine wave, square wave, pulse, and ramp-type signal generators
7. Ideal diode circuit
8. Signal compression
9. Absolute value circuit

2-1 INTRODUCTION

The applications for operational amplifiers seem endless. A glance at electronics articles will show many new applications. Most of these applications, however, are derived from a few basic op-amp circuits. It is the intent of this chapter to introduce these so-called basic configurations so that a reference can be established for understanding new circuits as they are encountered.

2-2 SUMMING, AVERAGING, AND SUBTRACTING CIRCUITS

Suppose we wish to add (mix) the signals together from two microphones to feed into a common amplifier. We could simply connect the microphones in parallel, but this would cause the impedance of one microphone to load down the other microphone, with a consequent reduction of signal level.

We can perform the adding operation with an op-amp without affecting the amplitude of each input signal if we use the summing circuit shown in Fig. 2-1. With this circuit, the virtual ground at the inverting input isolates the two signals V_A and V_B.

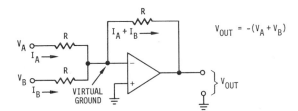

Figure 2-1   Inverting Summing Circuit

We can analyze the circuit operation by considering each input separately. Again, remembering the virtual ground at the inverting input, V_A causes a current to flow in the upper input resistor of:

$$I_A = \frac{V_A}{R} \tag{2-1}$$

and V_B causes a current to flow in the lower input resistor of:

$$I_B = \frac{V_B}{R} \tag{2-2}$$

Now I_A and I_B also flow through the feedback resistor, creating a voltage drop across this resistor of:

$$V_O = -(I_A + I_B)R \tag{2-3}$$

Since the left side of the resistor is at virtual ground and the right side is at the op-amp output, the voltage drop is the output voltage (V_O). Expressing the current in terms of voltage and resistance, we have:

$$V_O = - \left(\frac{V_A}{R} + \frac{V_B}{R} \right) R \qquad (2\text{-}4)$$

or

$$V_O = -(V_A + V_B)$$

Thus the output is the sum of the inputs with a sign change.

If we reduce the feedback resistor to $R/2$, then the output is:

$$V_O = - \left(\frac{V_A + V_B}{2} \right) \qquad (2\text{-}5)$$

This gives us the negative *average* of the two input signals. With three inputs, we would change the feedback resistor to $R/3$ to obtain the average at the output, and so on.

If we don't want the negative output for the sum and averaging circuits, we can use the non-inverting circuit shown in Fig. 2-2.

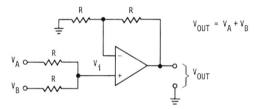

$$V_{OUT} = V_A + V_B$$

Figure 2-2   Non-Inverting Summing Circuit

In order to determine the voltage at the non-inverting input terminal, we need to convert the input circuit to what is called a Norton equivalent circuit, as shown in Fig. 2-3.

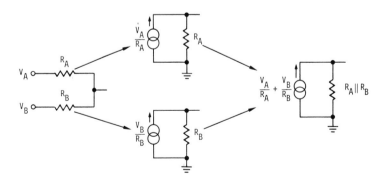

Figure 2-3   Norton Equivalent Circuit

The conversion is performed by shorting the non-inverting input terminal to ground and determining the current in each branch. For the upper branch, the current is V_A/R and for the lower branch V_B/R. These currents become the current generators indicated by double circles in Fig. 2-3.

Arrows at the current generators indicate the direction of current flow and are related to the polarity of the input voltage. In this case V_A and V_B have the same polarity. To complete the circuit, the branch resistors are then placed in parallel with the current generators.

This final circuit is the Norton equivalent for the input circuit. Note that the currents for the individual circuits are algebraically added (sign taken into account) to find the final circuit current generator value. The resistor value is determined by finding the parallel equivalent resistance of the two individual circuit resistors.

We can now determine the voltage at the non-inverting terminal, which is simply the voltage drop across the Norton circuit resistor caused by the flow of current from the current generator or:

$$V_i = \frac{(V_A + V_B)}{R} \frac{R}{2}$$

$$= \frac{V_A + V_B}{2} \tag{2-6}$$

This shows that the voltage at the non-inverting input is the average of the two inputs. If we had a follower circuit, (R_F shorted out), the output of the op-amp would be the average of the inputs. Using the value of R for R_1 and R_F as shown in Fig. 2-2, the gain is 2. The output voltage is then:

$$V_O = 2V_i$$

$$= 2 \left(\frac{V_A + V_B}{2} \right)$$

$$= V_A + V_B \tag{2-7}$$

which gives us the required sum of the two inputs.

If we had three inputs, the gain would have to be 3 in order to provide the sum of the output (make R_F twice R_1).

When using the summing and averaging circuits, we must take into account the source resistance of the driving circuit. Going back to the two-microphone situation, let us assume that the internal resistances are 10 kΩ and the signal voltage 10 mV. This resistance becomes part of the input circuit resistance and must be added to it. See Fig. 2-4.

Assuming a value of 1 KΩ for R and neglecting the 10-kΩ source resistance, we would have an input current of 10 μA, with the 10-mV input signal. This is not correct, because the actual input resistance is 11 kΩ and this yields a current of about 0.9 μA, which reduces the output voltage by a factor of 11.

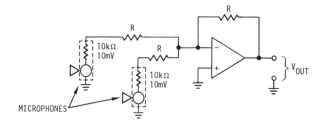

Figure 2-4 Summing Circuit with Input Resistance

For a 1% error, R should be 100 times greater than the source impedance of the driving device (microphone, transducer, etc.). If the input signals for a summing or averaging circuit are taken from op-amp outputs, then we can neglect the effect of source impedance (extremely low).

Different input resistors can be used for a summing circuit if we wish one input to have a greater effect on the output. This is called a *weighted input.*

Notice in Fig. 2-5 that the V_B input resistor is one half the value of the V_A resistor. Considering the circuit as an inverting amplifier with two inputs, the gain for V_A is -1 and the gain for V_B is -2. Thus, V_B has twice as much effect on the output signal.

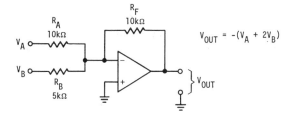

$$V_{OUT} = -(V_A + 2V_B)$$

Figure 2-5 Weighted Input Summer

We can build a subtracting circuit by feeding one signal into the inverting side of an op-amp and the other signal into the non-inverting side. But remember, the gain is greater by one from the non-inverting side. This can be taken care of by reducing the signal on the non-inverting side by using a voltage divider, as shown in Fig. 2-6.

The signal at the output caused by the inverting input signal is:

$$V_{OB} = -\frac{V_B R}{R}$$

$$= -V_B$$

From the non-inverting side we first must go through the voltage divider to find the signal at the non-inverting op-amp terminal. The signal at the non-inverting op-amp terminal is:

$$V_{ni} = \frac{V_A R}{2R}$$

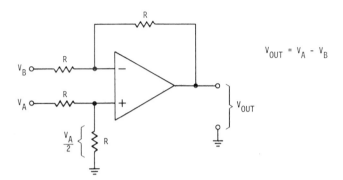

$$V_{OUT} = V_A - V_B$$

Figure 2-6 Subtracting Circuit

This signal will appear at the output as:

$$V_{OA} = V_{ni} \left(1 + \frac{R}{R} \right)$$

$$= 2V_{ni}$$

$$= 2V_A \left(\frac{R}{2R} \right)$$

$$= V_A$$

Finally, the total output is:

$$V_O = V_{OA} + V_{OB}$$

$$= V_A - V_B \qquad (2\text{-}8)$$

If we wish to have the *amplified* difference (subtraction) of the two signals, we can increase R_F by the required gain factor, but we must also increase the resistor at the non-inverting terminal to ground by the same factor. Figure 2-7 shows an example of a circuit with an output that is 10 times the difference of the input signals.

It must be stressed that the accuracy of the circuits described in this section is

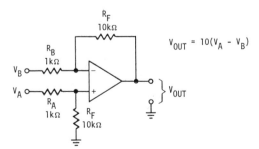

$$V_{OUT} = 10(V_A - V_B)$$

Figure 2-7 Amplified Difference Circuit

directly related to the accuracy of the circuit resistors. Typically, 1% resistors (or better) are used for these types of circuits.

2-3 INSTRUMENTATION AMPLIFIERS

Amplifiers used to increase the weak signal levels from transducers are called instrumentation amplifiers. Transducers, in turn, are devices that convert other forms of energy (optical, thermal, mechanical, etc.) into electrical signals. Typical transducers are:

- Photo devices: photo diodes, photo resistors
- Thermal devices: thermocouples, thermisters
- Mechanical: pressure, strain

These devices either generate an output voltage or change resistance in response to the non-electrical input.

The requirements for a good instrumentation amplifier are low internal noise so that weak signals will not be lost, high rejection of noise signals common to both inputs (common mode rejection), and high power supply change rejection.

It is not always possible to locate the amplifier with the transducer, so cabling between the two is sometimes required. Whenever we have cables and weak signals, we must be concerned with noise pickup. This pickup can be electrostatic (capacitive) or electromagnetic (induced). Figure 2-8 shows a transducer connected to a conventional (non-instrumentation) amplifier via a wire and indicates the noise coupling.

The electrostatic noise (V_e) is picked up by a capacitive "voltage divider" action between the capacitance from the noise source to the wire (C_S) and the capacitance from the wire to ground (C_W). Magnetic noise coupling (V_M) results from the magnetic noise source flux cutting the connecting wire and inducing a voltage in the wire.

In the circuit shown, the output voltage will be:

$$V_O = \left(\frac{R_F}{R_1} + 1 \right)(V_T + V_M + V_S) \tag{2-9}$$

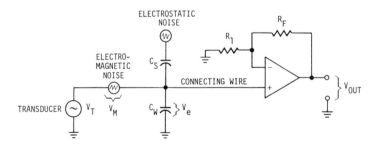

Figure 2-8 Noise Coupling

Thus, if the noise signals are greater than the transducer signal, it will be difficult to detect the transducer signal at the output of the amplifier.

By changing the circuit connections to the instrumentation amplifier circuit shown in Fig. 2-9, we can considerably improve the signal-to-noise ratio.

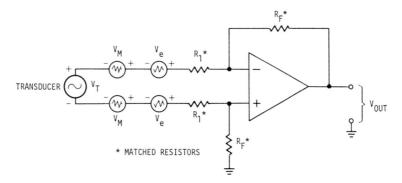

Figure 2-9   Instrumentation Amplifier (Balanced Input)

The major change in this circuit over the conventional amplifier previously discussed is the change to what is called a balanced input. The transducer is isolated from ground, and two wires are used to connect it to the amplifier. Each wire is balanced with respect to ground, thus each wire has the same resistance $(R_1 + R_F)$ to ground. This, plus the fact that the wires are twisted together, results in equal noise pickup in each wire.

If we refer back to the subtracting circuit discussed previously, we will find the noise output (V_{NO}) of the amplifier to be:

$$V_{NO} = \frac{R_F}{R_1} (V_e - V_e + V_M - V_M) \qquad (2\text{-}10)$$

$$= 0$$

This shows that the balanced configuration has eliminated the noise. In practice, the amount of noise reduction is a function of how close we can match the resistors, also the matching of the internal op-amp gains between the non-inverting and inverting inputs.

In order to determine the gain of the wanted transducer signal, we need to break the transducer generator into two generators, each with one half the transducer voltage, as shown in Fig. 2-10.

Notice that since the circuit is balanced with respect to ground, a point exists between the newly created signal generators which is at ground potential (virtual ground).

Considering the gain of the upper generator first, we have an inverted output signal (V_{Oi}) of:

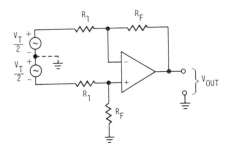

Figure 2-10 Equivalent Signal Circuit

$$V_{Oi} = -\frac{V_T}{2}\left(\frac{R_F}{R_1}\right)$$

The signal generator on the non-inverting side is 180° out of phase with the inverting side (note the plus and minus signs). This signal first goes through a voltage divider of R_1 and R_F, giving a voltage at the non-inverting input of:

$$V_{Oni} = -\frac{V_T}{2}\left(\frac{R_F}{R_F + R_1}\right)$$

This signal then goes through the amplifier and gives an output signal V_{Oni} of:

$$V_{Oni} = -\frac{V_T}{2}\left(\frac{R_F}{R_F + R_1}\right)\left(\frac{R_F}{R_1} + 1\right)$$

$$= -\frac{V_T}{2}\left(\frac{R_F}{R_F + R_1}\right)\left(\frac{R_F + R_1}{R_1}\right)$$

$$= -\frac{V_T}{2}\left(\frac{R_F}{R_1}\right)$$

If we add both outputs, we get:

$$V_O = V_{Oi} + V_{Oni}$$

$$= -\frac{V_T}{2}\left(\frac{R_F}{R_1}\right) - \frac{V_T}{2}\left(\frac{R_F}{R_1}\right)$$

$$= -V_T\left(\frac{R_F}{R_1}\right) \tag{2-11}$$

So the signal gain for the balanced input is simply the inverting gain, while the noise gain is very low.

When discussing the noise gain, the internal gain matching of the op-amp was mentioned. There can be a slight difference in gain for inverting and non-inverting signals because of internal component matching.

The manufacturer specifies this difference by using the term *common mode rejection ratio* (CMRR). For the 741 op-amp the minimum specified value is 70 dB, which is a ratio of about 3000. What this means is that if the *same* signal (common mode) is applied to both inputs of the op-amp, the output will be attenuated by a factor of 3000 rather than by the infinite factor (zero output) of the ideal op-amp. This concept can be further clarified by the example shown in Fig. 2-11.

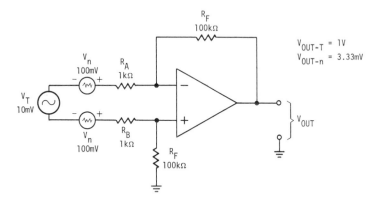

Figure 2-11   Practical Instrumentation Amplifier

With the 10-mV input signal, the output V_{OUT} is:

$$V_{\text{OUT}} = 10 \text{ mV} \left(-\frac{100 \text{ k}\Omega}{1 \text{ k}\Omega} \right)$$

$$= -1 \text{ V}$$

For the noise output (V_{NO}), we determine the output in the same manner as we did the signal, but we divide the result by the rejection ratio. Thus, V_{NO} is:

$$V_{NO} = \frac{100 \text{ mV} \left(\dfrac{100 \text{ k}\Omega}{1 \text{ k}\Omega} \right)}{3000}$$

$$= 3.33 \text{ mV}$$

We can see that even with the non-ideal op-amp, there is considerable reduction of the unwanted noise signal, even though in this example the input noise level was greater than the input signal level.

In order to further show the advantage of the balanced amplifier, let us work through an example using the unbalanced (single-ended) amplifier. The same signal and noise levels are assumed.

The signal output will be:

$$V_{OUT} = 10 \text{ mV} \left(-\frac{100 \text{ k}\Omega}{1 \text{ k}\Omega} \right)$$

$$= 1 \text{ V}$$

and the noise output:

$$V_{NO} = 100 \text{ mV} \left(-\frac{100 \text{ k}\Omega}{1 \text{ k}\Omega} \right)$$

$$= 10 \text{ V}$$

In this case the noise output is 10 times greater than the signal output, so any time we anticipate low signal input levels and high noise input levels, the *balanced input configuration* should be used.

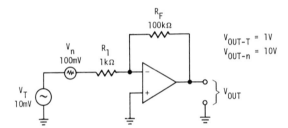

Figure 2-12 Single Ended Amplifier

If the transducer has a high internal resistance, the loading of $R_1 + R_F$ to ground in the balanced configuration could cause a drop in the transducer output voltage. To avoid this problem, the circuit shown in Fig. 2-13 can be used.

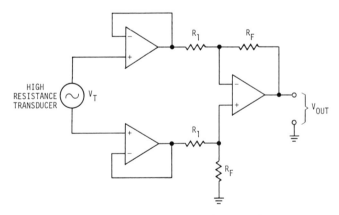

Figure 2-13 High Input Resistance Balanced Circuit

The two input followers represent essentially no load on the transducer, so the full transducer voltage is available. Voltage gain for this circuit is as before—R_F/R. If we wanted to vary the gain of the balanced circuits described, we would have to change both R_1 or R_F resistors by exactly the same amount. This creates a real tracking problem between the two potentiometers that would have to be used. We could gang the potentiometers on the same shaft, but then we have to worry about the linearity of the potentiometers. A simpler solution is to use the circuit shown in Fig. 2-14.

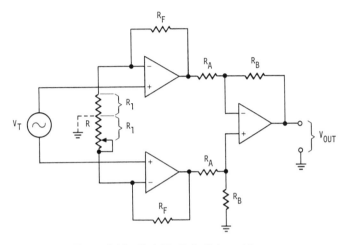

Figure 2-14   Variable Gain Balanced Input

In Fig. 2-14 the transducer feeds into the high impedance of a non-inverting amplifiers. The resistor R varies the gain. When part of this resistor is shorted out, the value of R is reduced. A virtual ground exists at the midpoint of the unshorted portion of R. This means that there would be no voltage difference between this point and ground. The virtual ground point divides the variable resistor R into two parts. The top portion is the R_1 resistor for the upper amplifier and the bottom portion is the R_1 resistor for the lower amplifier. If you break V_1 into two generators as we did before, you will find the overall gain is:

$$A_V = \left(\frac{R_F}{R_1} + 1\right)\left(-\frac{R_B}{R_A}\right) \qquad (2\text{-}12)$$

where again, R_1 is half of the unshorted value of R.

A numerical example will be used to clarify the configuration.

Let: $R_A = 1 \text{ k}\Omega$

$R_B = 50 \text{ k}\Omega$

$R_F = 100 \text{ k}\Omega$

$$R = 5 \text{ k}\Omega \text{ potentiometer}$$

Initially assume that R, the 5-kΩ potentiometer, has full resistance. The overall gain then is:

$$A_V = \left(\frac{100 \text{ k}\Omega}{\dfrac{5 \text{ k}\Omega}{2}} + 1 \right) \left(-\frac{50 \text{ k}\Omega}{1 \text{ k}\Omega} \right)$$

$$= (41)(-50)$$

$$= -2050$$

Reducing R to 1 kΩ gives an overall gain of:

$$A_V = \left(\frac{100 \text{ k}\Omega}{\dfrac{1 \text{ k}\Omega}{2}} + 1 \right) \left(-\frac{50 \text{ k}\Omega}{1 \text{ k}\Omega} \right)$$

$$= (201)(-50)$$

$$= -10{,}050$$

Reducing R to 0 Ω, the gain of the input op-amps becomes open loop, and the overall gain is the open loop gain multiplied by 50.

If we don't need the high impedance of the input inverting amplifiers, the variable gain circuit can be simplified as shown in Fig. 2-15.

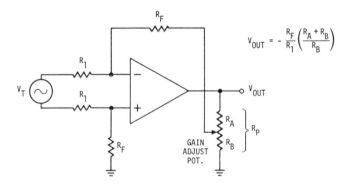

$$V_{OUT} = -\frac{R_F}{R_1} \left(\frac{R_A + R_B}{R_B} \right)$$

Figure 2-15   Simplified Variable Gain Balanced Circuit

The difference between the circuit and the original balanced input circuit described is that the feedback is taken from a voltage divider rather than directly from the output. The divider reduces the amount of negative feedback and therefore increases the gain. Remember, with no feedback we have open loop gain, and with full feedback (follower) we have unity gain. Thus, if the center tap of the potentiometer is moved closer to ground, there is less voltage feedback and the gain is increased. The gain increases by the reciprocal (α) of the attenuation factor, which is:

$$\alpha = \frac{R_A + R_B}{R_B} \qquad (2\text{-}13)$$

The overall gain (A_V) is then:

$$A_V = -\alpha \frac{R_F}{R_1}$$

$$= -\left(\frac{R_A + R_B}{R_B}\right)\frac{R_F}{R_1} \qquad (2\text{-}14)$$

Again, an example will help reinforce this concept.

Let:
$$R_1 = 10 \text{ k}\Omega$$
$$R_F = 100 \text{ k}\Omega$$
$$R_P = 2 \text{ k}\Omega$$

With the potentiometer set such that $R_A = 1.8 \text{ k}\Omega$, $R_B = 0.2 \text{ k}\Omega$, then the gain ($A_V$) is:

$$A_V = \left(\frac{1.8 \text{ k}\Omega + 0.2 \text{ k}\Omega}{0.2 \text{ k}\Omega}\right)\left(\frac{100 \text{ k}\Omega}{10 \text{ k}\Omega}\right) = 100$$

A slight unbalance of resistance seen by the transducer is created by the potentiometer resistance being in series with the feedback resistor. This effect is reduced if R_F is much greater than the potentiometer resistance.

Notice that the gain is increased by the factor α in the above equation. This means that the value of R_F can be reduced for a given gain. If the gain and R_1 requirements of an op-amp circuit cause the value of R_F to become too high (causing noise pickup), then a divider at the output will allow reduction of R_F for the same gain.

2-4 CURRENT BOOSTING

Earlier in Chapter 1 it was stated that the 741 op-amp can provide a maximum output current of 20 mA. But what if we need more current than this? Consider the +5-volt regulator circuit shown in Fig. 2-16.

This circuit provides a constant output voltage, which can be used for low voltage circuits (digital ICs). Without Q_1 the circuit can only provide 20 mA maximum to the load, but adding the transistor increases the current output by the beta (current gain) factor of the transistor. For the 2N5294 shown in the figure, the minimum beta is 30 with emitter currents in the 500 mA range. Multiplying the base

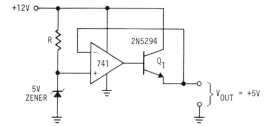

Figure 2-16 $+5$ Volt Regulator Circuit

current of 20 mA (the maximum available from the op-amp) by this beta gives a load (emitter) current capability of 600 mA.

If the load was a 50-Ω resistor, the load current would be:

$$I_L = \frac{5 \text{ V}}{50 \text{ }\Omega}$$

$$= 100 \text{ mA}$$

With this load current the required base current and, therefore, op-amp current is:

$$I_G = \frac{100 \text{ mA}}{30}$$

$$= 3.33 \text{ mA}$$

This shows that the op-amp current is reduced when the load current is reduced, since they are related by the beta factor. (In practice, beta usually decreases when the emitter current is reduced.)

Analyzing the complete circuit operation of Fig. 2-16, the non-inverting input is connected to the 5-V zener and, since the op-amp is in a follower configuration, the 5 V will appear at the emitter output. However, why is the inverting input fed back from the transistor emitter rather than the op-amp output? If we connected the inverting input from the op-amp output, the op-amp output would be at $+5$ V, but the emitter is always about 0.7 V less positive than the base for an NPN transistor. Thus, the output voltage would be only $+4.3$ V rather than $+5.0$ V.

We could try to find a 5.7-V zener which would then give us $+5.0$ output, but there are reasons for not doing this. It will be recalled from transistor theory that the base emitter diode drop changes with temperature at -2.2 mV/$^\circ$C for silicon. If we had a 50°C temperature increase, the new base emitter voltage drop would be:

$$V_{BE} = 0.7 \text{ V} - [(2.2 \text{ mV/}^\circ\text{C})(50^\circ\text{C})]$$

$$= 0.59 \text{ V}$$

Thus, the load voltage would change from $+5$ V to $+5.11$ V.

Including the transistor in the feedback loop by feeding back from its emitter will maintain +5 V at the output, since a difference cannot exist between the inverting and non-inverting terminals of the op-amp. As the temperature increases and the base emitter voltage drops, the op-amp output will drop to compensate for the change. Thus, the op-amp output will go from +5.7 V to +5.59 V to compensate for the transistor change, while the output will remain constant at 5 V.

Another reason for including the transistor in the feedback loop is to lower the output resistance of the circuit. The resistance seen looking back into the transistor emitter is approximately 26 mV/I_E. With a load current of 100 mA we have:

$$R_e = \frac{26 \text{ mV}}{100 \text{ mA}}$$

$$= 0.26 \ \Omega$$

If we take the feedback from the emitter, this resistance is reduced to an insignificant value because it is divided by the loop gain, which is the open loop gain (200,000 for a 741) for a follower since $A_V = 1$.

Op-amps are available that include the power transistor as part of the op-amp. The integrated circuit package has to be larger, and usually it costs more to have a single package rather than an op-amp and a separate transistor.

2-5 INTEGRATOR AND DIFFERENTIATOR

The terms *integrator* and *differentiator* are derived from calculus operations (see Appendix A). In op-amp circuits, an integrator is a circuit that has an output proportional to the product of the input voltage and time. Figure 2-17 shows the input-output relationship.

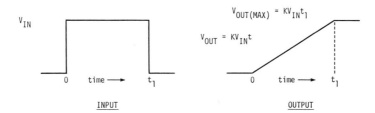

Figure 2-17   Input-Output for an Integrator

Notice that the output voltage reaches a maximum value of $V_{\text{OUT(MAX)}} = KV_{\text{IN}}t_1$ (a constant times the area of the input waveform) and holds at that value even though the input goes to zero.

A practical op-amp integrator is shown in Fig. 2-18.

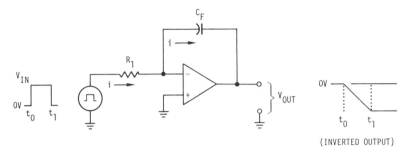

Figure 2-18   Op-Amp Integrator Circuit

If we assume a DC voltage V_{IN} is applied to the inverting input, and recognize that the op-amp inverting input is virtually grounded, then the current through R_1 is:

$$i = \frac{V_{IN}}{R_1}$$

Since no current flows into the op-amp terminals, this current must flow into and charge the feedback capacitor C_F. The voltage builds up on this capacitor in accordance with the relationship:

$$q = V_{C_F} C_F$$

but

$$q = it$$

and

$$V_{C_F} = \frac{it}{C_F}$$

$$V_{C_F} = -\frac{V_{IN}t}{R_1 C_F} \qquad (2\text{-}15)$$

With virtual ground at the op-amp inverting input, the capacitor voltage equals the output voltage ($V_{C_F} = V_{OUT}$). The last equation shows the integrator relationship between input and output, i.e., the output is the product of the input voltage V_{IN} and time (t) multiplied by a constant $1/R_1 C_F$. The minus sign in the equation reflects the inverting configuration. The slew rate of the op-amp must be considered if the output waveform has a steep slope. If V_{IN} varies with time, we would have to resort to calculus to predict the output waveform. (See Appendix A.)

Again looking back to Fig. 2-18, we see that after the positive input pulse has gone ($t > t_1$), the capacitor stays charged and the op-amp output remains at V_{CF}. If the capacitor had no leakage, the output would remain at this value forever. What happens if we now put in a negative pulse of the *same* amplitude and width? The output will ramp back to zero volts! Figure 2-19 shows this input/output relationship.

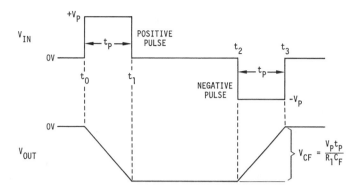

Figure 2-19   Integrator Input/Output Relationship

If the negative pulse was not identical to the positive pulse, the output would not go back to zero, i.e., if the negative pulse was the same width but a greater amplitude, the output would end up at a positive voltage level. This feature of the integrator makes it useful for a pulse comparison circuit.

With a series of positive pulses into an integrator, the output will go negative in a staircase fashion, as shown in Fig. 2-20. We can count the number of pulses in a given time interval by measuring the output voltage.

The output will keep going negative until limited by the op-amp negative power supply voltage. Actually, the limit will be one to two volts more positive than the supply. If it is desired to discharge the capacitor before the next pulse arrives, an FET can be connected across the capacitor terminals and turned on, when required, to provide the low resistance discharge path.

With a sine wave input to an integrator, the output is a cosine wave—or, in other words, we get a precise 90° phase shift between the input and output signals.

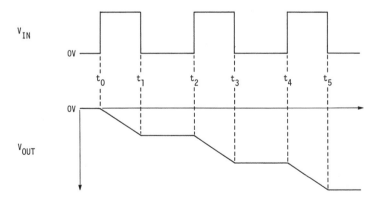

Figure 2-20   Integrator Response to a Pulse Train

However, we must operate at frequencies lower than the slew rate limited frequency.

Op-amp voltage and current offset must also be considered for an integrator. From a DC standpoint, the circuit is open loop. There is no DC feedback through the capacitor (C_F). Offset voltage will be amplified by the open loop gain, causing the op-amp output to saturate. The bias current (V_{B_1}) flowing through R_1 will cause the inverting terminal of the op-amp to go negative, which again will be amplified by the open loop gain. Solutions to these offset problems are to keep the value of R_1 as low as possible and to connect a resistor in parallel with the capacitor to reduce the DC gain. Use of an FET input op-amp will eliminate the bias current problem.

The differentiator circuit has an output that relates to the slope of the input signal, as shown in Fig. 2-21.

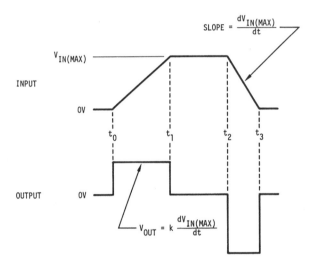

Figure 2-21   Differentiator Waveforms

The only time we have an output from the differentiator is when we have a slope on the input. The greater the slope, the greater the output voltage. Notice in Fig. 2-21 that there is no output from t_1 to t_2 (input slope is zero) and from t_2 to t_3 the output is negative (input has negative slope). The equation for the output is:

$$V_{\text{OUT}} = K \left(\frac{dV_{\text{IN}}}{dt} \right) \tag{2-16}$$

A practical differentiator circuit is shown in Fig. 2-22.

If we initially assume that R_1 has 0 Ω and if we apply a positive step input voltage to the circuit, capacitor C_1 will quickly charge to V_{IN}. All charging current for C_1 will flow through R_F, causing a voltage drop across the resistor. With the positive step input, the current direction is such that V_{OUT} will be negative (left side of R_F is at virtual ground). Thus, V_{OUT} is the result of C_1 charging current. If C_1 is not charging (or discharging), there is no output voltage.

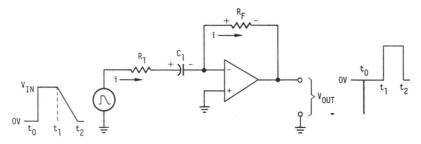

Figure 2-22   Practical Differentiator

The equation for V_{OUT} is:

$$V_{OUT} = - iR_F$$

but

$$q = C_1 V$$

and

$$i = C_1 \left(\frac{dV}{dt} \right)$$

thus

$$V_{OUT} = - R_F C_1 \left(\frac{dV}{dt} \right) \tag{2-17}$$

If dV/dt is a known linear slope, we can use this equation directly. But for any other type of input, we would have to use the expression for the input and take the derivative of the input using the tables for derivatives in Appendix A.

Without R_1 in the circuit, the circuit is unstable (could oscillate), since the intercept of the closed loop response with the open loop is at a slope greater than 20 dB per decade. The minimum value of R_1 for stability is determined from the equation:

$$R_{1(MIN)} = \sqrt{\frac{R_F}{2 \pi C_1 (GBP)}} \tag{2-18}$$

where GBP is the open loop gain bandwidth product for the particular op-amp. Adding R_1 will slow down the charging of C_1 and restrict the differentiator to slower-moving inputs. If R_1 is 100 Ω and C_1 is 0.1 μF, then the maximum charging slope of the capacitor is:

$$\frac{dV_{C_1}}{dt} = \frac{1 \text{ V}}{RC}$$

$$= \frac{1}{10^2 \times 10^{-7}}$$

$$= 0.1 \text{ V}/\mu S$$

In order not to be limited by the capacitor charging, a rule of thumb is to have the input signal slope less by a factor of 10—or in this example: 10 mV/μs. With a pulse or square wave input to the differentiator, the output will be as shown in Fig. 2-23. The differentiator in this case is used to form narrow pulses which can be used as trigger pulses.

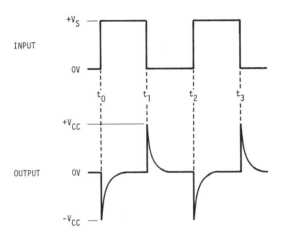

Figure 2-23   Differentiator Waveforms

With R_1 in the circuit, the output waveform is not a sharp spike but has an exponential trailing edge as determined by time constant $R_1 C_1$. Peak output voltage is determined from the maximum initial current through R_F, as follows:

$$V_{\text{OUT}} = iR_F$$

However, with no charge on C_1, the initial current is:

$$i = \frac{V_S}{R_1}$$

thus
$$V_{\text{OUT}} = \frac{V_s R_F}{R_1} \tag{2-19}$$

It can be seen from Fig. 2-23 that in order to maintain a good differentiated output, the minimum pulse width should be greater than 10 times $R_1 C_1$.

2-6 BASIC FILTER CIRCUITS

A filter circuit is used to pass certain sine wave frequencies and reject unwanted frequencies. This is important when we want to detect a signal when noise or unwanted signals are present. If we wish to receive a radio signal, we tune in the station and, in the process, we set up a filter circuit which passes the desired signal and rejects all unwanted stations. It should be understood that noise within the pass range of the filter will not be rejected.

Filter circuits at radio frequencies consist of resonant circuits using inductors and capacitors, but at low frequencies the inductors become too large and expensive, so resistors and capacitors are used. We will discuss only the RC filters in this section. Capacitors and resistors used in these circuits have to be of high quality, which means close tolerance and small value change with temperature and time— otherwise, wanted signals may be rejected and unwanted signals passed.

Actual filter circuits can be of four basic types:

1. Low pass
2. High pass
3. Band pass
4. Band stop (notch)

We will look at each of these RC filters and then describe practical filter circuits using op-amps.

2-7 LOW PASS FILTER

Low pass filters allow low frequencies to pass while rejecting high frequencies, as shown in Fig. 2-24.

The ideal filter graph would have a sharp step between those signals that are passed and those that are not passed—but in the real world, filters have a more gradual transition, as indicated in Fig. 1-26. With the gradual slope it was decided that those signals passed are those with an amplitude greater than 0.707 of the peak value, while those with an amplitude less than 0.707 of the flat portion are considered rejected. The frequency at which the amplitude is reduced to 0.707 is called the *cutoff* frequency.

With V_{IN} a constant voltage from a sine wave generator, we find that as frequency increases, the reactance of the capacitor decreases and when the capacitor reactance approaches the resistor value, the output signal (V_{OUT}) will decrease. At

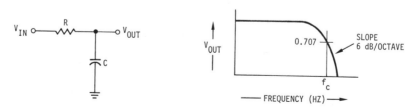

Figure 2-24   Low Pass Filter

the cutoff frequency $V_{OUT} = 0.707\ V_{IN}$ and the resistor value is equal to the reactance of the capacitor or:

$$R = \frac{1}{2\ \pi f_c C}$$

and
$$f_c = \frac{1}{2\ \pi\ RC}$$

At the cutoff frequency, there is a 45° phase shift between the input and output (output lags input). This phase shift increases as we increase the input frequency beyond the cutoff frequency and finally reaches a maximum value of 90°, while the slope of the output voltage versus frequency curve reaches a maximum value of 6 dB per octave (output voltage halves when the frequency is doubled).

When using filters, we must consider the source resistance (r_s) of the circuit that is driving the filter, and also the loading effect of the circuit connected to the output. Resistance of the driving circuit effectively adds to R and will reduce the cutoff frequency of the filter. Loading resistance across the output will appear in parallel with R (Thevenin's equivalent) and will raise the cutoff frequency as well as lower the output voltage at all frequencies.

Ideally, we want to drive a filter from a zero-ohms source and connect the output to an infinite-ohms load. We can get close to the ideal configuration if we drive the filter from an op-amp follower (very low output resistance) and connect the output of the filter to a second op-amp follower (very high input resistance) as shown in Fig. 2-25.

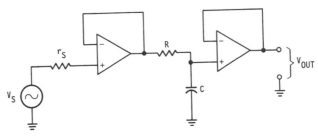

Figure 2-25 Buffered Low Pass Filter

If we would like to have greater rejection of frequencies beyond the cutoff frequency, we can add another low pass filter section. This will give us twice the slope of a single filter section for 12 dB per octave (double the frequency and the amplitude drops to one fourth). A dual-section filter is called a *second-order filter*.

Connecting the second filter at the output of the first filter can create loading problems. This can be prevented by making the resistance of the second filter 100 times larger and the capacitor smaller by a factor of 100. However, we can use the same components for both filter sections by inserting an op-amp between the two filters as shown in Fig. 2-26.

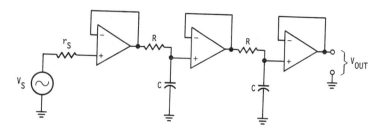

Figure 2-26 Two Section Low Pass Filter

The op-amp input resistance being very high doesn't load the filter, and the low output impedance of the op-amp is a good driving source for the filter.

A simple first-order low pass filter can be generated by simply connecting a capacitor across the feedback resistor R_F of an op-amp amplifier as indicated in Fig. 2-27.

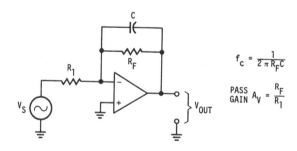

$$f_c = \frac{1}{2\pi R_F C}$$

$$\text{PASS GAIN } A_V = \frac{R_F}{R_1}$$

Figure 2-27 Low Pass Filter—First Order

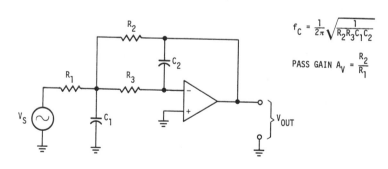

$$f_c = \frac{1}{2\pi}\sqrt{\frac{1}{R_2 R_3 C_1 C_2}}$$

$$\text{PASS GAIN } A_V = \frac{R_2}{R_1}$$

a) MULTIPLE FEEDBACK LOW-PASS FILTER

Figure 2-28 Second Order Low Pass Filters

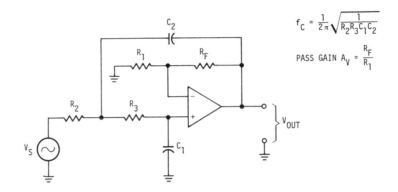

$$f_c = \frac{1}{2\pi}\sqrt{\frac{1}{R_2 R_3 C_1 C_2}}$$

$$\text{PASS GAIN } A_V = \frac{R_F}{R_1}$$

b) VOLTAGE CONTROLLED VOLTAGE SOURCE (VCVS) LOW-PASS FILTER

Figure 2-28 (cont'd.)

This type of filter is called an *active filter* because the op-amp is part of the filter circuit. As indicated, the cutoff frequency is determined by R_F and C, while the gain in the pass band is simply R_F/R_1.

There are several types of second-order (12 dB per octave slope) low pass filters. Two of the more common types, along with their equations, are presented in Fig. 2-28.

2-8 HIGH PASS FILTERS

By reversing the position of the resistor and capacitor in the simple low pass filter, we can create a high pass filter. This type of filter passes all signals *above* the cutoff frequency up to the frequency response of the op-amp. The configuration for the first-order active filter is shown in Fig. 2-29, along with the frequency response.

The capacitor blocks DC, so there is no DC at the output. As the frequency increases, the reactance of the capacitor decreases, which raises the gain. At still higher frequencies the capacitor is essentially a short circuit and the gain is simply R_F/R_1.

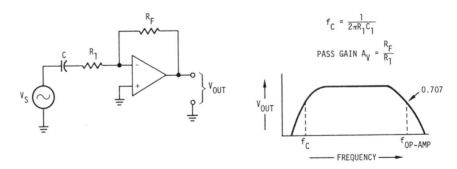

$$f_c = \frac{1}{2\pi R_1 C_1}$$

$$\text{PASS GAIN } A_V = \frac{R_F}{R_1}$$

Figure 2-29   High Pass Active Filter—First Order

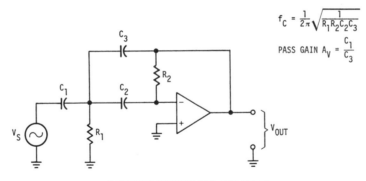

$$f_C = \frac{1}{2\pi}\sqrt{\frac{1}{R_1 R_2 C_2 C_3}}$$

$$\text{PASS GAIN } A_V = \frac{C_1}{C_3}$$

a) MULTIPLE FEEDBACK HIGH-PASS FILTER

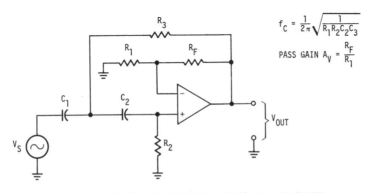

$$f_C = \frac{1}{2\pi}\sqrt{\frac{1}{R_1 R_2 C_2 C_3}}$$

$$\text{PASS GAIN } A_V = \frac{R_F}{R_1}$$

b) VOLTAGE CONTROLLED VOLTAGE SOURCE (VCVS) HIGH-PASS FILTER

Figure 2-30   Second Order High Pass Filters

Circuit diagrams and equations for second-order high pass filters are given in Fig. 2-30.

If you compare these circuits with the low pass versions, it is apparent that the positions of the capacitors and resistors have been interchanged to achieve the high pass feature.

2-9 BAND PASS FILTERS

Passing a narrow band of frequencies about some center frequency is the characteristic of a band pass filter. The simplest form of band pass filter is a combination of a low pass and a high pass filter (see Fig. 2-31).

The filter is defined in terms of the center frequency (f_c) and the bandwidth ($f_2 - f_1$). Q is the ratio of center frequency over the bandwidth. The higher the Q, the narrower the bandwidth.

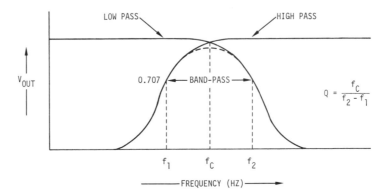

Figure 2-31 Band pass Filter Derived from High and Low Pass

We can pass a signal through an op-amp low pass filter and then through an op-amp high pass filter (or vice versa) to achieve the band pass.

Op-amp circuits using a single op-amp are shown in Fig. 2-32 (see p. 63).

It is obvious from looking at the equations that the band pass circuits are much more complex than the high or low pass circuits. In practice, the required center frequency and bandwidth (or Q) is known, and the capacitor values are first selected because the range of available capacitors is much more limited than resistors. Both the resistors and capacitors used must be close tolerance (1% or less) and stable with temperature. This means metal film resistors and mica, ceramic, polystyrene, or Teflon capacitors.

Op-amps used in filter circuits should have an open loop gain at least 100 times greater than the filter circuit gain at the circuit's highest frequency, i.e., if the center frequency of a band pass filter is at 1 kHz and the gain is 20, the op-amp should have an open loop gain of at least 2000 at 1 kHz. This example requires an op-amp with a minimum open loop gain bandwidth product of 2000×1 kHz or 2 MHz.

2-10 SIGNAL GENERATORS

Signal generators fall into several different categories:

1. Sine wave
2. Square wave
3. Pulse
4. Ramp (triangle wave, sawtooth, etc.)

With op-amps we can generate all of these output waveforms. Let us first look at a sine wave generator, shown in Fig. 2-33 (see p. 64). This generator is a Wien bridge-type oscillator.

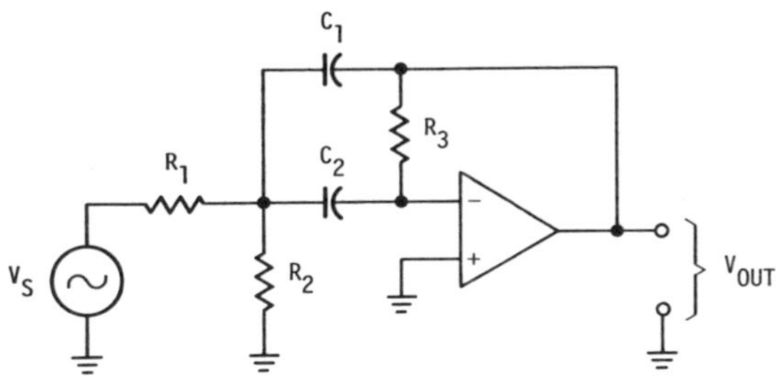

$$f_C = \frac{1}{2\pi}\sqrt{\frac{1}{R_3 C_1 C_2}\left(\frac{1}{R_1}+\frac{1}{R_2}\right)}$$

PASS BAND GAIN
$$A_V = \frac{1}{\left(\dfrac{R_1}{R_3}\right)\left(1+\dfrac{C_1}{C_2}\right)}$$

$$Q = \sqrt{R_3\left(\frac{1}{R_1}+\frac{1}{R_2}\right)}\bigg/\left[\sqrt{\frac{C_1}{C_2}}+\sqrt{\frac{C_2}{C_1}}\right]$$

a) MULTIPLE FEEDBACK BAND-PASS FILTER

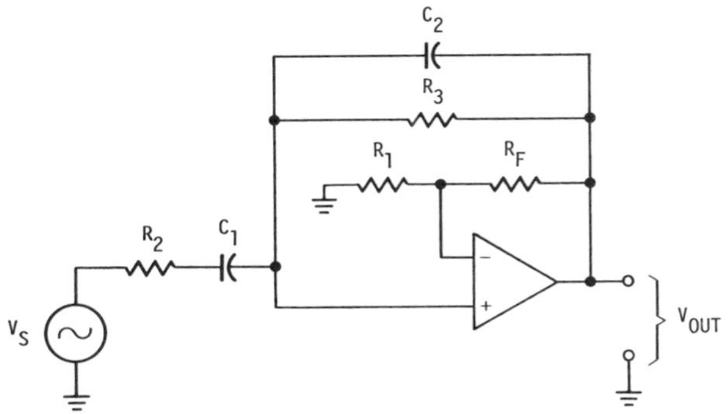

$$f_C = \frac{1}{2\pi}\sqrt{\frac{1}{R_2 R_3 C_1 C_2}}$$

PASS BAND GAIN
$$A_V = \frac{\dfrac{R_F}{R_1}}{\left(1-\dfrac{R_F}{R_1}\right)\left(\dfrac{R_2}{R_3}+\dfrac{C_2}{C_1}\right)+1}$$

$$Q = \frac{1}{\sqrt{\dfrac{R_2 C_1}{R_3 C_2}}+\sqrt{\dfrac{R_3 C_2}{R_2 C_1}}-\left(1-\dfrac{R_F}{R_1}\right)\sqrt{\dfrac{R_3 C_1}{R_3 C_2}}}$$

b) VOLTAGE CONTROLLED VOLTAGE SOURCE (VCVS) BAND-PASS FILTER

Figure 2-32 Voltage Controlled Voltage Source (VCVS) Bandpass Filter

63

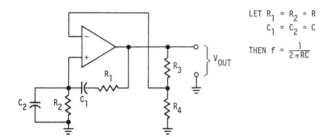

$$\text{LET } R_1 = R_2 = R$$
$$C_1 = C_2 = C$$

$$\text{THEN } f = \frac{1}{2\pi RC}$$

Figure 2-33   Wein Bridge Sine Wave
Oscillator

The conditions for an oscillator to oscillate are that the feedback signal must be in phase (positive feedback) with the input and have a gain of one or greater from output back to input. For the Wien bridge oscillator shown, the positive feedback is achieved by the voltage divider consisting of $R_1 C_1$ and $R_2 C_2$. Maximum positive feedback occurs at the one frequency where there is no phase shift between the input to the divider and output of the divider (non-inverting input). At this frequency we have oscillation.

The negative feedback divider R_3 and R_4 serves to control the amount of positive feedback to maintain a sine wave output. Without the negative feedback the output would be clipped. R_4 is usually chosen to be about one half of R_3. (With $R_1 = R_2$ and $C_1 = C_2$)

Actually, the circuit shown in Fig. 2-33 is quite unstable; any drift will cause the output to either saturate or disappear. Figure 2-34 shows a more practical oscillator circuit, with R_4 made variable and diode-connected across R_3 to increase stability. If the output signal gets too high, the diodes conduct applying heavier negative feedback across R_4.

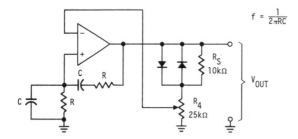

$$f = \frac{1}{2\pi RC}$$

Figure 2-34   Stabilized Wein Bridge
Oscillator

The maximum frequency of oscillation for a given op-amp should be about one thousandth of the open loop gain bandwidth product (1 MHz for a 741 op-amp for a maximum oscillator frequency of 1 kHz).

Square wave generators are simple in design, as indicated in Fig. 2-35.

The output switches between the positive and negative saturation levels (about ± 10 V) and the divider feeds back one tenth of the output to the non-inverting input. Assume the output is at $+10$ V, then $+1$ V is fed back to the non-inverting input. The capacitor (C) will start charging to the positive output voltage via R_1. But when the

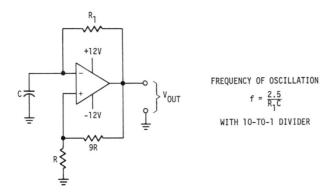

FREQUENCY OF OSCILLATION

$$f = \frac{2.5}{R_1 C}$$

WITH 10-TO-1 DIVIDER

Figure 2-35 Square Wave Generator

capacitor voltage just exceeds (by less than a millivolt) $+1$ V, the inverting input is now more positive than the non-inverting input and the op-amp output swings to -10 V. The voltage at the non-inverting input is now -1 V and the capacitor at the inverting input starts charging toward the -10 V output. As soon as the capacitor voltage is a little more negative than -1 V, the output goes back to $+10$ V and the cycle continues. The waveforms at inverting input, non-inverting input, and the output are shown in Fig. 2-36.

The voltage on the capacitor during charging is

$$e_c = E \left(1 - \varepsilon^{-\frac{t}{RC}} \right) \tag{2-20}$$

but the capacitor starts charging from -1 V, which makes $E = 11$ volts; so

$$e_c = 11 \text{ V} \left(1 - \varepsilon^{-\frac{t}{RC}} \right)$$

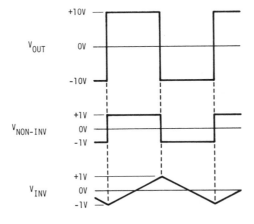

Figure 2-36 Square Wave Generator Waveforms

The inverting input will exceed the non-inverting input when e_C is just greater than $+1$ V. The charging time (t_C) then is determined by making $e_C = 2$ V and substituting in the equation:

$$+2\text{ V} = 11\text{ V} \left(1 - \varepsilon^{-\frac{t}{RC}}\right)$$

Solving for $\varepsilon^{-\frac{t}{RC}}$: $\varepsilon^{-\frac{t}{RC}} = 9/11$

Taking the natural log of both sides

$$-\frac{t_c}{RC} = ln\ 9/11$$

$$t_c \sim 0.2RC$$

but the total period for the square wave is $T = t_C + t_D = 2t_C$ where t_D is the capacitor discharge time and is equal to t_C. The frequency of square wave oscillation is then:

$$f = 1/T = 1/2t_C = 2.5\ RC$$

The square wave output rise and fall times are controlled by the slew rate of the op-amp.

A slight change in the square wave generator circuit converts it into a pulse generator. With reference to Fig. 2-37, a resistor (R_2) and a diode (D) are connected across R_1. Now the positive charging time constant is shorter than the negative charging time constant because when the output is positive the diode conducts, which puts R_2 in parallel with R_1.

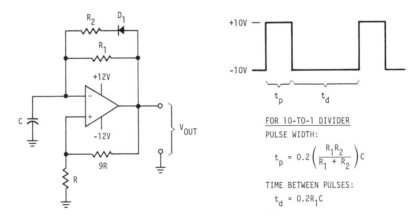

FOR 10-TO-1 DIVIDER
PULSE WIDTH:

$$t_p = 0.2\left(\frac{R_1 R_2}{R_1 + R_2}\right)C$$

TIME BETWEEN PULSES:

$$t_d = 0.2R_1 C$$

Figure 2-37   Positive Pulse Generator

Neglecting the diode drop, the capacitor charges positive with a time constant of $CR_1R_2/(R_1 + R_2)$ and negative with a time constant of R_1C. The positive charging time of the capacitor determines the output pulse width. R_2 can be made variable to control the pulse width and R_1 variable to change the time between pulses (assuming R_2 is less than R_1). Reversing the diode provides a negative output pulse.

Two op-amps can be used to make a circuit that will provide a triangular wave output. Figure 2-38 shows such a circuit.

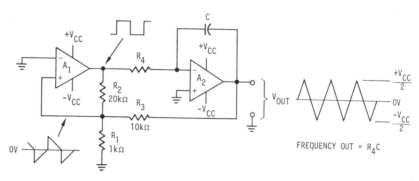

Figure 2-38   Triangle Wave Generator

Assume the output of amplifier A_1 is in positive saturation. This will provide a positive input to the non-inverting terminal via R_2 and R_1. Amplifier A_2 is connected as an "integrator circuit" and with a positive input level to the left of R_4, the output will start to ramp in a negative direction. Because of the divider action of R_3 and R_1, the non-inverting input of A_1 will also start to become less positive. When the non-inverting input of A_1 goes a little below ground, A_1 output will go to negative saturation, forcing the non-inverting input further negative through the R_2 and R_1 divider. The input to the integrator A_2 is now a negative level, and its output starts to ramp positive. Again, the non-inverting input of A_1 follows the output of A_2 and starts going positive. With the non-inverting a little above ground, the output of A_1 shifts back to positive saturation and the oscillation continues. With the values shown for R_1, R_2, and R_3, the output voltage swing will be about one half of the total supply (since R_2 is twice R_3). The frequency of the triangle wave is determined by R_4 and C, and is simply:

$$f = \frac{1}{R_4C} \tag{2-21}$$

The maximum frequency is limited by the slew rate of the op-amp. It is a good practice to keep the maximum triangle wave slope to less than a tenth of the slew rate.

Connecting a diode and resistor in parallel with R_4 converts the triangle wave generator into a sawtooth generator. The modified circuit is shown in Fig. 2-39.

With R_5 much less than R_4, the negative going output of the integrator will have a steeper slope (C charges quicker).

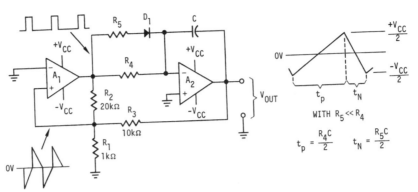

Figure 2-39   Sawtooth Generator

2-11 DIODE CIRCUITS

We have seen the use of the diode in the pulse generator and sawtooth generator circuits. In this section we will consider op-amp circuits where the "main element" is a diode. Let us first review the basic silicon diode curve as shown in Fig. 2-40.

Notice the curve is non-linear in the region below a diode drop of 0.7 V. At 25°C the diode equation is:

$$I_F = I_S \varepsilon^{\frac{V_F}{26 \text{ mV}}} \tag{2-22}$$

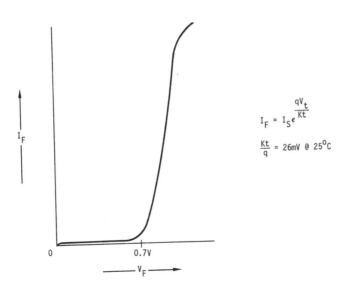

$$I_F = I_S \varepsilon^{\frac{qV_t}{Kt}}$$

$$\frac{Kt}{q} = 26\text{mV} @ 25°C$$

Figure 2-40   Basic Diode Curve

Where I_S is the reverse leakage current (picoamps for a silicon diode). If we solve for the forward voltage drop V_F:

$$V_F = -26 \left(\ln \frac{I_F}{I_S} \right) mV \qquad (2\text{-}23)$$

we see that the voltage drop across the diode is related to the natural logarithm of the current (I_F). This means that the voltage across the diode increases little when the input current is increased. For example, if the current went from 20 μA to 400 μA, the voltage across the diode would go from 0.437 V to 0.514 V (assumes I_S of 1 pA).

This feature of the diode can be used to build a signal compression circuit as shown in Fig. 2-41.

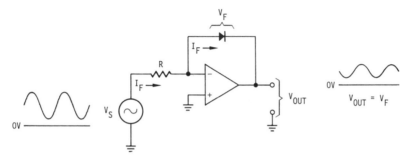

Figure 2-41   Signal Compression Circuit

The output voltage from the op-amp is simply the diode voltage V_F (left side of the diode is virtually grounded), while the input voltage generates the diode current (I_F) through the resistor (R). This circuit can be used to compress speech prior to transmission to allow more efficient operation of the transmitter device (higher average power level). The output voltage peaks at about 0.7 V, so amplification is usually required following the compressor stage.

By reversing the position of the diode and the resistor, we have an "expander" circuit which can be used in a receiver to "restore" compressed speech.

Suppose we wish to convert a small sine wave voltage to DC in order to read the voltage on a conventional meter movement. Our first approach would be to feed the AC through a diode to rectify the signal and then filter the half-wave signal with a capacitor to provide a DC level. But as we saw above, the diode is very non-linear, particularly at low signal levels. There would be very little voltage output from the circuit with low input signals. A solution to the problem is to use the diode with an op-amp as shown in Fig. 2-42.

The circuit uses the basic op-amp characteristic—that the inverting and non-inverting signals must be the same. Therefore, the op-amp output rises for positive input signals to overcome the diode drop. In other words, as the diode drop increases, the op-amp output rises by the same amount so the input and output signal are the

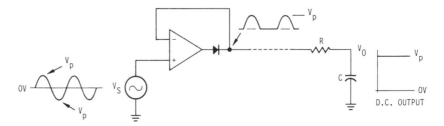

Figure 2-42   Diode Linearizer Circuit

same. The diode cannot pass the negative half cycle, thus the op-amp output goes into negative saturation, but this does not appear at the output. We then have an output signal consisting of the positive alternations of the input signal, and the circuit acts like an ideal diode. Gain can be added to the circuit by simply adding an R_F and R_1 resistor. The RC network at the output is used to convert the alternation to DC.

Connecting the capacitor directly to the output of the op-amp converts the circuit into a positive peak detector. This can be used, for example, to detect the maximum transient voltage on an AC line. The capacitor charges to the peak and maintains this charge over the specified time period.

A further modification of the diode linearizing circuit is the precision absolute value circuit of Fig. 2-43.

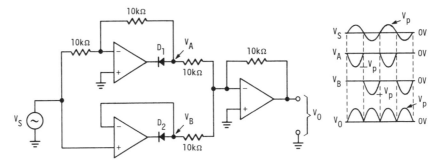

Figure 2-43   Precision Absolute Value Circuit

This circuit provides a positive output regardless of the sign of the input signal—like a full-wave rectifier circuit. The circuit will work for low-level signals because we don't have to worry about voltage drops across the diodes. Changing the polarity of D_1 and D_2 will give a "negative only" output. For ultimate precision, the diodes should be matched along with the 10 kΩ resistors.

2-12 PRACTICAL EXAMPLES

EXAMPLE 2-1

For the circuit shown in Fig. 2-44, find the following with $V_{IN_1} = +1$ V and $V_{IN_2} = +2$ V:

a) Output voltage.

b) Value of R_2.

c) Maximum input voltages.

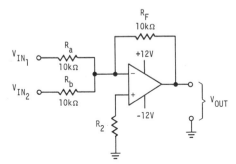

Figure 2-44   Inverting Summing Circuit

Solution:

a) Since the op-amp inverting terminal is at virtual ground, each input signal is amplified as if the other signal were not present. The circuit gain is −1 for each input, and the output signal is:

$$V_{OUT} = 1\text{ V}(-1) + 2\text{ V}(-1)$$

$$= -3\text{ V}$$

b) The value of R_2 is selected for offset compensation and is the parallel combination of the R_F resistor and the two input resistors (R_a and R_b), or:

$$R_2 = \frac{10\text{ k}\Omega}{3}$$

$$= 3.3\text{ k}\Omega$$

c) The maximum output signal from an op-amp is about 1 V less than the supply voltage, or for the example circuit, ±11 V. Thus, the input signals added together must not exceed 11 V.

EXAMPLE 2-2

With the circuit shown in Fig. 2-45, find the following:

a) Determine the output waveform with a +1-V input step at $t = 0$ and $t = 1.5$ s.

b) Determine the maximum error in part (a) due to bias current.

c) Determine the output with a 2-V peak input sine wave.

Solution:

a) Using the following equations:

$$V_{OUT} = \frac{-(1\,V)t}{0.1}$$

$$= -10t \; V$$

at $t = 0$ s,

$$V_{OUT} = 0 \; V$$

at $t = 1.5$ s,

$$V_{OUT} = -15 \; V$$

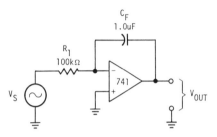

Figure 2-45   Integrator Circuit Example

b) The maximum bias current for a 741 is 600 nA (0.6 μA) and this current is supplied by the positive input voltage. The total current through the 100 kΩ resistor R_1 is:

$$i = \frac{1\;V}{100\;k\Omega} = 10 \; \mu A$$

Therefore, 0.6 μA of this current does not flow into the capacitor as was assumed in part (a). The actual current is 9.4 μA and the output voltage after 1.5 s is:

$$V_{OUT} = - \frac{(15\;V)9.4\;\mu A}{10}$$

$$= - 14.1 \; V$$

Thus, part (a) was in error by 0.9 V. It should be noted that this error can be reduced by reducing the value of R_1.

c) With a sine wave input, we need to use the following equations:

$$V_{OUT} = \frac{1}{R_1 C_F} \int V_{IN} dt$$

$$= 10 \; V \int (2V) \sin 2\,\pi f t dt$$

From the tables in Appendix A:

$$\int (2\;V) \sin 2\,\pi f t dt = - \frac{(2\;V) \cos 2\,\pi f t}{2\,\pi f}$$

$$V_{OUT} = \frac{(1\ V)\cos 2\pi ft}{\pi f}$$

The output is a cosine wave which is 90° out of phase with the input, and the output amplitude decreases as the input frequency increases ($1/f$ relationship).

EXAMPLE 2-3

For the circuit shown in Fig. 2-46, find:

a) The minimum value of R_1.

b) The maximum input square wave frequency.

c) The output with a 2-V peak input sine wave.

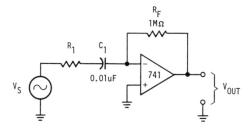

Figure 2-46   Differentiator Circuit Example

Solution:

a) Using the rule that for a 741 the value of R_1 should be not less than 0.001 R:

$$R_1 = 0.001(1\ M\Omega)$$

$$= 1\ k\Omega$$

b) The maximum input frequency can be determined by using the rule of thumb that a minimum pulse width (or half square wave period) of 10 times the RC time constant. Half the square period should be no less than:

$$T/2 = 10\ R_1 C$$

$$T = 20\ R_1 C$$

$$= (20)10^3 10^{-8}\text{s}$$

$$= 0.2\ \text{ms}$$

or $$f = 5000\ \text{Hz}$$

c) The general expression for the output from the differentiator is from the following equation.

$$V_{OUT} = \frac{(R_F C_1)\ dV}{dt}$$

With a sine wave input we use the Table of Differentials in Appendix A:

$$\frac{d(\sin 2\pi ft)}{dt} = 2\pi f \cos 2\pi ft$$

thus

$$V_{OUT} = 2\pi f C_1 \cos 2\pi ft$$

This result indicates that the output is 90° shifted from the input sine wave and the amplitude is increased as the frequency increases.

EXAMPLE 2-4

Design a circuit to amplify a voltage of 10 mV from a 10 kΩ resistance transducer to a level of 1 V. The transducer is located in an electrical noisy environment.

Solution. Since we are dealing with noise pickup, the balanced differential circuit should be used to cancel out common mode noise. We need a circuit gain of 100; we could use the circuit of Fig. 2-11 but the source resistance of 10 kΩ will lower the gain (R_1 would effectively become 6 kΩ). Raising R_1 will make the R_F values too high. A better approach is the circuit of Figure 2-47.

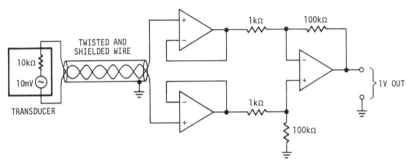

Figure 2-47   Circuit Solution

EXAMPLE 2-5

A burglar alarm circuit uses a frequency varying square wave (siren) to drive a loudspeaker. The power supply is 12 V, and a 10-W, 8-Ω speaker is to be used. Design the output drive circuit starting with the op-amp output stage.

Solution. The speaker will require a current drive of:

$$I = \sqrt{\frac{10\ W}{8\ \Omega}}$$

$$= \quad 1.12\ A$$

If we are using a 741-op-amp, the maximum output current is about 20 mA. We need to boost this current by about 60. This could be done if we connected a transistor at the output of the op-amp, but a beta of 60 is a little high for a single power transistor. So it is better to use two transistors in a Darlington connection as shown in Fig. 2-48.

EXAMPLE 2-6

A band pass filter is required with a center frequency of 10 kHz, a Q of 20, and a gain of 10.

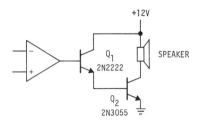

+12V

Q_1
2N2222

SPEAKER

Q_2
2N3055

Figure 2-48   Siren Output Stage

Solution.   If we use the multiple feedback filter, the circuit is shown in Fig. 2-49.

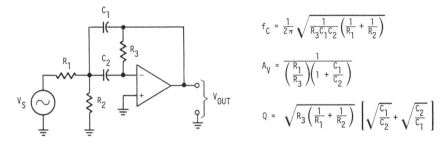

$$f_C = \frac{1}{2\pi} \sqrt{\frac{1}{R_3 C_1 C_2} \left(\frac{1}{R_1} + \frac{1}{R_2} \right)}$$

$$A_V = \frac{1}{\left(\frac{R_1}{R_3} \right) \left(1 + \frac{C_1}{C_2} \right)}$$

$$Q = \sqrt{R_3 \left(\frac{1}{R_1} + \frac{1}{R_2} \right)} \left[\sqrt{\frac{C_1}{C_2}} + \sqrt{\frac{C_2}{C_1}} \right]$$

Figure 2-49   Bandpass Filter Circuit

Let us make $C_1 = C_2 = C$ and pick a value for C of 0.01 μF.
Using the known values in the gain equation and solving for R_3:

$$R_3 = 20 R_1$$

From the equation for Q:

$$R_2 = R_1 / 4$$

Finally, from the equation for center frequency:

$$R_1 = 8 \text{ k}\Omega$$

To complete the selection of parts: $R_1 = 8 \, \Omega$, $R_2 = 2 \text{ k}\Omega$, $R_3 = 160 \text{ k}\Omega$ and $C = 0.01 \, \mu$F.

Using the earlier stated criteria for op-amp selection, we need an op-amp with an open loop gain bandwidth product of $(10 \times 100)(10 \text{ kHz}) = 10 \text{ MHz}$.

EXAMPLE 2-7

Design a circuit to provide a sawtooth waveform for a CRT deflection circuit that will increase in voltage by 10 V in 1 ms and retrace (drop) by 10 V in 10 μs.

We can use the sawtooth generator previously discussed and shown below:

Solution.   Since R_3 is one half of R_2, the output swing will be one half of the saturated output of A_1. With ± 12-V power supplies, the output of A_1 will be ± 10 V (may need diode clamping), and A_2 output will be the required ± 5 V. The positive slope time is:

$$t_p = \frac{R_4 C}{2}$$

If we let $C = 0.01 \ \mu F$, then:

$$R_4 = \frac{(2)10^{-3}S}{10^{-8}F}$$

$$= 200 \ k\Omega$$

For the negative slope:

$$t_N = \frac{R_5 C}{2}$$

or:

$$R_5 = \frac{(2)10^{-5}S}{10^{-8}F}$$

$$= 2 \ k\Omega$$

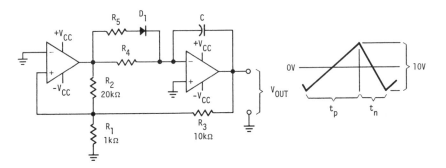

Figure 2-50   Sawtooth Generator Circuit

The diode will create a slight non-linearity for the negative going retrace, but this is not important since the CRT beam is only active during the positive slope.

Things to consider with the op-amp used is a slew rate faster than the required slopes (1 V/μs) and keeping low bias currents through R_4 (FET type op-amp).

2-13 CHAPTER 2—SUMMARY

- We can add any number of signals together using an op-amp summing circuit. The output signal can be the positive sum or the negative sum.

- By modifying the summing circuit, we can generate an op-amp averaging or subtracting circuit.

- Amplifiers used to amplify very weak signals are called instrumentation amplifiers. These amplifiers use low-noise op-amps and balanced circuit configurations to obtain maximum common mode rejection.

- The available current output from an op-amp can be boosted by using a transistor at the output.
- Op-amps can be used to perform the mathematical operations of integration and differentiation. The output from an integrator is the area under the curve of the input signal, while the output of the differentiator relates to the slope of the input signal.
- Because of their high input impedance and low output impedance, op-amps are a good choice for active filter circuits.
- We must consider the frequency response of the op-amp used in filter circuits because it can change the overall filter response.
- The most common types of filters are the multiple feedback and the VCVS versions.
- Using op-amps, we can generate sine wave, square wave, pulse, or ramp-type waveforms. Higher-frequency waveforms are limited by the frequency response or slew rate of the op-amp.
- Diodes can be used with op-amps to provide the ideal diode characteristic (no forward drop) or for signal compression.

2-14 CHAPTER 2—EXERCISE PROBLEMS

1. Determine the output voltage for the inverting summing circuit in Fig. 2-51.

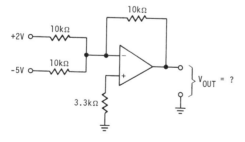

Figure 2-51 Inverting Summing Circuit

2. Show a circuit that will provide an output $V_o = 3A + 2B - C$ where A, B, C are the inputs and can be positive or negative.

3. If the common mode rejection for the instrumentation amplifier circuit in Fig. 2-52 is 94 dB, find the output signal and output noise levels.

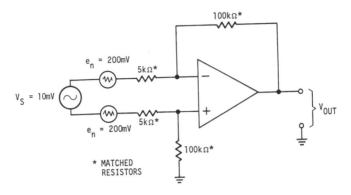

Figure 2-52   Instrumentation Amplifier (Balanced Input)

4. It is required to boost the output current from an op-amp from 50 mA to 1 A. Show the circuit and select components.

5. Determine the voltage at the output of the integrator circuit shown in Fig. 2-53 *after* the second pulse input has passed.

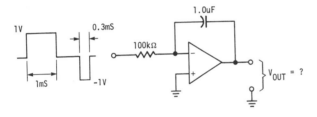

Figure 2-53   Integrator Circuit

6. Determine the output voltage from an op-amp differentiator circuit with $C_1 = 0.1$ nf and $R_F = 100$ kΩ if the slope of the input signal is:

   a) $\dfrac{0.1 \text{ V}}{\mu s}$

   b) $\dfrac{-20 \text{ mV}}{\mu s}$

7. Determine the component values for a low pass VCVS filter with a cutoff frequency of 5 kHz and an inband gain of 20. Also specify the op-amp requirements. Assume $R_2 = R_3$ and $C_1 = C_2 = 10$ nF.

8. Specify the component values for a Wien bridge oscillator circuit to generate a sine wave of 500 Hz. (Let $C_1 = C_2 = 0.01 \mu F$ and $R_3 = 10$ kΩ.)

9. It is desired to have a pulse generator with a pulse width of 50 μs and a frequency of 1 kHz. Show the complete circuit and specify all values. (Assume $R_1 = 10$ kΩ.)

10. A sine wave has a peak-to-peak amplitude of 30 mV. Provide a circuit that will rectify the waveform and provide a positive peak output of 1.2 V.

2-15 CHAPTER 2—LABORATORY EXPERIMENTS

Experiment 2A

Purpose: To demonstrate summing and averaging circuits.

Requirements: Connect up the summing circuit shown in Fig. 2-54, matching up the 20 kΩ resistors within 1%.

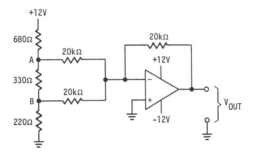

Figure 2-54 Experiment 2A—Summing Circuit

Measure:

1. Voltage from point A to ground and voltage from point B to ground, and compare with (−) sum at the output of the op-amp.
2. Change the value of R_F to convert the circuit into an averaging circuit and verify that the output is the average of A and B.

Experiment 2B

Purpose: To show how a balanced input amplifier rejects common mode noise.

Requirements: Connect up the balanced input circuit shown in Fig. 2-55, using resistors matched to within 1%.

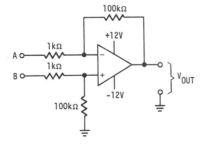

Figure 2-55 Experiment 2B—Balanced Input Circuit

Measure:

1. Isolate the sine wave generator from ground and connect between points A and B and measure the circuit gain. Compare with the computed value.
2. Connect points A and B together and connect the generator between this point and ground (common mode input). Measure the output signal voltage and compare with the computed value.

Experiment 2C

Purpose: To demonstrate operation of a band pass filter circuit.

Requirements: Select components for a VCVS filter circuit with the following specifications:

    a) Center frequency $f_c = 200$ Hz
    b) Bandwidth 20 Hz ($Q = 10$)
    c) Gain $= 10$

Measure:

1. Center frequency
2. Bandwidth
3. Gain

Experiment 2D

Purpose: To show operation of a square wave generator and a pulse generator.

Requirements: Connect up the square wave generator shown in Fig. 2-56, selecting components for a frequency of 500 Hz.

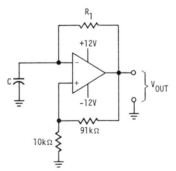

Figure 2-56 Experiment 2D—Square Wave Generator

Measure:

1. The square wave output frequency and compare with the predicted value.
2. Convert the circuit into a *positive* pulse generator by connecting a resistor diode combination across R_1. Select components to make a 10% duty cycle $t_p/(t_d + t_p)$.

3

Linear Voltage Regulators

OBJECTIVES

This chapter will cover the following basic concepts of the voltage regulator:
1. Need for a regulator
2. Fixed and variable regulators
3. Short circuit current protection
4. Power dissipation calculations
5. Ripple rejection
6. Fold-back current limiting
7. Thermal protection
8. Regulation computation
9. Tracking regulators

3-1 INTRODUCTION

Voltage regulators are used to maintain a predetermined output voltage of a power supply despite variations in the input AC line voltage and in the output load. The voltage regulators are inserted between the output of an unregulated power supply and the input to the load as shown in Fig. 3-1. Integrated circuit (IC) voltage regulators considerably simplify power supply design by replacing discrete components such as transistors and vacuum tubes, which were used in early supplies. However, power transistors are still sometimes used in conjunction with the linear voltage regulators to control the higher currents associated with the larger power supplies.

We will consider the two main types of voltage regulators, the linear voltage regulator and the switching voltage regulator. The linear voltage regulator will be described in this chapter and the switching voltage regulator in Chapter 4.

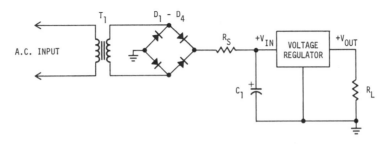

Figure 3-1   Complete Power Supply Circuit Configuration

3-2 LINEAR VOLTAGE REGULATORS

The block diagram for the LM723 variable output voltage regulator is shown in Fig. 3-2. (See Appendix E for detailed specifications.)

The LM723 includes a +7-V reference source which can be used directly, or attenuated, to provide an input to the non-inverting side of an error amplifier. A sample of the power supply output voltage is applied to the inverting input of this amplifier. These two voltages are compared by the error amplifier to produce a voltage that is used to control the current through a series pass transistor. The series pass transistor is connected between the filter capacitor and the load; it provides current to the load resistor. This current through the load resistor develops the output voltage.

Current limiting is achieved by sampling the output current through a resistor connected between the output of the regulator and the load. If the current is too high, the current limiting transistor is turned on, drawing current from the error amplifier, which reduces current to the series pass transistor and prevents a further increase in the load current.

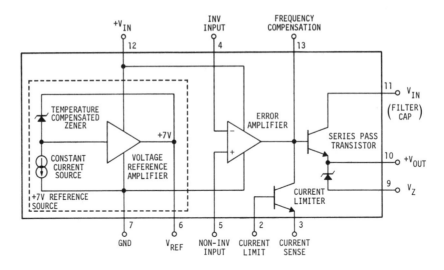

Figure 3-2   LM 723 Block Diagram

3-3 TYPICAL LINEAR VOLTAGE REGULATOR APPLICATIONS

A typical application is shown in Fig. 3-3 for a variable $+2$-VDC to $+7$-VDC variable power supply. The unregulated input voltage is $+12$ V and current limiting is set at 50 mA:

The error amplifier is connected as a voltage follower and, therefore, its output voltage is the same as the voltage at the center of the "output adjust" potentiometer R_2. Resistor R_1, in series with the amplifier inverting input, is used to minimize bias current offset. Ideally, the resistance of R_1 should equal the equivalent resistance of the potentiometer R_2. The current limiting resistor R_{CL} is determined by dividing 0.6 V (current limit transistor turn-on voltage) by the specified limiting current (I_{CL}) of 50 mA.

An example of a variable $+7$-V to 34-V power supply is shown in Fig. 3-4.

This power supply has an external transistor Q_1 to boost the output current up to 400 mA. The error amplifier is connected as a non-inverting amplifier with a gain A_V:

$$A_V = \frac{R_2}{R_1} + 1 \tag{3-1}$$

Resistors R_1 and R_2 are determined by the wiper position of the output adjust potentiometer as indicated in Fig. 3-4. The output voltage V_{OUT} is simply the reference source voltage V_{REF} multiplied by the gain A_V. The required value of the current limiting resistor R_{CL} is:

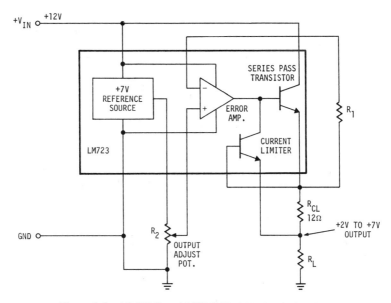

Figure 3-3   +2-VDC to +7-VDC Variable Power Supply

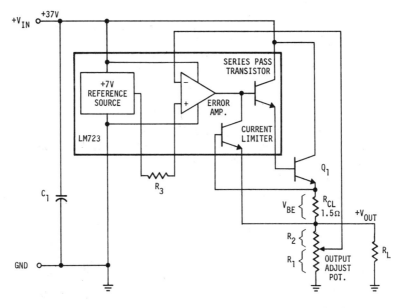

Figure 3-4   High Current Variable Power Supply (+7 VDC to +34 VDC)

$$R_{CL} = \frac{V_{BE}}{I_{CL}} \tag{3-2}$$

$$= \frac{0.6 \text{ V}}{0.4 \text{ A}}$$

$$= 1.5 \ \Omega$$

In order to determine the requirements for the external transistor Q_1, we assume that we want to limit the power dissipation P_{ID} in the regulator to 500 mW, which is the 723 rating at 50°C. If the regulator input voltage is +37 V and the output is shorted to ground (worst case load), then the voltage drop across the regulator is 37 V.

This gives a maximum regulator current $I_{R(max)}$ of:

$$I_R(\text{MAX}) = \frac{P_{ID}(\text{MAX})}{V_R} \tag{3-3}$$

$$= \frac{500 \text{ mW}}{37 \text{ V}}$$

$$= 13.5 \text{ mA}$$

We can determine the minimum beta (B) for Q_1 by assuming that $I_C = I_E = I_{OUT}$ and then:

$$B = \frac{I_{OUT}(\text{MAX})}{I_R(\text{MAX})} \tag{3-4}$$

$$= \frac{400 \text{ mA}}{13.5 \text{ mA}}$$

$$\sim 30$$

Q_1 also requires a voltage rating of at least 37 V. If the beta of the Q_1 is higher than 30 at 400 mA, then less current is drawn from the regulator and the regulator dissipation drops. We shall now consider the requirements for the regulator input power source.

The limits of the input voltage are determined by the internal circuitry of the chip. If the input voltage is raised, the internal dissipation increases since power dissipated P_{ID} is:

$$P_{ID} = (V_{IN} - V_{OUT})I_{OUT} \tag{3-5}$$

where V_{OUT} is the regulator output voltage and I_{OUT} is the load current. Raising the input voltage too high will cause the internal semiconductors of the regulator to break down and the chip will be damaged. If the input voltage is too low, the semiconductors within the regulator have too little voltage across them to operate correctly.

The LM723 regulator has a maximum input voltage of 40 V and a minimum input of 9.5 V. A minimum voltage of 3 V required across the regulator in order to stay in regulation.

3-4 DETERMINING POWER SUPPLY COMPONENT VALUES

The selection of the power supply transformer, diodes, and filter capacitor is determined by the load voltage and current, and the regulator characteristics. This can best be explained by an example. (*Note:* The example is in reference to Fig. 3-1.)

Assume the following requirements:

—Load voltage, V_{OUT}, 10 V $\pm$ 0.5% under line and load variations.

—Load current max, $I_{OUT(max)}$, is 50 mA.

—Max peak-to-peak output ripple voltage (V_R) at full load is 10 mV.

—AC input line voltage variation is $\pm$ 10%.

—Maximum operating temperature 50°C.

To meet these requirements, a LM723 voltage regulator is selected. This device has the following specified characteristics:

1. Line regulation—0.1%
2. Load regulation—0.2%
3. Power supply ripple rejection—74 dB
4. Minimum input-output voltage differential—3 V (min. input voltage 9.5 V)
5. Maximum input voltage differential—38 V
6. Maximum power dissipation 660 mW at 25°C (500 mV at 50°C)

The regulator input voltage at the filter capacitor can range from 10 V plus the minimum output-input voltage differential of 3 V or 13 V, to a maximum value of 40 V. With a load current of 50 mA, plus 4 mA idle current for the regulator, the maximum allowable voltage across the regulator V_{RM} with 500 mW (50°C) power dissipation is:

$$V_{RM} = \frac{P_{ID}}{I_{OUT}} \tag{3-6}$$

$$= \frac{500 \text{ mW}}{54 \text{ mA}}$$

$$= 9.2 \text{ V}$$

With 10 V at the output, the maximum regulator input voltage, then, is 9.2 V + 10 V or 19.2 V.

If a transformer with a secondary voltage of 12 V rms is available, the peak voltage across the secondary is approximately 17 V ($12 \times \sqrt{2}$). The filter capacitor voltage is then 15.8 V (17 V minus two 0.6-V diode drops). Under 10% high input line voltage conditions, we have 17.5 V across the capacitor, and with 10% low line, 14 V. Note that the high line condition of 17.5 V is less than the maximum input of 19.2 V (determined above, from the maximum power dissipation).

In determining the filter capacitor, the minimum voltage on the capacitor for correct regulator action is 13 V. Depending upon the size of the capacitor, a certain amount of ripple voltage ($\triangle V_{p-p}$) appears across it as shown in Fig. 3-5.

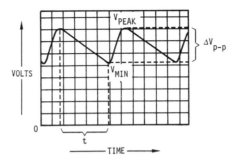

Figure 3-5 Power Supply Output Ripple Voltage Waveform

Under low input AC line voltage conditions, the voltage across the capacitor is 14 V. Remembering that the minimum input to the regulator is 13 V for 10 V out, then the allowable ripple voltage $V = (14 \text{ V} - 13 \text{ V}) = 1$ V. We can determine the required size of the filter capacitor from:

$$C = \frac{I_L t}{\triangle V} \qquad (3\text{-}7)$$

$$= \frac{54 \text{ mA} \times 8 \text{ ms}}{1 \text{ V}}$$

$$= 432 \ \mu\text{F}$$

where I_L is the load current of 50 mA plus idle current of 4 mA, and t is the capacitor discharge time (approximately 8 ms for full-wave rectification and 16 ms for half-wave rectification on a 60-hz line). A practical selection for the capacitor would be 500 μF at 25 V.

The surge resistor R_S in series with the filter capacitor along with the transformer DC resistance R_T, limits the initial turn-on current (I_O). When power is first turned on, the capacitor appears as a short circuit, and the current must be limited to below the diode maximum one cycle surge rating of typically 30 A for a 1-A diode (about the smallest size available). We must assume that the power is turned on at the worst possible time—a high line voltage and at the peak of the input sine wave. Thus:

$$R_S + R_T = R_{\text{TOTAL}} = \frac{V_{\text{IN}}(\text{AC MAX})}{I_o} \qquad (3\text{-}8)$$

$$= \frac{17.5 \text{ V}}{30 \text{ A}}$$

$$= 0.58 \ \Omega$$

If the transformer-measured DC resistance (R_T) is 0.08Ω, then R_S must be 0.5Ω. This completes the selection of components.

The question now is whether the required output characteristics have been met. With 1-V peak-to-peak ripple voltage (V_{CR}) on the capacitor, the 74 dB (ratio 5000) ripple rejection yields:

$$V_R = \frac{V_{CR}}{\text{Ripple Res. Ratio}} \qquad (3\text{-}9)$$

$$= \frac{1 \text{ V}}{5000}$$

$$= 0.2 \text{ mV}$$

This is well within our desired 10-mV value.

Going back to the 723 specifications, a 10% change in the AC line voltage yields a 0.01% change in output voltage (10% × 0.1%). A load change from 5 mA to 50 mA will give a 0.2% change or a total of 0.21%, which again is well within the ± 0.5% or 1% total required.

3-5 FOLD-BACK CURRENT LIMITING

The method of current limiting previously described limits the maximum current that can be drawn from the power supply to that value which develops 0.6 V across the current-sensing resistor R_{CL}. If the output of the power supply is shorted to ground, then the full input voltage is dropped across the regulator, and the regulator power dissipation is the product of this voltage and the short circuit current. With high input voltages, this could result in too much device dissipation and burnout of the regulator.

A technique to avoid this situation is to use a fold-back current limiting technique (shown in Fig. 3-6), which reduces the current drawn from the regulator when a short is applied across the output.

This circuit differs from the fixed current limiter in that the base of the current limit transistor Q_2 is connected to a voltage divider (R_1 and R_2) rather than to the output of the series regulator transistor Q_1.

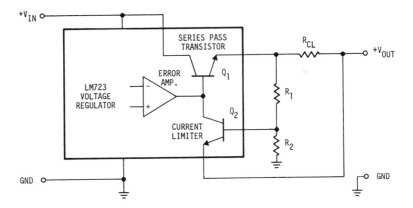

Figure 3-6 Fold-Back Current Limiting Circuit Configuration

Let us consider the circuit action with the voltage divider and the current-sensing resistor R_{SC} shown in Fig. 3-7.

Referring to Fig. 3-7, the voltage from the base-to-emitter of Q_2 is $V_{BE} = (V_{CL} - V_1)$. V_1 is negative because V_B is closer to ground than V_{ST}. If no current is being drawn from the regulator (no load), then $V_{CL} = 0$ and the base-to-emitter junction of Q_2 is reversed biased. With an increasing current from the regulator, the voltage V_{CL} increases and the negative bias on Q_2 is reduced. Further increase in load current will cause V_{CL} to cancel V_1, resulting in a positive voltage across the current limit transistor base-to-emitter junction. When this voltage V_{BE} reaches 0.6 V, Q_2 turns on and the output load current will not be able to increase beyond the value that developed this voltage. This is referred to as the *current limit point.*

The operation described to this point is similar to that of a linear regulator. However, if we decrease the load resistor below the value that caused current limiting, the output voltage will drop. When V_{OUT} drops, V_{ST} also drops and therefore V_1 is reduced. This results in less load current required through R_{CL} in order to maintain V_{BE} at 0.6 V.

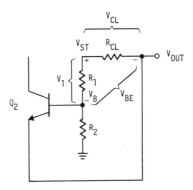

Figure 3-7 Fold-Back Resistor Network

The relationship of the output current and voltage versus load resistance is shown in Fig. 3-8. When the resistance drops below R_{MIN}, the output voltage starts to drop and current foldback starts to occur.

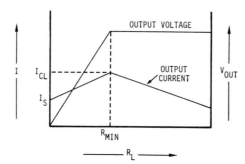

Figure 3-8 Fold-Back Current, Voltage + Load Resistance

With a short across the output, V_{OUT} is zero and $V_{ST} = V_{CL}$. Therefore,

$$V_{BE} = V_{ST} - V_1 \qquad (3\text{-}10)$$

Looking at Fig. 3-7 with a short on the output,

$$V_1 = \frac{V_{ST} R_1}{R_1 + R_2} \qquad (3\text{-}11)$$

and at the current limit point $V_{BE} = 0.6$ V, thus from Eqs. 3-10 and 3-11,

$$0.6 \text{ V} = V_{ST} - \left(\frac{V_{ST} R_1}{R_1 + R_2}\right) \qquad (3\text{-}12)$$

Solving for V_{CL} (since $V_{ST} = V_{CL}$):

$$V_{CL} = 0.6 \text{ V} \left(\frac{R_1 + R_2}{R_2}\right) \qquad (3\text{-}13)$$

but V_{CL} is also equal to $I_S R_{CL}$ where I_S is the load current with a short on the output, substituting this relationship in Eq. 3-13.

$$R_{CL} = \frac{0.6 \text{ V}}{I_S} \left(\frac{R_1 + R_2}{R_2}\right) \qquad (3\text{-}14)$$

Returning to the situation with a load resistor that just causes current limiting, then:

$$V_{BE} = 0.6 \text{ V} = V_{CL} - V_1 \qquad (3\text{-}15)$$

and

$$V_1 = \frac{V_{ST} R_1}{R_1 + R_2} \qquad (3\text{-}16)$$

from Fig.3-7

$$V_{ST} = V_{OUT} + V_{CL} \qquad (3\text{-}17)$$

Substituting Eqs. 3-16 and 3-17 into Eq. 3-15,

$$0.6 \text{ V} = V_{\text{CL}} - \frac{(V_{\text{OUT}} + V_{\text{CL}})R_1}{R_1 + R_2} \tag{3-18}$$

Solving for V_{CL}:

$$V_{\text{CL}} = \frac{0.6 \text{ V}(R_1 + R_2) + V_{\text{OUT}} R_1}{R_2} \tag{3-19}$$

But V_{CL} is also equal to $I_{\text{CL}}R_{\text{CL}}$.

With I_{CL} the required non-shorted current limit (maximum current with no reduction in the output voltage), and substituting this relationship into Eq. 3-19 and solving for I_{CL}:

$$I_{\text{CL}} = \frac{0.6 \text{ V}(R_1 + R_2) + V_{\text{OUT}} R_1}{R_{\text{CL}} R_2} \tag{3-20}$$

Let us now define the fold-back factor K as the ratio of I_{CL} divided by I_S.

$$K = \frac{I_{\text{CL}}}{I_S} \tag{3-21}$$

$$= \frac{0.6 \text{ V}(R_1 + R_2) + V_{\text{OUT}} R_1}{0.6 \text{ V}\left(\dfrac{R_1 + R_2}{R_{\text{CL}} R_2}\right) R_{\text{CL}} R_2}$$

$$= \frac{0.6 \text{ V}(R_1 + R_2) + V_{\text{OUT}} R_1}{0.6 \text{ V}(R_1 + R_2)}$$

$$= 1 + \frac{V_{\text{OUT}} R_1}{0.6 \text{ V} (R_1 + R_2)} \tag{3-22}$$

The fold-back factor K indicates how much the current will be reduced when the output is shorted.

A practical example will reinforce understanding of these concepts. Let us assume the following requirements:

$$V_{\text{OUT}} = 5 \text{ V}$$

$$I_{\text{CL}} = 80 \text{ mA}$$

$$I_S = 40 \text{ mA}$$

These requirements yield a K factor of 2. Let $R_1 = 10 \text{ k}\Omega$, then R_2 can be determined by rearranging Eq. 3-22:

$$R_2 = \frac{V_{\text{OUT}} R_1}{0.6 \text{ V}(K - 1)} - R_1 \tag{3-23}$$

$$= \frac{5\ \text{V} \times 10\ \text{K}}{0.6\ \text{V}} - 10\ \text{k}$$

$$= 73\ \text{k}\Omega$$

The 72-kΩ resistor would be selected for R_2 since it is the closest standard 5% resistor value to the calculated value.

Resistor R_{CL} can be determined from Eq. 3-14 as follows:

$$R_{\text{CL}} = \frac{0.6\text{V}}{I_S} \left(\frac{R_1 + R_2}{R_2} \right)$$

$$= \frac{0.6\text{V}}{40\ \text{mA}} \left(\frac{10\text{k} + 72\text{k}}{72\text{k}} \right)$$

$$= 17\ \Omega$$

A practical selection for R_{CL} would be a 18-Ω resistor.

We can check this result by determining the values of V_{CL} and V_1 at the start of the current limit as follows:

From Eq. 3-16
$$V_1 = \frac{V_{\text{ST}} R_1}{R_1 + R_2}$$

$$= 0.12\ V_{\text{ST}}$$

From Eq. 3-17
$$= 0.12(V_{\text{OUT}} + V_{\text{SC}})$$

and
$$0.6\ \text{V} = V_{\text{SC}} - V_1$$

Substitute for V_1
$$= V_{\text{SC}} - 0.12(V_{\text{OUT}} + V_{\text{SC}})$$

$$= V_{\text{SC}} - 0.12(5\ \text{V} + V_{\text{SC}})$$

$$= V_{\text{SC}} - 0.6\ \text{V} - 0.12\ V_{\text{SC}}$$

$$1.2\ \text{V} = 0.88\ V_{\text{SC}}$$

$$V_{\text{SC}} = 1.36\ \text{V}$$

$$V_1 = 0.12(5\ \text{V} + 1.36\ \text{V})$$

$$= 0.76\ \text{V}$$

The difference between V_{CL} and V_1 is the desired 0.6 V, which turns on the current limit transistor. Note that the voltage at the emitter of the series transistor V_{ST} is 5 V + 1.36 V = 6.36 V.

With a short on the output, the following conditions exist.

$$V_{\text{OUT}} = 0$$

$$V_1 = 0.12\ V_{\text{CL}}$$

$$0.6 \text{ V} = V_{SC} - 0.12 \ V_{CL}$$

$$V_{CL} = 0.72 \text{ V}$$

and $$V_{ST} = 0.72 \text{ V}$$

If the input to the example regulator is $+12$ V, the internal power dissipation of the regulator, with a short on the output and with no fold-back current limiting, would be $P_L = (12 \text{ V} - 0.6 \text{ V})80 \text{ mA} = 920 \text{ mW}$. This is beyond the maximum allowable power dissipation (660 mW at 25°C) for the device. With the fold-back current limiting used, $K = 2$, the dissipation with a shorted output is $P_{FB} = (12 \text{ V} - 0.782)40 \text{ mA} = 449 \text{ mW}$, which is below the 660 mW maximum rating.

3-6 THERMALLY PROTECTED REGULATORS

Some regulators, like the LM140 series, have built-in thermal overload protection circuits which will shut down the regulator if the internal dissipation becomes too great. This prevents the regulator from being damaged from a combination of too large a voltage drop across it and too much current through it. Figure 3-9 shows a simplified typical circuit used within the regulators for overload protection.

Referring to Fig. 3-9, the combination of the voltage reference source and resistors, R_1 and R_2, sets up a fixed voltage on the base of Q_1, which is below the transistor "turn-on" voltage. Q_1 base-to-emitter junction is temperature sensitive (typically -2.2 mV/°C), and when the temperature of the base-to-emitter junction increases, the "turn-on" voltage decreases. Transistor Q_1 is normally physically located within the device next to the output regulator transistor (Q_2) for maximum thermal conduction. As power is dissipated by Q_2, Q_1 heats up and its "turn-on" voltage decreases. When the power dissipated by Q_2 is great enough, Q_1 will turn on and Q_2 base current I_B will be shunted to ground ($I_S = I_B$), causing Q_2 to turn off. This causes the regulator output voltage and current to drop to zero, thus preventing the regulator from being damaged. Thermal turnoff usually occurs when the junction temperature is in the range of 160–190°C.

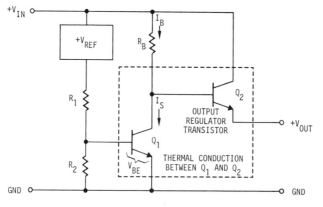

Figure 3-9 Thermal Overload Protection Circuit

Regulators with built-in thermal protection do not need fold-back current limiting for protection against a short circuit on the regulator output. However, conventional current limiting is sometimes still provided to protect against too much dissipation in the load circuits.

Since thermally protected regulators operate on the basis of device temperature and not on the power controlled by the device, the amount of output current is a function of the device's heat sinking. The better the regulator is heat-sink mounted, the greater the possible output current.

3-7 HIGHER CURRENT ADJUSTABLE REGULATORS

The 723 regulator previously described was one of the first variable-voltage-type linear regulators. It has been replaced in new applications requiring more current by the LM117 types (LM217, LM317). The LM 317K can, for example, provide in excess of 2A and dissipate 20 W in its TO-3 package. It also has internal thermal overload protection as described in the previous section.

Output voltage on the LM317 is controlled by the resistors and an internal 1.25-V reference as shown in Fig. 3-10.

Regulation is achieved by the LM317 maintaining the 1.25-V reference voltage across R_1 under all possible normal load conditions. The output voltage is:

$$V_{\text{OUT}} = V_{\text{REF}} + (I_A + I_1)R_2$$

where $I_1 = V_{\text{REF}}/R_1$ and I_A is the bias current from the adjust terminal and has a maximum value of 100 μA.

If we neglect I_A, then:

$$V_{\text{OUT}} = V_{\text{REF}} + V_{\text{REF}} R_2/R_1$$
$$= V_{\text{REF}}(1 + R_2/R_1) \tag{3-24}$$

This equation shows that if we decrease R_2 to zero, the output voltage is simply the internal reference voltage. Increasing R_2 can raise the output voltage to a desired maximum value of about 2 V less than the regulator input voltage.

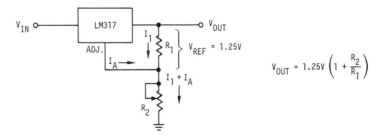

Figure 3-10   Output Voltage Adjust Current for LM 317

With a maximum input voltage specification of 40 V, the maximum output then is 38 V. It is not suggested, however, that the regulator be operated at the absolute maximum input level of 40 V, since an increase in power line voltage caused by transients could destroy the device. A maximum input voltage of 35 V is more conservative and yields an output voltage as high as 33 V.

Let us look at an example using the LM317. Assume we have the following requirements:

Output voltage range 2–20 V

Output current capability 1.0 A

Line voltage variation $= 10\%$

Maximum output ripple 10 mV

For the output voltage of 20 V we need to determine values for R_1 and the potentiometer used for R_2. Let us first find the value of R_1. Remember, we neglected the 100 μA of bias current when we derived Eq. 2-24. In order to do this we should make the current (I_1) through R_1 a hundred times larger than I_A, or:

$$I_1 = 100\,I_A$$
$$= 10\,\text{mA}$$

From Ohm's law:

$$R_1 = V_{\text{REF}}/I_1$$
$$= 1.25\ \text{V}/10\ \text{mA}$$
$$= 125\ \Omega$$

Let us use a standard value of 120 Ω (5%).

V_{REF} has a nominal value of 1.25 V, but it can be anywhere in the range of 1.2 to 1.3 V for a given device. The maximum value of R_2 occurs with a reference voltage of 1.2 V. Using this value and solving for R_2 in Eq. 3-24:

$$R_2 = R_1\left(\frac{V_{\text{OUT}} - V_{\text{REF}}}{V_{\text{REF}}}\right)$$
$$= 120\,\frac{(20 - 1.2)\text{V}}{1.2\ \text{V}}$$
$$= 1880\ \Omega$$

If we pick a standard 2 Ω pot, this value will accommodate the tolerance variations of R_1.

For 20 V out, we must have an input voltage under low line voltage conditions of no less than 22 V. Selecting a standard 20-V (RMS) transformer provides about 24 V (V_{IN}) at the input to the regulator with 10% low line.

$$V_{\text{IN}} = [(\sqrt{2} \times 20 \times 0.9) - 1.4] \text{ V}$$

$$\sim 24 \text{ V (low line)}$$

The 1.4 V in the equation corresponds to the diode drops in a bridge rectifier, and the 0.9 factor reduces the nominal voltage by 10%. Under high line voltage conditions, the voltage input to the regulator will be:

$$V_{\text{IN}} = [(\sqrt{2} \times 20 \times 1.1) - 1.4] \text{ V}$$

$$\sim 30 \text{ V (high line)}$$

This is all right because it is below our chosen maximum value of 35 V.

Since the LM317 is thermally protected, the available current from the device is a function of the:

1. Device internal current limiting
2. Voltage across the regulator (difference between input and output)
3. Heat sinking

With the LM317, the internal current limiting is set at 2.2 A with up to 12 V across the regulator and decreases to 0.4 A with 40 V differential between the input voltage and output voltage. (See Fig. 3-11.)

Setting the power supply for an output voltage of 2 V, the voltage at high line across the regulator is:

$$V_{\text{REF}} = (30 \text{ V} - 2 \text{ V})$$

$$= 28 \text{ V}$$

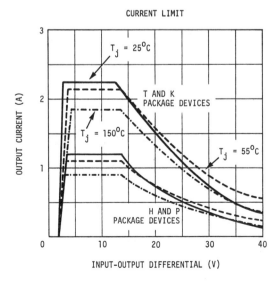

Figure 3-11 Current Limit vs Input-Output Differential

Referring to Fig. 3-11 and assuming a T or K package, we find that the maximum current out with this voltage across the regulator is about 0.75 A. But we wanted a 1.0-amp power supply! This points up an important limitation of variable power supplies—the current capability reduces as we lower the output voltage because the voltage drop across the regulator increases and the dissipation in the regulator increases. Actually, we can get 1.0-A output when the output voltage is increased to 5 V. If we must have the 1.0 A with 2-V output, we have two choices—we can add a current-boosting transistor to the output of the LM317 or we can use a higher power regulator like the LM350. This regulator has a maximum current output of 4.5 A and drops to 1.2 A with 28 V across the device.

Heat sinking affects the current capability of a regulator because the more heat we can remove from the device, the higher the possible internal dissipation. Let us go over some of the basics of heat transfer. It is helpful to relate the flow of heat (watts) to current flow in an electrical circuit. The temperature is analogous to voltage, and thermal resistance is like circuit resistance, but is measured in degrees per watt. From these relationships, we can formulate a Thermal Ohm's Law where temperature (T) is the product of watts (P_D) times thermal resistance (θ).

A thermal equivalent circuit is shown in Fig. 3-12 for a regulator mounted on a heat sink.

In the equivalent circuit, the watts dissipated within the device are handled like a constant current generator—constant because the power loss in the regulator is independent of temperature and is simply the product of the voltage drop across the regulator times the load current.

If we know the thermal resistances of each part of the thermal circuits, shown in Fig. 3-12, along with the surrounding air temperature (T_A) and the maximum junction temperature (T_J), we can determine the maximum allowable power dissipation since:

$$P_D = \frac{T_J - T_A}{\theta_{JC} + \theta_{CS} + \theta_{SA}} \tag{3-25}$$

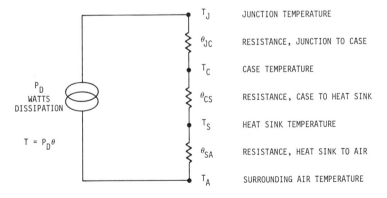

Figure 3-12 Thermal Equivalent Circuit

The LM317 will go into thermal shutdown when the junction temperature exceeds 170°C. But for more reliable device operation, we will restrict the junction temperature to 150°C. Thermal shutdown can still occur with a short circuit on the regulator output. The thermal resistance of the junction to case (θ_{JC}) specified for the LM317 TO-3 package is 3°C/W. If we use a mica insulator coated with silicon grease between the case and the heat sink, the thermal resistance of this mounting is (from Table 3-1) 0.4°C/W. To find the thermal resistance of the heat sink, we initially pick the dimensions and use the nomograph of Fig. 3-13.

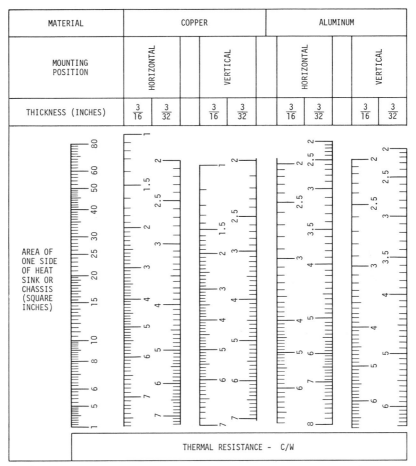

INSTRUCTION FOR USE: SELECT THE HEAT-SINK AREA AT LEFT AND DRAW A HORIZONTAL LINE ACROSS THE CHART FROM THIS VALUE. READ THE VALUE OF MAXIMUM THERMAL RESISTANCE DEPENDING ON THE THICKNESS OF THE MATERIAL, TYPE OF MATERIAL, AND MOUNTING POSITION.

Figure 3-13 Heat Sink Thermal Resistance

Assume a 16-in.^{2} (4-in.-square) heat sink of 3/16-in. aluminum mounted vertically. From the chart the thermal resistance is 3.4°C/W. Finally, if we assume

Table 3-1   Mounting Thermal Resistance θ_{cs} (°C/W)

Package	Direct Contact	With Silicon Grease	With Grease and Mica Insulator
TO-3	0.5–0.7	0.3–0.5	0.4–0.6
TO-202	1.5–2.0	0.9–1.2	1.2–1.7
TO-220	1.0–1.3	0.6–0.8	0.8–1.1

an ambient (room) temperature of 25°C (77°F), we can calculate the maximum power dissipation for the regulator using Eq. 3-25.

$$P_D = \frac{(150 - 25)°C}{(3 + 0.4 + 3.5)°C/W}$$

$$= 18.4 \text{ W}$$

Going back to our original requirement for a regulator that can provide 1-A output with a voltage range of 2 to 20 V, we can determine the load current for the allowable 18.4 W power dissipation by dividing by the maximum voltage drop across the device. The maximum voltage drop of 28 V occurs when we adjust the power supply for 2-V output, at high line (input 30 V). For this condition the maximum allowable current (I_M) is

$$I_M = \frac{18.4 \text{ W}}{28 \text{ V}}$$

$$= 0.66 \text{ A}$$

This value is less than the 0.75-A value determined from the data sheet curve of Fig. 3-11—but the heat sink dimensions used for the manufacturers' data were not specified. It had to have a lower thermal resistance (larger heat sink) than the one used in our example!

The analysis we have conducted on regulator heat sinking applies to any type of semiconductor. All we need to know about the device is the maximum junction temperature and the thermal resistance from the junction to the case.

Commercial heat sinks are available with finned structures and mounting provisions for different devices' packages (TO-3, etc.). Their main advantage is that they are smaller for a given thermal resistance than the sheet metal heat sinks.

3-8 FIXED VOLTAGE REGULATORS

Regulators built into electronic equipment are mostly the fixed voltage type. TTL logic, for example, requires a fixed 5-V supply and op-amps typically ± 12 V.

Table 3-2 Common Voltage Regulators

Type	Output voltage (V)	Max input voltage (V)	Max current (A)	Line regulation (%)	Load regulation (%)	Ripple rejection typical (db)	Max junction temperature (°C)	Thermal resistance junction to case (°C/WATT)
LM340	+5, +12 or +15	+35	2.4	1	1	72	150	4
LM79XX	−5, −12 or −15	−35	2.2	1	1	70	150	5
LM341	+5, +12 or +15	+35	1.0	1	2	71	125	12
LM79MXX	−5, −12 or −15	−35 (−25 for −5 reg.)	0.8	1	2	70	125	12
LM323	+5	+20	4.2	0.5	2	60	125	2
LM723C	+2 to +37	+40	0.15	0.1	0.2	74	125 (NO THERMAL PROTECTION)	140 (DIP PACKAGE)
LM317	+1.2 to +37	+40	2.2	0.2 (+10V OUT)	0.3	65	125	3 (TO-3 PACKAGE)
LM317HV	+1.2 to +57	+60	2.2	0.2 (+10V OUT)	0.3	65	125	3 (TO-3 PACKAGE)
LM337	−1.2 to −37	−40	2.2	0.2 (−10V OUT)	0.5	60	125	3 (TO-3 PACKAGE)
LM337HV	−1.2 to −47	−50	2.2	0.2 (−10V OUT)	0.5	60	125	3 (TO-3 PACKAGE)
LM350	+1.2 to +33	+35	4.5	0.3 (+10V OUT)	0.5	65	125	1.5 (TO-3 PACKAGE)
LM338	+1.2 to +32	+35	9.5	0.3 (+10V OUT)	0.5	60	125	1.0 (TO-3 PACKAGE)

Fixed voltage regulators are obviously simpler to use since they are usually three-terminal devices. The input terminal is connected to the power supply filter capacitor, the common terminal is connected to ground, and the output terminal to the load. The same adjustable regulator features of internal current limiting and thermal overload protection are common to fixed regulators.

A popular fixed regulator is the LM340. This device comes in three versions, with output voltages of $+5$, $+12$, or $+15$ V and maximum output currents from 0.1 to 1.5 A (depending on package configuration). Line regulation ranges from 0.01% for the higher quality units to 1% for the commercial versions. Load regulation can be as low as 0.3% or as high as 1%, and ripple rejection ranges from 54 dB to 80 dB. As always, the higher the price you are willing to pay, the better the regulator.

Table 3-2 lists the key characteristics for several commercial types of fixed regulators, with both positive and negative output voltages. Also listed are some variable regulators.

The maximum current given for these regulators is at a power dissipation level below thermal shutoff, and as we noted in the design example for the LM317 regulator, these current values are considerably lower at the higher power dissipation levels. Note that both fixed and variable negative voltage regulators are available that complement the positive voltage devices. Also see from the table that the maximum output voltages are quite optimistic for the variable regulators when one considers the maximum possible input voltages and allow for line voltage variations. The LM317HV and the LM337HV provide for more conservative designs if higher output voltages are required.

All of the regulators described in this chapter use high gain amplifiers to maintain good regulation. This high gain can cause oscillations to appear at the output of the regulator. These oscillations can be reduced by adding a high frequency bypass capacitor at the input and also at the output of the regulator. A good choice is a 0.01-μF ceramic or mica capacitor connected as close as possible to the regulator terminals.

3-9 TRACKING POWER SUPPLIES

When both positive and negative power supplies are used, it is sometimes required that if one of the supplies has a change in output voltage, the other supply also must change by the same amount. This is called *power supply tracking* and is used to make the operation of circuits using dual supplies more predictable. One circuit used to accomplish tracking is shown in Fig. 3-14.

The input voltages to the regulators are chosen to exceed the minimum voltage drop across the regulators (2.5 V for the LM341 and 1.1 V for the LM79). As described previously, the capacitors must keep the 120 Hz input ripple from dropping the regulator input voltages below the minimum value.

Tracking is accomplished by the op-amp and the resistive divider. If both outputs are exactly 12 V, the center tap of the divider (Pin 2 of op-amp) will be at the

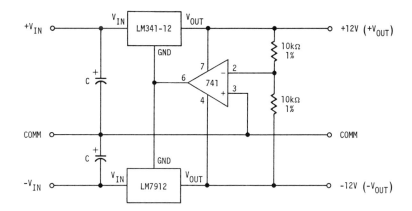

Figure 3-14 Dual Tracking ± 12V Supplies

common (ground) potential. This means there will be no difference in voltage between the op-amp input leads, and the op-amp output will also be at the common potential. Let us now assume that a load on the +12-V supply causes the supply to drop by 100 mV. The voltage across the divider is now 23.9 V rather than 24 V, and the voltage drop across each resistor will be 11.95 V. This means the voltage at the center of the divider will be −50 mV. The inverting input to the op-amp is now more negative than the non-inverting input, so the op-amp output goes positive. Since the positive regulator has a fixed voltage of 11.9 V between the output and common terminal, raising the voltage on the common terminal will cause the output voltage to increase. The positive output voltage will increase until the center tap of the divider is back to the common potential. This occurs when the positive output voltage is increased by 50 mV. To achieve this, the op-amp output had to increase by this amount. But since the common terminal of the negative regulator is also raised, the negative supply decreases by 50 mV. The final result is output voltages of ±11.95 V. We must have a slight difference in the op-amp inputs in order to have +50 mV at the output. However, with an open loop gain of 100,000, the center tap of the resistors (inverting input) will only be $0.5 \mu V$ above the common potential. This represents the error in tracking of the output voltages.

A similar analysis will show that increases or decreases in both the positive and negative supplies will be tracked.

3-10 EXAMPLE PROBLEMS

The concepts presented in this chapter will be further illustrated by additional example problems.

EXAMPLE 3-1
We require a power supply with the following specifications:

a) Output voltage $+9$ V $\pm$ 2%

b) Maximum load current 30 mA

c) Maximum output ripple 5 mV (peak to peak)

d) Regulation 1% total (line and load)

e) Operation to 35°C

Solution. Since we need only 30 mA, a low-current regulator like the LM723 can be used. The complete external circuit using this chip is shown in Fig. 3-15.

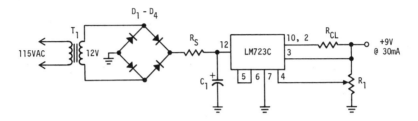

Figure 3-15   9 Volt 30 mA Power Supply

Selecting a transformer with a 12-V secondary voltage gives us a DC voltage (V_C) on the filter capacitor of:

$$V_C = (12 \text{ V} \times \sqrt{2} \text{ V} - 1.4)$$

$$= 15.6 \text{ V}$$

At 10% high line this voltage increases to $(13.2 \text{ V} \times \sqrt{2} - 1.4) = 17.3$ V and at low line decreases to $(10.8 \text{ V} \times \sqrt{2} - 1.4 \text{ V}) = 13.9$ V.

If we set the current limit for 30 mA, then the maximum regulator dissipation (P_M) with a short on the output is:

$$P_M = 17.3 \text{ V} \times 30 \text{ mA}$$

$$= 518 \text{ mW}$$

The maximum specified dissipation at 25°C is 660 mW and 500 mW at 50°C. At our maximum required temperature of 35°C and using straight-line derating from 25°C, we have a maximum dissipation of 616 mW. Since the computed dissipation of 518 mW is less than this value, we do not need to use fold-back current limiting.

The pot R_1 is used to adjust the gain of the error amplifier and set the output voltage at precisely $+9$ V. Fixed resistors could be used in place of R_1 if the output voltage is not critical.

We can now determine the remaining circuit values:

a) Use 1-A diode bridge (30-A surge)

b) $R_S = 17.3 \text{ V}/30 \text{ A} = 0.6 \ \Omega$ (includes transformer resistance)

c) $C_1 = 30 \text{ mA} \times 8 \text{ mA}/1.9 \text{ V} = 126\mu\text{F}$ (1.9 V is the permissible ripple at low line—allowing 3 V across the regulator). Use a standard 150 μF 25 V capacitor.

d) $R_{CL} = 0.6 \text{ V}/30 \text{ mA} = 20$ ohms.

e) Allowing about 1 mA to flow through R_1, we get a resistance of 9 V/1 mA $= 9K$. Use a standard 10-kΩ pot for the output voltage adjustment control. A complete circuit with the selected parts can now be shown in Fig. 3-16. The 0.01-μF capacitors are added to prevent oscillations.

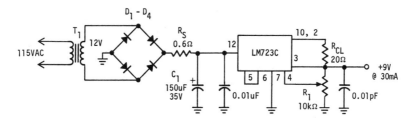

Figure 3-16   9 Volt Power Supply with Circuit Values

The output ripple (V_R) is determined by dividing the input ripple (across C_1) by the ripple rejection from Table 3-2 of 74 dB (5000).

$$V_R = 1.9 \text{ V}/5000$$
$$= 0.38 \text{ mV}$$

which is much less than the specified 5 mV.

Again from Table 3-2, we add the line and load regulation for a maximum change of 0.3% (0.1% + 0.2%), which is better than the required 1%.

EXAMPLE 3-2

We require a power supply for op-amps to provide ± 12 V at 1 A up to a temperature of 50°C. Output ripple should be no greater than 10 mV and the total regulation (line and load) should be less than 3% (0.36 V).

Solution.   A conservative selection would be the LM340-12 for the positive regulator and LM7912 for the negative regulator. Both regulators can provide in excess of the required 1 A and have a combined load-line regulation of 2%.

The circuit schematic is shown below.

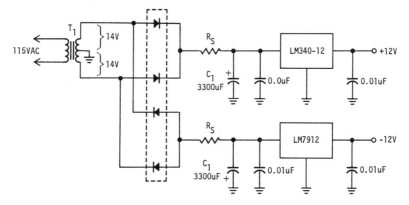

Figure 3-17 $\pm$ 12 Volt, 1 Amp Supply

Note that we need a center-tapped transformer in order to operate off a single winding. A standard 28 V, C.T. 2-A transformer is selected which inputs to the regulator $(14 \text{ V} \times 0.9 \times \sqrt{2} - 0.7)\text{V}$ or 17.1 V at 10% low line voltage and 21 V at 10% high line.

The minimum specified input voltage for the LM340 is 14.6 V and 14 V for the LM7912. This allows a maximum ripple on the filter capacitors of $(17.1 - 14.6)$ V $= 2.5$ V at low line voltage. From this we can size the filter capacitors (C_1):

$$C_1 = (1 \text{ A} \times 8 \text{ ms})/2.5 \text{ V}$$

$$= 3200 \ \mu\text{F}$$

Let us use $3300\text{-}\mu\text{F}$ 25-V capacitors. The regulator output ripple will be the capacitor ripple divided by the ripple rejection of 70 dB (3000) from Table 3-2, or 0.8 mV, which is less than the required 10 mV. For the rectifiers, we can use individual 1-A, 100-V PRV diodes or a 1-A, 100-V PRV bridge rectifier. The required surge resistance is 21 V/30 A $= 0.7 \ \Omega$—which is one half the transformer resistance plus R_S.

This completes the selection of parts for the power supply, but what about the heat sinking of the regulators? The maximum power dissipation occurs in the regulators at high line with 1 A of current flowing and is $(21 - 12)$ V $\times 1$ A or 9 W. With the maximum junction temperature of 150°C for both regulators, and a maximum air temperature of 50°C, the maximum total thermal resistance (θ_T) is:

$$\theta_T = (150 - 50)°\text{C}/9 \text{ W}$$

$$= 11.1°\text{C/W}$$

Thermal resistance junction to case for the 7912 is 5°C/W; and if we mount the regulator to the heat sink with a silicon-grease-covered mica washer, the thermal resistance case to heat sink is 0.4°C/W. This requires a heat sink thermal resistance of:

$$\theta_{HS} = (11.1 - 5 - 0.4)°\text{C/W}$$

$$= 5.7°\text{C/W}$$

If we select a heat sink of 3/16-inch aluminum-mounted horizontal, then, from Fig. 3-13, we read a required area of about 7 in.[2]. For both regulators on the same heat sink, we require 14 in.[2].

EXAMPLE 3-3

Add a +5-V, 1-A TTL supply capability to the power supply in Example 3-2. Output ripple is to be less than 10 mV and combined line-load regulation less than 3%. As before, the maximum input voltage to the regulator is 21 V, while the minimum is 17.1 V.

Solution. For the new power supply we must add two more 1-A rectifiers of 100-V PRV, and the same 0.7-Ω surge resistor used for the $\pm$ 12-V supplies.

The minimum specified input voltage to the regulator is 7.3 V, so the allowable ripple is $(17.1 - 17.3)$ V or 9.8 V. The filter capacitor C_1 is then:

$$C_1 = (1 \text{ A} \times 8 \text{ ms})/9.8 \text{ V}$$

$$= 816 \ \mu\text{F}$$

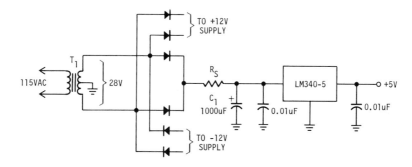

Figure 3.18   Addition of +5V Supply

Use a 1000 μF at 25 V. Notice that the capacitor for this supply is about one third the capacitance of the other two supplies. The reason is with the filter capacitor at 17.1 V (low line) and a minimum regulator input voltage of 7.3 V, we can allow more ripple on the filter capacitor. But do we still meet the output ripple requirements? Dividing the capacitor ripple of 9.8 V by the regulation rejection from Table 3-2 of 72 dB (4000), we have 2.5 mV, which is less than the specified 10 mV.

Using the higher-input voltage with the low voltage supply reduced the size of the filter capacitor, but it does increase the regulator power dissipation P_D:

$$P_D = (21 - 5) \text{ V} \times 1 \text{ A}$$

$$= 16 \text{ W}$$

The required thermal resistance is:

$$\theta_T = (150 - 50)°\text{C}/16 \text{ W}$$
$$= 6.25°\text{C}/\text{W}$$

Which requires a heat sink resistance of:

$$\theta_{HS} = (6.25 - 5 - 0.4)°\text{C}/\text{W}$$

$$= 0.85°\text{C}/\text{W}$$

Referring to Figure 3-13 for the heat sink size, we find we need a 3/16-inch copper heat sink mounted vertically, but with an area of about 70 in.[2]! This is not very practical, so we need to lower the regulator power dissipation. This could be accomplished by using a separate transformer with a lower voltage winding for the +5-V regulator or putting a resistor in series with the regulator input to lower the regulator dissipation. Let us use this latter approach with the series resistor (R_1), as shown in Fig. 3-19.

If we choose a 6-Ω resistor for R_1, this will drop 6 V at 1 A and reduce the dissipation in the regulator from 16 W to $(21 - 5 - 6)$ V $\times$ 1 A or 10 W. Recalculating the total thermal resistance:

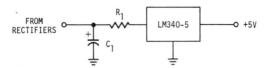

Figure 3-19   Resistor Added to Reduce Regulator Dissipation

$$\theta_T = (150 - 50)°C/10\ W$$
$$= 10°C/W$$

And solving for θ_{HS}:

$$\theta_{HS} = (10 - 5 - 0.4)°C/W$$
$$= 4.6°C/W$$

We now find we can use a more reasonable size heat sink of 9 in.² of 3/16-inch aluminum mounted vertically.

With R_1 added, we must allow 6 V less ripple on the filter capacitor (3.8 V), so the new value is:

$$C_1 = (1\ A + 8\ mS)/3.8\ V$$
$$= 2200\ \mu F$$

The resistor added will dissipate 6 W at full load—so a derated 10-W resistor should be used (resistor power ratings are always at 25°C). Mounting the resistor away from the heat sinks will prevent additional heat loading of the regulators.

3-11 CHAPTER 3—SUMMARY

- A regulator is used between the power supply filter capacitor and the load to keep the load voltage constant with variations in the input line voltage and load current.
- All integrated circuit regulators contain the following internal elements:

  a) A reference voltage
  b) An error amplifier
  c) A means for protecting against excessive current flow

- The LM723 regulator is a low-power adjustable regulator that provides load currents up to 150 mA and has an output voltage range from +2 V to about +35 V (conservative).
- Current limiting can protect both the regulator and the load under normal operating conditions.

- Fold-back current limiting reduces the current as the load resistance decreases to abnormally low values (possible short circuit) and protects the regulator from excessive power dissipation.

- The fold-back factor K is found by dividing the maximum output current, with full output voltage, by the current obtained with a short on the output.

- Regulators with internal thermal overload protection do not require fold-back current limiting.

- The LM317 regulator is a high-current (2.2 A) regulator with an output voltage range from $+2$ V to $+35$ V.

- Since there is less voltage drop across an adjustable regulator when the output voltage is high, the output current can be greater for the same power dissipation.

- Allowable ripple on a filter capacitor is the difference between the low line input voltage and the minimum permissible input voltage to the regulator. The larger the allowable ripple, the smaller the capacitance value.

- Heat sinking a voltage regulator reduces the internal transistor's junction temperature and allows greater output current for a given voltage drop across the regulator.

- We can determine the required heat sink for a regulator by using the Thermal Ohm's Law to find the heat sink thermal resistance.

3-12 CHAPTER 3—EXERCISE PROBLEMS

1. Find the power dissipation in the transistor Q_1 in Fig. 3-4 with a short on the regulator output.

2. If the output voltage of a 723 regulator is 10 V and the input voltage to the regulator is 15 V at low line, find the maximum allowable ripple on the filter capacitor.

3. Design a regulated power supply using an LM 723 voltage regulator to provide an output of $+9$ V at 20 mA(CL point) with an output ripple voltage of less than 10 mV peak to peak. Specify all component values and power ratings, from transformer to output. Use 12V transformer.

4. Design a regulated power supply to provide an output of $+5$ V at 20 mA with less than 25 mV peak-to-peak ripple. Use an LM723 voltage regulator and specify all component values from transformer to output. Use 8V transformer.

5. Provide a circuit and specify parts for a regulated power supply with an output voltage variable from $+7$ to $+28$ V with a maximum load current of 10 mA using a LM723 regulator. Use 24V transformer.

6. A power supply is required with an output of $+15$ V at 150 mA. Input voltage to the regulator is $+20$ V. Specify all component values for the regulator circuit. (Assume maximum temperature of 50°C and use an LM723 regulator.)

7. Design a fold-back current limiting circuit for an LM723 regulator with an input voltage of $+15$ V and output of $+5$ V at 50 mA. Select a K factor to protect the device at 50°C when the output is shorted. (Let $R_1 = 10k\Omega$.)

8. Specify the output transistor and design a fold-back current limiting circuit for a +10-V, 1-A power supply using a LM723. Input voltage to the LM723 regulator is +15 V. Circuit should be able to sustain a continuous short on the output at 50°C. (Let $R_1 = 10$ kΩ.)

9. Using an LM317K regulator, show the complete circuit (from transformer to output) for a variable power supply capable of providing 0.5 A from 1.25 V to 18 V. Maximum output ripple to be 10 mV. Use 18V transformer.

10. Find the area of a 3/16-in. aluminum heat mounted horizontally with grease and insulator for the regulator of Problem 9 if the maximum ambient temperature is 45°C.

11. Select a regulator from Table 3-2 and determine all other power supply components to provide an output voltage of +5 V with 1.5-A current capability. Output ripple to be no greater than 5 mV. (Use a 10-V RMS transformer).

12. Determine the area of a 3/16-in. aluminum heat sink mounted vertically for the regulator in Problem 11 if the maximum ambient temperature is 40°C.

3-13 CHAPTER 3—LABORATORY EXPERIMENTS

Experiment 3A

Purpose: To design and test a LM723 power supply voltage regulator circuit.

Requirements: The power supply is to have an output voltage of +9 V at 20 mA (current limit point) (Reference Problem 3-3).

Measure:

1. Output voltage
2. Line regulation (± 10% line variation)
3. Load regulation (1 mA to 20 mA)
4. Ripple rejection
5. Short circuit current limit
6. Reference voltage
7. Standby current drain (no load)

Compare the above measurements with the data sheet values in Appendix E.

Experiment 3B

Purpose: To design and test a variable voltage power supply using a LM317K regulator.

Requirements: The power supply is to be variable from 1.25 V to 18 V with an output current of 0.5 A (Reference Problem 3-9).

Measure:

1. Voltage range
2. Line regulation ($\pm$10% line variation)
3. Load regulation (1 mA to 500 mA)
4. Ripple rejection
5. Short circuit current limit
6. Standby current drain (no load)

Compare the above measured values with the data sheet values in Appendix E.

4

Switching Regulators

OBJECTIVES

At the completion of this chapter you should know the following concepts:
1. Advantages of a switching regulator
2. Block diagram for a typical regulator
3. Step-down switching regulator operation
4. Step-up switching regulator operation
5. Motor control using a switching regulator

4-1 INTRODUCTION

The linear voltage regulator can be considered as a variable dropping resistor in series with the load. All load current, therefore, flows through it, resulting in a power (I^2R) loss. This power loss can be rather high, and power supply efficiencies of less than 50% are common with the linear-type regulator.

The switching regulator improves power supply efficiency at the expense of increased complexity. An ideal switch has no power loss associated with its operation; there is either full voltage across the switch and no current, or full current and no voltage, thus no power (EI) loss. The switching regulator uses this concept to reduce power loss.

We can use a switching regulator to step up or step down the DC input voltage to the regulator just as a transformer changes AC voltages. Let us first look at the step-down regulator.

4-2 STEP-DOWN SWITCHING POWER SUPPLY

A step-down switching power supply reduces the input voltage to a lower value in a much more efficient manner than a voltage divider.

A fundamental step-down switching power supply circuit is shown in Fig. 4-1. Conventional current is indicated, and a switching regulator controls the external power switching transistor Q_1.

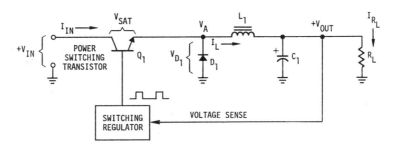

Figure 4-1   Step-Down Switching Regulator

Circuit operation is as follows: Q_1 is a switch that has "on" and "off" times (t_{OFF} and t_{ON}) controlled by the switching regulator. With Q_1 "on," current flows from V_{IN} and is supplied to the load (R_L) via L_1, CR_1 is reverse biased, and C_1 is charging. With Q_1 turned off, L_1 forces V_A negative to keep current flowing in the inductor. CR_1 now starts conducting and the load current flows through CR_1 and L_1. The current flow during the "on" and "off" times is indicated in Fig. 4-2.

The longer the switch is closed, the more energy stored in the inductor and the greater voltage buildup on the output capacitor (C_1), which means a higher output voltage.

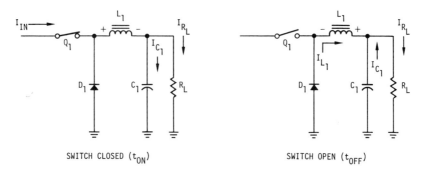

Figure 4-2 Current Flow for Step-Down Switching Regulator

Looking at the equations for the switching regulator which are derived in Appendix B, we see that the output voltage V_{OUT} is:

$$V_{OUT} = V_{IN} \left(\frac{t_{ON}}{t_{OFF} + t_{ON}} \right)$$

$$= V_{IN} \left(\frac{t_{ON}}{T} \right) \tag{4-1}$$

where T is the period of the pulse from the switching regulator and t_{ON} the pulse width.

By increasing the "on" time, we can increase the output voltage, and as expected, if Q_1 is left "on," the output voltage equals the input voltage ($t_{ON} = T$).

The input current I_{IN} is:

$$I_{IN} = I_{OUT} \left(\frac{t_{ON}}{T} \right) \tag{4-2}$$

Therefore, increasing the "on" time causes the input current to increase.

Equations 4-1 and 4-2 resemble the equation for a step-down transformer with the factor t_{ON}/T similar to the turns ratio. The lower the turns ratio, the lower the secondary output voltage.

If we neglect switching losses, efficiency (η) for the step-down switching regulator is approximated by the equation:

$$\eta \sim \frac{V_{OUT}}{V_{OUT} + 1} \tag{4-3}$$

which shows that the higher the output voltage, the greater the efficiency.

Let us now look at the detailed operation of the switching regulator integrated circuit.

4-3 A TYPICAL SWITCHING REGULATOR DEVICE

A typical switching regulator device is the LM3524 regulating pulse width modulator-integrated circuit. Figure 4-3 illustrates the device in block diagram form within a basic switching regulator circuit configuration. Q_A and Q_B act as the switches that control the voltage to the output circuit; the rest of the internal circuitry varies the "on" time of these transistors in response to load voltage changes.

The positive input voltage $+V_{IN}$ is stepped down to the output voltage and is also used to power the internal reference regulator of the LM3524. This reference regulator is a voltage regulator capable of providing a $+5$-VDC voltage (V_{REF}) output at 50 mA to internal or external circuitry. Externally, V_{REF} is normally applied to the non-inverting input of an internal error amplifier via a 2:1 voltage divider consisting of resistors R_3 and R_4. A sample of the switching regulator circuit output voltage ($+V_{OUT}$) is fed back to the inverting input of the error amplifier via a voltage divider consisting of resistors R_1 and R_2. Any error between the divided-down $+V_{OUT}$ and $V_{REF}/2$ at the non-inverting input of the error amplifier is amplified and applied to the inverting input of an internal voltage comparator. A sawtooth signal from an internal free-running oscillator is connected to the comparator non-inverting input.

During circuit operation, if the switching regulator output voltage is too "high," the output voltage of the error amplifier will go "low." With the applied sawtooth waveform, this will cause the output of the comparator to be in the "high" state for a longer period of time, and in the "low" state for a shorter period of time. The output of the voltage comparator is applied as an input to two NOR gates U_1 and U_2. Output of either NOR gate will be "low" when any one of its inputs is "high." Conversely, if all inputs are "low," the outputs are "high." The outputs of the NOR gates, when high, turn "on" the two output transistors Q_A and Q_B.

With the output of the comparator in the "high" state for a longer period of time, the NOR gates outputs are low for a greater period of time. This causes the output transistors to be off longer, resulting in a decrease in the average current from $+V_{IN}$ to the load. Allowing less current to flow to the load decreases the voltage across the output filter capacitor C_1, thus lowering the average output DC voltage. The internal toggle flip-flop Q and $\bar{Q}$ outputs serve to alternately turn on Q_A and Q_B. A short oscillator pulse is coupled to the NOR gates to disable the gates and prevent the output transistors from both being on during the switching time. The oscillator output pulse frequency is determined by timing capacitor C_T and timing resistor R_T.

Power supply current limiting is achieved by sampling the voltage across a resistor in series with the load. Inputs to the internal current limit amplifier are connected across this resistor (R_{CL})—inverting lead to the most positive side of the resistor and non-inverting lead to the most negative side. If the current through this resistor is great enough to develop 200 mV across it, the current limit amplifier output goes "low." This forces the comparator output to go "low," which reduces the output voltage from the regulator as previously described.

Compensation capacitor C_{COMP} and resistor R_{COMP} are used to set, and roll off, the gain of the error amplifier at high frequencies to prevent oscillations from the

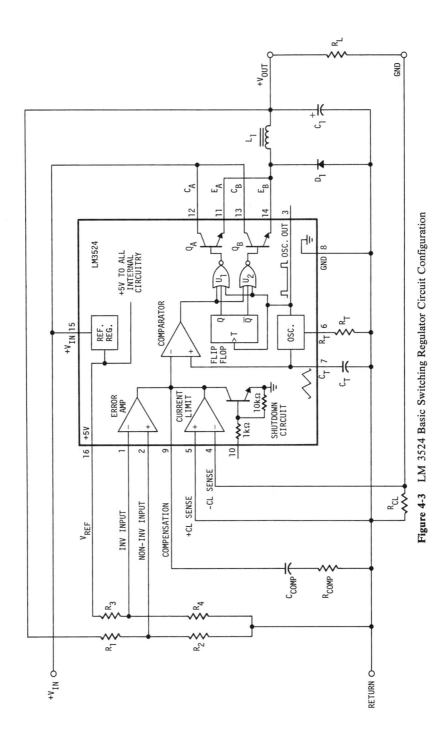

Figure 4-3 LM 3524 Basic Switching Regulator Circuit Configuration

phase shift provided by L_1 in much the same way as the compensation capacitor is used with an op-amp.

A shutdown circuit, consisting of a NPN transistor and a 1 k and 10 kΩ resistor, turns "off" the output transistors via the comparator and NOR gates when a positive biasing voltage is applied to the input. This allows for external deactivation of the switching regulator.

4-4 A PRACTICAL STEP-DOWN SWITCHING REGULATOR

A complete schematic for a step-down switching regulator using the LM3524 is shown in Fig. 4-4. This regulator has an input voltage of 12 V and an output voltage of 5 V, with a load current of 1 A and maximum ripple of 10 mV. The current limit is to be set at 1.3 A and the desired oscillator frequency is 30 KHz.

We shall use these requirements to determine the values of the circuit components. Design equations for the step-down switching regulator (derived from prior equations and Appendix B) are grouped together below for convenience.

$$R_1 = 5 \ k \left(\frac{V_{OUT}}{2.5} - 1 \right) \text{ assumes } R_2, R_3 \text{ and } R_4 = 5 \ k\Omega \qquad (4\text{-}4)$$

$$R_{CL} = \frac{200 \text{ mV}}{I_{OUT}(\text{MAX})} \qquad (4\text{-}5)$$

$$f_{OSC} = \frac{1}{R_T C_T} \qquad (4\text{-}6)$$

$$L_1 = \frac{2.5 \ V_{OUT} \ (V_{IN} - V_{OUT})}{I_{OUT} \ V_{IN} \ f_{OSC}} \qquad (4\text{-}7)$$

$$C_O = \frac{V_{OUT}(V_{IN} - V_{OUT})}{8\Delta V_{OUT} \ V_{IN} \ L_1 \ f^2_{OSC}} \qquad (4\text{-}8)$$

where ΔV_{OUT} is the ripple across the load resistor.

In Fig. 4-4, the output capacitance C_O is the "combination" of C_2 and C_3. Ceramic capacitor C_2 provides a low impedance at high frequencies where the electrolytic capacitor C_3 becomes a high impedance.

To use the equations, we shall again list the requirements.

1. Input voltage $(V_{IN}) = +12$ V
2. Output voltage $(V_{OUT}) = +5$ V
3. Output ripple $(V) = 10$ mV
4. Current limit $(I_{OUT\text{-max.}}) = 1.3$ A
5. Oscillator frequency $(f_{OSC}) = 30$ kHz

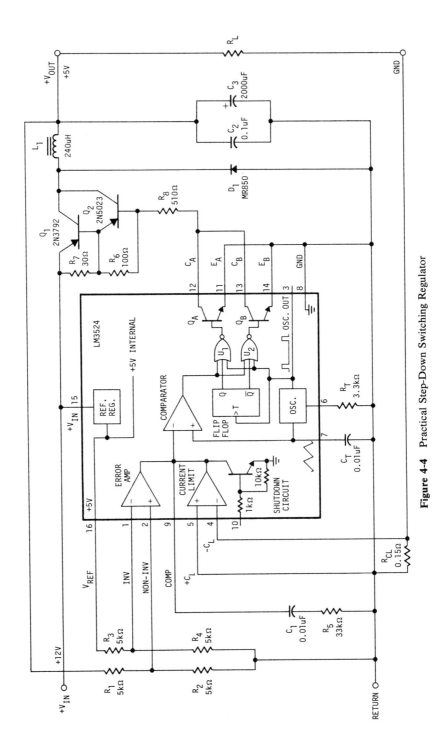

Figure 4-4 Practical Step-Down Switching Regulator

6. Selected value for $C_T = 0.01 \, \mu\text{F}$

The internal $+5$-VDC reference regulator output has been divided in half to bias the error amplifier inputs to the middle of its operating range by using two 5-kΩ resistors.

Transistors Q_1 and Q_2 have been added to boost the output current capability to 1 A. Since each LM3524 output transistor is "on" for half the period, actually 45%, they have been paralleled to provide for a longer possible duty cycle (up to 90%). This allows a lower input voltage for a given output voltage.

Equation 4-4 for R_1 is derived from the equation for a non-inverting amplifier. Solving for R_1,

$$R_1 = 5 \, k \left(\frac{5 \, \text{V}}{2.5} - 1 \right) \Omega \qquad (4\text{-}9)$$

$$= 5 \, k\Omega$$

The current limit resistor R_{CL} is determined from Eq. 4-5:

$$R_{\text{CL}} = \frac{200 \, \text{mV}}{1.3 \, \text{A}}$$

$$= 0.15 \, \Omega$$

With a switching frequency for the regulator of 30 kHz and C_T of 0.01 μF, the value of R_T is:

$$R_T = \frac{1}{30 \, \text{kHz} \times 0.01 \, \mu\text{F}}$$

$$= 3.3 \, k\Omega$$

The value of the inductor (L_1) is:

$$L_1 = \frac{2.5 \times 5 \, \text{V}(12 \, \text{V} - 5 \, \text{V})}{1 \, \text{A} \times 12 \, \text{V} \times 30 \, \text{kHz}}$$

$$= 240 \, \mu\text{H}$$

Now we can determine the value of the output capacitor (C_O):

$$C_O = \frac{5 \, \text{V}(12 \, \text{V} - 5 \, \text{V})}{8 \times 10 \, \text{mV} \times 2.4 \times 10^{-4} \text{H}(3 \times 10^4 \, \text{kHz})^2}$$

$$= 2025 \, \mu\text{F}$$

Notice that the higher the oscillator frequency, the lower the value of L_1 and C_O required. The maximum oscillator frequency is limited by the switching time of Q_1 and Q_2. As a rule of thumb, to maximize efficiency, the minimum oscillator period should be about 100 times longer than the transistor switching time, e.g., if a transistor has a 1-μs switching time, the oscillator period would be 100 μs, which gives a maximum oscillator frequency of 10 kHz. This is due to the power losses in Q_1 and Q_2 during their switching time. The more often they are switched, the higher the losses and the lower the overall circuit efficiency. Another factor that limits high-frequency operation is the core losses in L_1. This requires the use of ferrite or molypermally core materials. Consideration of the transistor and inductor losses usually limits the maximum switching frequency to less than 50 kHz.

At the required switching frequency of 30 kHz, for our example, we have a period of 33.3 μs. A 2N3792 transistor has a switching time of less than 0.3 μs, which satisfies the criteria for keeping the switching losses low. This transistor can also switch currents up to 10 A and therefore would be a good choice for Q_1. The 2N5023 transistor, used for Q_2, switches in less than 0.1 μs. Its dissipation is lower because it only has to supply the base current for Q_1 along with the current for the R_7 stabilizing resistor.

Remember that the compensation components R_5 and C_1 are used to roll off the gain of the error amplifier at high frequencies to avoid oscillations. The components chosen in this example provide a 3-dB down frequency of 480 Hz ($X_C = 33$ k).

4-5 STEP-UP SWITCHING REGULATOR

A step-up switching regulator provides a higher voltage at the output of the regulator than is applied to the input. Figure 4-5 shows the basic circuit for step-up operation.

Referring to Fig. 4-5, transistor Q_1 acts as a switch to control the anode voltage of diode CR_1. During the "on" time (t_{ON}), Q_1 is turned on and current flows through inductor L_1, which stores energy. Diode CR_1 is reversed biased by an anode voltage equal to V_{SAT} (typically less than 0.2 VDC). Output load current I_{RL} is supplied by

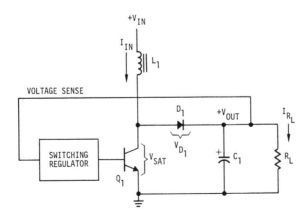

Figure 4-5   Step-Up Switching Regulator

the charge stored in the filter capacitor C_1—since C_1 was charged during the previous "off" time.

At the start of the "off" time (t_{OFF}), Q_1 is turned off and L_1 maintains current flow by raising the anode voltage of CR_1 to the point where CR_1 is forward biased. The output current is now supplied through L_1 and CR_1 to the load resistor. Any charge lost from C_1 during t_{ON} is now replenished. The current flow during the "on" and "off" times is shown in Fig. 4-6.

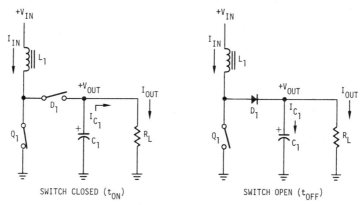

Figure 4-6   Current Flow for Step-Up Switching Regulator

The relationships of voltage, current, and efficiency for the step-up switching regulator are derived in Appendix B and are as follows:

$$V_{OUT} = V_{IN} \left(\frac{T}{t_{OFF}} \right) \tag{4-10}$$

This equation shows that the output voltage is increased if we reduce the Q_1 "off" time (or increase the "on" time).

$$I_{IN} = I_{OUT} \left(\frac{T}{t_{OFF}} \right) \tag{4-11}$$

Equation 4-11 indicates that decreasing the "off" time increases the input current requirements—higher output voltage.

Efficiency for the step-up regulator is related to the "off" time and the output voltage.

$$\eta = \frac{V_{OUT}}{V_{OUT} + \left(\dfrac{T}{t_{OFF}} \right)} \tag{4-12}$$

Equation 4-12 shows that the efficiency increases when a higher output voltage is required and also increases if the "off" time is increased (less "on" time). If we substitute Eq. 4-10 into Eq. 4-12, we get:

$$\eta = \frac{V_{IN}}{V_{IN} + 1} \tag{4-13}$$

Equation 4-13 means that for a given output voltage, raising the input voltage raises the efficiency (the less step-up, the higher the efficiency).

4-6 A PRACTICAL STEP-UP SWITCHING REGULATOR

Fig. 4-7 shows a practical step-up switching regulator using a LM3524 integrated circuit to provide an output of +15 V at 0.5 A with an input of +5 V. The desired oscillator frequency is 30 kHz and the maximum ripple at the output is to be 5 mV.

Transistors Q_1 and Q_2 are added to boost the current capability up to 0.5 A. The network D_1, C_1 forms a slow-start circuit which holds the output of the error amplifier initially low to reduce the duty cycle to a minimum. Without this circuit, the inductor may saturate at turn-on because of the high initial current required to charge the output capacitor to the required output voltage.

Notice that since the input voltage is less than 8 V, the internal reference source cannot be used. This means that there is no line voltage regulation, and that any variations on the 5 V input will appear amplified by a factor of 3 at the output. This variation is reduced by using a voltage reference diode as shown in Fig. 4–8 (see p. 124). (see p. 124)

With the reference voltage of 2.49 V at the non-inverting terminal, V_{OUT} is:

$$V_{OUT} = 2.49 \text{ V} \left(\frac{R_1}{R_2} + 1 \right)$$

and

$$R_1 = R_2 \frac{(V_{OUT} - 2.49 \text{ V})}{2.49 \text{ V}}$$

The design equations, previously derived and in Appendix B for the step-up voltage switching regulator are grouped below:

$$R_1 = \frac{2.4 \text{ K}(V_{OUT} - 2.49 \text{ V})}{2.49 \text{ V}} \quad \text{assumes 2.49 V REF and } R_2 = 2.4 \text{ k}\Omega \tag{4-14}$$

$$f_{OSC} = \frac{1}{R_T C_T} \tag{4-15}$$

$$L_1 = \frac{2.5 \, V_{IN}^2 (V_{OUT} - V_{IN})}{I_{OUT}(V_{OUT})^2 \, f_{OSC}} \quad \text{in henrys} \tag{4-16}$$

$$C_4 = \frac{I_{OUT}(V_{OUT} - V_{IN})}{\Delta V_{OUT} \, f_{OSC} \, V_{OUT}} \quad \text{in farads} \tag{4-17}$$

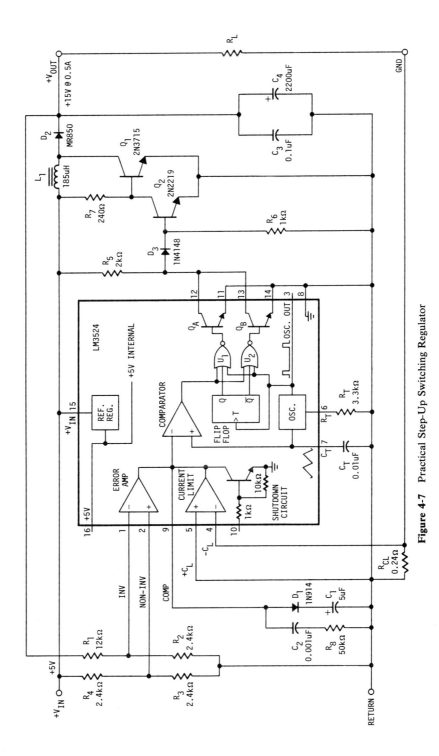

Figure 4-7 Practical Step-Up Switching Regulator

123

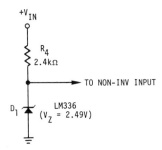

Figure 4-8 External Reference

The values of the output filter components L_1 and C_4 can be determined from these equations. As with the step-down switching regulator, the selection of the oscillator frequency is determined from the switching time of Q_1 and Q_2. We want the maximum switching frequency to reduce the values of L_1 and C_4, but we will lower the efficiency of the regulator if the switching time of the transistors (Q_1 and Q_2) is not much shorter than one hundredth the switching frequency period. Since the 2N3715 transistor used for Q_1 has a switching time less than 1 μs, we can neglect the power loss in this transistor.

Let us again list the circuit requirements:

1. Output current $(I_{OUT}) = 0.5$ A
2. Input voltage $(V_{IN}) = +5$ V
3. Output voltage $(V_{OUT}) = +15$ V
4. Output ripple $(V) = 5$ mV
5. Oscillator frequency $(f_{OSC}) = 30$ kHz
6. Select $C_T = 0.01$ μF

From these requirements we can first determine the value of R_1:

$$R_1 = \frac{2.4 \text{ k}(15 \text{ V} - 2.49 \text{ V})}{2.49 \text{ V}}$$

$$= 12 \text{ k}\Omega$$

With a required oscillator frequency of 30 kHz:

$$R_T = \frac{1}{C_T f_{OSC}}$$

$$= \frac{1}{10^{-8}\text{F} \times 3 \times 10^4 \text{Hz}}$$

$$= 3.3 \text{ K}$$

For the inductor L_1:

$$L_1 = \frac{2.5(5 \text{ V})^2(15 \text{ V} - 5 \text{ V})}{0.5 \text{ A}(15 \text{ V})^2 \, 3 \times 10^4 \text{Hz}}$$

$$= 185 \ \mu\text{H}$$

Using this value of L_1, we can now determine the value for C_4:

$$C_4 = \frac{0.5 \text{ A} \times (15 \text{ V} - 5 \text{ V})}{5 \times 10^{-3} \text{ V} \times 3 \times 10^4 \text{ Hz} \times 15 \text{ V}}$$

$$= 2200 \ \mu\text{F}$$

This completes the selection of components for the step-up regulator. Let us now look at a different application of a switching regulator—that of motor speed control.

4-7 MOTOR SPEED CONTROL

We can control the speed of a DC motor by simply putting a resistor in series with the motor. Increasing the resistance will lower the motor current and cause the motor to slow down. This form of speed control, however, has the same disadvantage as the linear regulator: There is power loss in the controlling element (resistor). For battery-powered motors, we need to minimize losses, since any lost power shortens the time between recharging. Using a switching regulator as the control element considerably reduces losses.

For large battery-powered motors in such applications as fork-lift trucks and people movers, a silicon controlled rectifier (SCR) controller is used. A simplified schematic is shown in Fig. 4-9.

It will be recalled that an SCR turns on and effectively becomes a short circuit when a trigger signal is applied to the gate, so it behaves like a switch and is shown as such in Fig. 4-9.

When the SCR is turned on, current flows through the motor windings and the motor rotates. The SCR is now turned off, and the interruption of current causes the

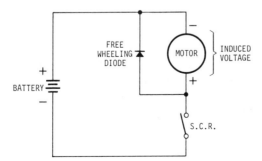

Figure 4-9 S.C.R. Speed Control

energy stored in the field of the motor windings to induce a voltage across the motor which is positive at the SCR end and negative at the positive battery terminal. This induced voltage forces current through the freewheeling diode, and motor current continues to flow even with the switch (SCR) open. What happens then is that when the SCR is on we have motor current flow from the positive terminal of the battery, through the motor and the SCR, and back to the negative terminal of the battery. When the SCR is turned off, motor current flows back around through the freewheeling diode. The net result is that current flows *continuously* through the motor. But how do we control the speed? The average current in the motor is a function of how long the SCR is turned on, because the longer current "on" time allows more energy buildup in the magnetic field. This results in greater current flow when the field collapses. Speed control is achieved by varying the "on" time of the SCR or by varying how often we turn on the SCR.

4-8 A MOTOR CONTROL REGULATOR

At low motor voltages and currents, we can use a switching regulator for speed control. A possible regulator for this application is the LH1605, which can output voltages from 3 to 30 V and currents up to 5 A. The internal block diagram for this regulator is shown in Fig. 4-10.

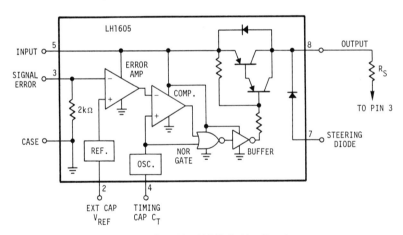

Figure 4-10   LH 1605 Switching Regulator

Internal operation of the LH1605 is very similar to that of the LM3524. Pulse width modulation of the output results from a comparison of the error input signal and a sawtooth waveform from the oscillator. The output voltage is selected by a single resistor (R_S) in accordance with the equation

$$V_{OUT} = \frac{2.5 \text{ V}(R_S + 2 \text{ k})}{2 \text{ k}}$$

(4-18)

Oscillation frequency is determined by the capacitor C_T:

$$f_{OSC} = \frac{1}{4 \times 10^4 \, C_T} \qquad (4\text{-}19)$$

The other equations required are the same as provided earlier for the step-up and step-down power supply circuits.

Since the power transistors and steering diode (freewheeling) are included in the regulator chip, the external circuit using the LN1605 is considerably simpler than that of the LM3524.

Let us look at a motor speed control circuit using this device. Assume the following design requirements:

$$V_{IN} = 24 \text{ V}$$

$$V_{MOTOR} = 12 \text{ V}$$

$$I_{MAX} = 2 \text{ A (current limit)}$$

$$\text{Motor Speed} = 1500 \text{ RPM}$$

The motor control circuit is shown in Fig. 4-11.

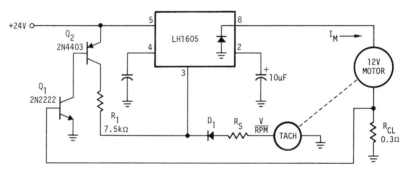

Figure 4-11   Motor Speed Regulator

The prime difference between this circuit and the power supply circuits previously examined is that the required output is not a voltage but current (I_M). Filtering is accomplished by the inductance of motor rather than a combination of inductance and capacitance.

Since we wish to control speed, the tachometer (DC generator) is used to provide a DC feedback voltage proportional to the motor speed. We can determine the value of R_S by knowing the output voltage from the tach at the required speed. Let us assume this value is 5 V at the required 1500 RPM, then rewriting equation (4-18):

$$R_S = \frac{2 \text{ k}(5 \text{ V} - 2.5 \text{ V})}{2.5 \text{ V}}$$

$$= 2 \, k\Omega$$

The control system will now maintain the speed by insuring that the tach output stays at 5 V—even with mechanical load changes on the motor.

If we refer back to Fig. 4-10, we find that there is no provision for an overcurrent limit on the regulator. But what if the motor stalls? The tach output will drop to zero and the regulator will turn full on to supply current to the motor. This could damage both the regulator and the motor. A current limit circuit is included which consists of R_{CL}, R_1, Q_1, Q_2, and D_1. If the motor current reaches 2 A, the voltage drop across R_{CL} is 0.6 V, which will turn on Q_1; this causes Q_2 to conduct. With Q_2 in saturation, the top end of R_1 is close to 24 V and the error amp input rises toward +5 V because of the internal 2 K resistor (D_1 blocks the loading of the tach). Having +5 V at the error amp input causes the regulator output transistor to approach cutoff, which prevents further increase in motor current. In actual operation, the voltage at pin 3 is just enough to allow 2 A of motor current.

4-9 POWER DISSIPATION

Thermal considerations for the switching regulator discussed in this chapter are the same as for the linear regulators. However, the power dissipation (P_D) is considerably lower than the linear regulator and is determined from the product of the internal switching transistor "on" voltage drop (V_S), current (I_S), and the duty cycle (t_{ON}/T); or

$$P_D = \frac{V_S I_S t_{ON}}{T} \tag{4-20}$$

The worst-case duty cycle (highest value) should be used in this equation.

4-10 CHAPTER 4—SUMMARY

- There is less power loss in a switching regulator than in a linear regulator because the control device is either full on or off.
- Switching regulators use an inductor to change the input DC voltage to the regulator to a higher or lower output voltage—just like a transformer does with AC.
- The combination of an inductor, diode, and capacitor converts the chopped regulator output voltage into DC.
- The LM3524 regulator converts an error voltage into a pulse width modulated (PWM) output signal. If the input error voltage indicates that the output voltage is too low, the output pulse will widen.
- Operating the LM3524 at a higher switching frequency reduces the size of the inductor and capacitor. However, too high a frequency can lower the efficiency because of transistor switching losses and inductor core losses.

- The efficiency of a switching regulator power supply increases as the output voltage is raised.
- Motor speed control with minimum power loss can be achieved with a switching regulator.
- The LM1605 switching regulator is a higher current device, but it does not have the internal protection features of the LM3524.

4-11 CHAPTER 4—EXERCISE PROBLEMS

1. If the output transistor for a switching regulator switches in 0.5 μs, determine the maximum switching oscillator frequency.
2. The input voltage to a step-down switching regulator is 12 V. If an external switching transistor is on 40% of the time, find the output voltage.
3. Find the output voltage from a step-up switching regulator if the conditions are the same as Problem 2.
4. Calculate the efficiency of a switching regulator with 5 V input and 15 V output. Neglect switching losses.
5. Select the value of the R_S resistor in order to have an output voltage from the LH1605 regulator of 7.5 V.
6. Compute the power dissipation for the internal switching transistor if the "on" voltage is 0.2 V, the current 2 A and the duty cycle 60%.
7. With reference to Fig. 4-9, find the average motor current if the SCR current is 10 A and the SCR is on 40% of the time.
8. Determine the relative efficiency of a switching regulator and a linear regulator—both with input voltage of 25 V and 15 V output. Assume the same load current and neglect switching losses.
9. With reference to Fig. 4-4, determine the value of R_1 if the output voltage is 10 V. Determine R_T and C_T if the frequency of oscillation is 10 kHz.
10. Find the value of the filter inductor and filter capacitor for a switching regulator with following characteristics:

$$V_{\text{in}} = 20 \text{ V}$$

$$V_{\text{out}} = 10 \text{ V}$$

$$I_{\text{out}} = 100 \text{ mA}$$

$$f_{\text{osc}} = 10 \text{ kHz}$$

$$\text{Max ripple} = 10 \text{ mV}$$

$$\text{Current limit} = 140 \text{ mA}$$

11. Design a switching power supply to meet the following requirements:

$$V_{\text{in}} = 10 \text{ V}$$

$$V_{out} = 15 \text{ V}$$
$$I_{out} = 1 \text{ A}$$
$$f_{osc} = 20 \text{ kHz}$$
$$\text{Max ripple} = 5 \text{ mV}$$
$$\text{Current limit} = 1.4 \text{ A}$$

Specify all component values.

12. It is required to maintain the speed of a 12-V DC motor at 1000 RPM using a switching regulator. A tach is available with an output of 1 V/100 RPM and the motor draws 2 A of current at the maximum load. If the input voltage is 16 V, specify all required component values. Restrict the maximum output current to 4 A.

4-12 CHAPTER 4—LABORATORY EXPERIMENTS

Experiment 4A

Purpose: To design and test a step-down switching regulator circuit.

Requirements: Power supply to have an output voltage of 14 V at 0.5 A. Input voltage to be 16 V DC, and the switching frequency of the regulator to be around 30 kHz. Output ripple is to be less than 100 mV peak to peak and the current limit to be set at 0.8 A.

Measure:

1. Output voltage
2. Load regulation
3. Output ripple
4. Current limit point
5. Efficiency

Compare with predicted values.

Experiment 4B

Purpose: To design and test a step-up switching regulator circuit.

Requirements: The output voltage is to be 18 V and the maximum current limit set at 80 mA. Input voltage of 12 V should be used and a switching frequency of 20 kHz. Maximum ripple to be 50 mV peak to peak.

Measure:

1. Output voltage
2. Load regulator
3. Output ripple
4. Current limit point
5. Efficiency

Compare with predicted values.

5

Voltage References and Current Sources

OBJECTIVES

At the completion of this chapter you should know the following:
1. Purpose of a voltage reference
2. Advantages of a voltage reference over a zener diode
3. Applications of a voltage reference
4. Generation of a current source
5. Applications for a current source

5-1 INTRODUCTION

Whenever there is a need for a voltage or current to some prescribed accuracy, there has to be some absolute reference device that can be used for comparison. With the 723 regulator, the output voltage was compared with an internal reference voltage of 7 V. In this chapter we will look at integrated circuit references, which are available for use in equipment where a standard voltage reference or current source is required. But first we will consider the zener diode.

5-2 ZENER DIODES AS VOLTAGE REFERENCES

The most common reference voltage source is a zener diode, a device that relies on the constant voltage characteristic of reverse breakdown. However, variations in the zener voltage can occur when current is changed through the diode. This is due to the internal resistance (zener impedance) of the diode.

For example, a particular zener has a breakdown voltage of 12 V at 50 mA and a zener impedance of 10 Ω. If the current through the zener is changed to 100 mA, the new zener voltage is:

$$V_Z = 12 \text{ V} + (100 \text{ mA} - 50 \text{ mA}) \times 10 = 12.5 \text{ V} \qquad (5\text{-}1)$$

We see, then, that the zener voltage has changed by 0.5 V with current through the device. The higher the zener impedance, the greater the voltage change for a given current. With a given zener, we can eliminate the voltage change by keeping the current constant through the device.

5-3 VOLTAGE REFERENCE DEVICES

A voltage reference semiconductor consists of an integrated circuit with a zener diode and active circuits that keep the current through the zener relatively constant. This gives a very small voltage change with load current changes, and therefore the voltage reference device has a low effective zener impedance—typically less than 1 Ω. Figure 5-1 shows the typical internal circuit of the device.

In operation, the + input terminal is equivalent to the cathode terminal on a zener diode and is connected in series, with a current limiting resistor R_L.

If the + input terminal goes more positive by an amount Δ V, the change is transferred by the zener diode to the base of Q_1. This causes a more positive voltage on the base of Q_1 and this transistor is turned on more, resulting in more collector current and a drop in collector voltage. Since the collector of Q_1 is connected directly to the base of Q_2 (PNP), this transistor is also turned on more, causing greater current to flow through the external resistor R_L. The net effect is that the original positive increase is negated by the additional volt drop across R_L caused by the increased current through Q_2. The gain of Q_1 and Q_2 enables the circuit to correct for changes in the input voltage, and this lowers the effective "zener impedance" of the device. Since

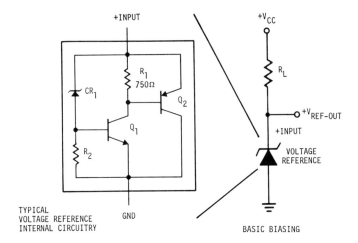

Figure 5-1   Voltage Reference Equivalent Circuit

the components are on one IC chip, greater temperature stability is achieved with the voltage reference by use of internal compensation.

Being a two-terminal device like a zener, the voltage reference is used in circuit applications the same way a zener is used. However, the range of available voltages for the voltage references devices is not as extensive as can be obtained with zener diodes.

5-4 TYPICAL VOLTAGE REFERENCES

One particular voltage reference is the LM329. This device has a nominal voltage of 6.9 V and has an operating current range of 1 mA to 15 mA. The zener impedance is about 1 Ω, so that we can expect only 14 mV change in the reference voltage over the complete current excursion.

The temperature coefficient for the LM329C is a maximum of 20 parts per million (2×10^{-5}) per degree centigrade. This means that if the ambient temperature changes from 25°C to 50°C, the new reference voltage will be:

$$V_{50} = 6.9 \text{ V} + 6.9 \text{ V}(50 - 25)2 \times 10^{-5}$$
$$= 6.9 \text{ V} + 0.00345 \text{ V}$$
$$= 6.90345 \text{ V} \tag{5-2}$$

or a change of 3.45 mV over the temperature range.

Other voltage references are shown in Table 5-1.

The table indicates a reference voltage range from 1.22 V to 15 V and initial tolerances from 0.05% to 5%. Dynamic resistances are from 0.6 Ω to 2.2 Ω, and

Table 5-1 Some available voltage references

Reverse Breakdown Voltage V_R at I_R	Device	Voltage Tolerance Max, $T_A = 25°C$	Voltage Temperature Drift-ppm/°C Max or mV Max Change Over Temperature Range		Current Range, I_R	Output Dynamic Impedance (Max)
			Drift (Max)	Temperature Range		
1.22	LM113	±5%	100 ppm typ	−55°C to +125°C	500μA to 20 mA	0.8Ω
1.22	LM313	±5%	100 ppm typ	0°C to 70°C	500 μA to 20 mA	0.8Ω
1.22	LM113-1	±1%	50 ppm typ	−55°C to +125°C	500 μA to 20 mA	0.8Ω
1.22	LM113-2	±2%	50 ppm typ	−55°C to +125°C	500 μA to 20 mA	0.8Ω
1.235	LM185	±1%	20 ppm typ	−55°C to +125°C	1 mA to 20 mA	0.6Ω
1.235	LM285	±1%	20 ppm typ	−25°C to +85°C	1 mA to 20 mA	0.6Ω
1.235	LM385B	±1%	20 ppm typ	0°C to +70°C	1 mA to 20 mA	1Ω
1.235	LM385	−2.5, +2	20 ppm typ	0°C to +70°C	1 mA to 20 mA	1Ω
2.49	LM136	±2%	18 mV	−55°C to +125°C	400 μA to 10 mA	0.6Ω
2.49	LM136A	±1%	18 mV	−55% to +125°C	400 μA to 10 mA	0.6Ω
2.49	LM236	±2%	9 mV	−25°C to +85°C	400 μA to 10 mA	0.6Ω
2.49	LM236A	±1%	9 mV	−25°C to +85°C	400 μA to 10 mA	0.6Ω
2.49	LM336	±4%	6 mV	0°C to +70°C	400 μA to 10 mA	1Ω
2.49	LM336B	±2%	6 mV	0°C to +70°C	400 μA to 10 mA	1Ω
2.5	LM185-2.5	±1.5%	20 ppm typ	−55°C to +125°C	20 μA to 20 mA	0.6Ω
2.5	LM285-2.5	±1.5%	20 ppm typ	−25°C to +85°C	20 μA to 20 mA	0.6Ω
2.5	LM385-2.5	±3%	20 ppm typ	0°C to +70°C	20 μA to 20 mA	1Ω
2.5	LM385B-2.5	±1.5%	20 ppm typ	0°C to +70°C	20 μA to 20 mA	1Ω
5.0	LM136-5.0	±2%	36 mV	−55°C to +125°C	400 μA to 10 mA	0.6Ω
5.0	LM136A-5.0	±1%	36 mV	−55°C to +125°C	400 μA to 10 mA	0.6Ω
5.0	LM236-5.0	±2%	18 mV	−25°C to +85°C	400 μA to 10 mA	0.6Ω
5.0	LM236A-5.0	±1%	18 mV	−25°C to +85°C	400 μA to 10 mA	0.6Ω
5.0	LM336-5.0	±4%	12 mV	0°C to +70°C	400 μA to 10 mA	1Ω
5.0	LM336B-5.0	±2%	12 mV	0°C to +70°C	400 μA to 10 mA	1Ω
6.90	LM129A	+3%, −2%	10 ppm	−55°C to +125°C	0.6 mA to 15 mA	1Ω
6.90	LM129B	+3%, −2%	20 ppm	−55°C to +125°C	0.6 mA to 15 mA	1Ω
6.90	LM129C	+3%, −2%	50 ppm	−55°C to +125°C	0.6 mA to 15 mA	1Ω
6.90	LM329B	±5%	50 ppm	0°C to +70°C	0.6 mA to 15 mA	2Ω
6.90	LM329C	±5%	20 ppm	0°C to +70°C	0.6 mA to 15 mA	2Ω
6.90	LM329D	±5%	100 ppm	0°C to +70°C	0.6 mA to 15 mA	2Ω
6.95	LM199A	+1%, −2%	0.5 ppm	−55°C to 125°C	0.5 mA to 10 mA	1Ω
6.95	LM199A	+1%, −2%	10 ppm	85°C to +125°C	0.5 mA to 10 mA	1Ω
6.95	LM199	+1%, −2%	1 ppm	−55°C to +85°C	0.5 mA to 10 mA	1Ω
6.95	LM199	+1%, −2%	15 ppm	85°C to +125°C	0.5 mA to 10 mA	1Ω
6.95	LM299A	+1%, −2%	0.5 ppm	−25°C to +85°C	0.5 mA to 10 mA	1Ω
6.95	LM299	+1%, −2%	1 ppm	−25°C to +85°C	0.5 mA to 10 mA	1Ω
6.95	LM399A	±5%	1 ppm	0°C to +70°C	0.5 mA to 10 mA	1.5Ω
6.95	LM399	±5%	2 ppm	0°C to +70°C	0.5 mA to 10 mA	1.5Ω
6.95	LM3999	±5%	5 ppm	0°C to +70°C	0.6 mA to 10 mA	2.2Ω
10.00	LH0070-0	0.1%	20 mV	−25°C to +85°C	0 mA to 20 mA	1Ω
10.00	LH0070-1	0.1%	10 mV	−25°C to +85°C	0 mA to 20 mA	1Ω
10.00	LH0070-2	0.05%	4 mV	−25°C to +85°C	0 mA to 20 mA	1Ω
10.24	LH0071-0	0.1%	20 mV	−25°C to +85°C	0 mA to 20 mA	1Ω
10.24	LH0071-1	0.1%	10mV	−25°C to +85°C	0 mA to 20 mA	1Ω
10.24	LH0071-2	0.05%	4 mV	−25°C to +85°C	0 mA to 20 mA	1Ω
Adjustable— 5V, 6V, 10V, 12V, 15V	LH0075	±0.5%	0.003%/°C typ	−55°C to +125°C	1 mA to 200 mA	1Ω
Adjustable— 5V, 6V, 10V, 12V, 15V	LH0075C	±1%	0.003%/°C typ	0°C to +70°C	1 mA to 200 mA	1Ω
Adjustable— −5V, −6V, −10V, −12V, −15V	LH0076	±0.5%	0.003%/°C typ	−55°C to +125°C	1 mA to 200 mA	1Ω
Adjustable— −5V, −6V, −10V, −12V, −15V	LH0076C	±1%	0.003%/°C typ	0°C to +70°C	1 mA to 200 mA	1Ω

temperature drift is from 100 parts per million to 4 mV over the full temperature range.

The LH0075 and LH0076 devices are really adjustable voltage precision regulators (the LH0076 has a negative output voltage) which can provide currents up to 200 mA. Both devices can output up to 27 V by using one external resistor; or by just arranging connections, 5 V, 6 V, 10 V, 12 V, or 15 V can be selected.

5-5 VOLTAGE REFERENCE APPLICATION

In operation, the two terminals of the voltage reference are used like a zener diode. An application example is shown in Fig. 5-2.

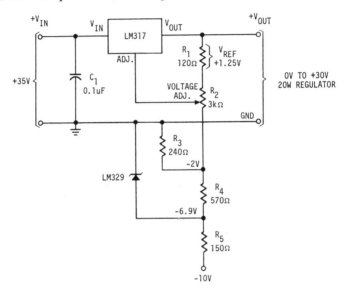

Figure 5-2 0–30 VDC 20 Watt Power Supply

Remember that the LM317 regulator has a minimum output voltage of +1.25 V. But, by biasing the lower terminal of the voltage adjust potentiometer negative, we are able to provide a zero output voltage capability.

As the circuit shows, the LM329 provides a stable −6.9 V at the junction of R_4 and R_5. This voltage is reduced to −2 V by the voltage divider action of R_4 and R_3. Using −2 V rather than −1.25 V insures that the output will reach zero volts. Since the output voltage (and current flow) is taken from the V_{OUT} terminal to the GND terminal, the output voltage can't go negative.

Another application of the voltage reference is to improve the regulation of an adjustable regulator like the LM317. The circuit is shown in Fig. 5-3.

It will be recalled in the chapter on linear regulators that the LM317 has an internal reference of 1.25 V and the device regulates by maintaining this voltage

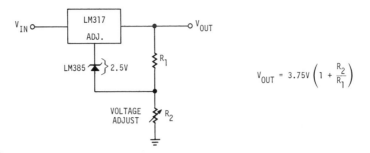

Figure 5-3   Improved LM317 Regulator

$$V_{OUT} = 3.75V \left(1 + \frac{R_2}{R_1}\right)$$

across R_1. Adding the LM385 in series with the adjustment terminal effectively increases the reference to $(1.25 \text{ V} + 2.5 \text{ V})$ or 3.75 V, which means a greater portion of the output voltage is regulated. Actually, the regulator is improved by the factor of 3 that the reference voltage is increased. The disadvantage of this circuit is that the minimum output voltage is now 3.75 V rather than 1.25 V. However, by adding a negative reference similar to that shown in Fig. 5-2, the output voltage can be brought back to zero.

Let us now look at an application of a voltage reference as a calibration standard. A much-used standard for voltage measurements is a standard cell, which is a battery with a precise output voltage. This standard can be replaced, however, with the circuit given in Fig. 5-4, which uses a LM199 precision reference.

If we want extreme (standard cell) accuracy with a voltage reference, we must keep the temperature of the device constant. The LM199 (299, 399) comes in a package that contains a small oven and a built-in, temperature-controlled heater. By maintaining the temperature of the reference at a value greater than the ambient (surrounding) temperature, the temperature of the reference is essentially independent of ambient temperature changes. That is, if the ambient temperature rises, the internal heater output is reduced—which maintains a constant temperature at the reference. Going back to Fig. 5-4, a standard cell has an output voltage of 1.0192 V with a

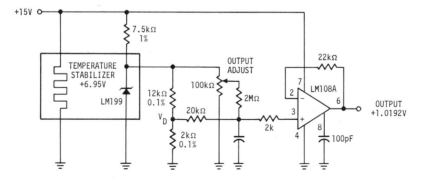

Figure 5-4   Standard Cell Replacement Circuit

temperature coefficient of $-10 \, \mu\text{V/}^\circ\text{C}$. In the schematic, the 6.95-V reference is divided down by the 12 k and 2 k to provide a voltage V_D.

$$V_D = \frac{6.95 \text{ V} \times 2 \, k}{14 \, k}$$

$$= 0.9929 \text{ V}$$

The 100-k pot in series with the 2-MΩ resistor forms another voltage divider with the 20-k resistor to provide the input to the non-inverting side of the follower. Let us assume that the wiper on the pot is all the way to the top, then the input voltage (V_{NI}) to the follower is:

$$\text{max } V_{\text{NI}} = \frac{(6.95 - 0.9929)\text{V} \times 20 \, k}{2.02 \text{ M}} + 0.9929 \text{ V}$$

$$= 1.0519 \text{ V}$$

With the wiper on the pot at the bottom of its range, the top end of the 2-MΩ resistor goes to ground and forms a voltage divider with the 20-Ω resistor which divides down the 0.9929 V to:

$$\text{min } V_{\text{NI}} = \frac{0.9929 \text{ V} \times 2 \text{ M}}{2.02 \text{ M}}$$

$$= 0.9831 \text{ V}$$

We can see, then, with the pot, we have the adjustment capability to set the voltage at the required 1.0192 volts.

The LM108A op-amp used in this circuit has a maximum 1-mV offset over a temperature range of -55° to $+125^\circ\text{C}$. Since this can be adjusted out with the pot, the critical factor becomes changes of the offset voltage and offset current with a temperature. The maximum offset voltage variation with temperature of 5 $\mu\text{V/}^\circ\text{C}$ for this op-amp compares favorably with the standard cells value of 10 μV. Considering the offset current change over the temperature range, the coefficient is 2.5 pA/$^\circ\text{C}$. The change in output voltage (ΔV_{00}) is:

$$\Delta V_{00} = 2.5 \times 10^{-12} \text{A/}^\circ\text{C} \times 2.2 \times 10^4 \, \Omega \times 180^\circ\text{C}$$

$$= 9.9 \, \mu\text{V} \text{ (which is negligible!)}$$

But what about the voltage reference? The temperature coefficient for the LM199 over a temperature range of -55°C to $+85^\circ\text{C}$ is 0.0001%/$^\circ\text{C}$. This is equivalent to a coefficient of 7 $\mu\text{V/}^\circ\text{C}$. Combining this with the op-amp coefficient, we get a total of 13 $\mu\text{V/}^\circ\text{C}$, which is slightly over the 10 $\mu\text{V/}^\circ\text{C}$ stated for the standard cell.

Over the temperature range from 85°C to 125°C, the LM199 degrades to 15 $\mu\text{V/}^\circ\text{C}$, which takes it beyond the standard cell variation. However, since standard cells are mainly used in calibration labs, which are temperature controlled at around 25°C, this should not create a problem.

A further application for a voltage reference is to expand the scale of a voltmeter. Let us assume that we are required to measure the load regulation of a 5-V power supply. If the specified regulation is 0.1%, then we need to be able to measure a 5-mV change in the 5-V supply. A 3½-digit digital voltmeter (DVM) will measure only down to about 10 mV. But let us assume the meter has a more sensitive scale— like a 100-mV full scale. We can set up the expanded scale measuring circuit shown in Fig. 5-5.

With the 10-K pot across the voltage reference we can adjust the voltage at the wiper from 0 V to 6.9 V.

Using the circuit to measure the load regulation of the power supply, we go through the following steps:

1. Set the load resistor (R_L) to a value that will cause the minimum load current to flow (usually about 1 mA).

2. Set the voltmeter to the 5-V scale and connect it between the power supply output and the wiper of the pot, as shown in the figure.

3. Adjust the 10-turn pot until the DVM reads approximately 0 V.

4. Now set the DVM to the 100-mV scale and readjust the pot for 0 V.

5. Lower R_L to provide the full load current from the power supply and read the DVM voltage. This voltage is the change in the output voltage (ΔV).

6. The percent load regulation is determined by the equation:

$$\% \text{ regulation} = \frac{\Delta V \times 100}{5 \text{ V}}$$

What the reference circuit has done is to null out the 5-volt output so we are able to detect only the changes (ΔV) in the output voltage on a more sensitive scale of the DVM. The applications described in this section for the voltage reference are

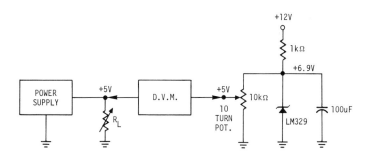

Figure 5-5   Expanded Scale Measuring Circuit

just a few of many uses for this device. We shall now see that precise constant current sources can also be generated by using a voltage reference.

5-6 CURRENT SOURCES

A current source is generated by putting a very high resistance (R_S) in series with the output terminals of a voltage source. If R_S is at least 100 times greater than the maximum load resistor, we can say that the current is constant regardless of changes in the load resistor.

Current sources are useful for calibrating meters or for generating linear-ramp-type waveforms. The equation for charging a capacitor is:

$$i = C \frac{\Delta V}{\Delta t}$$

If the current is kept constant, the slope $\Delta V/\Delta t$ is constant. A linear ramp generated this way can be used for the horizontal and vertical deflection circuits used in a cathode ray tube (CRT).

A precision 1-mA current "source" is shown in Fig. 5-6:

The voltage at the non-inverting input is the supply voltage of 15 V minus the 2.5-V reference, or 12.5 V. Since there can be no difference in the voltage between the two inputs of an op-amp, the voltage at the bottom of the 2.5-k resistor (R_S) must also be at 12.5 V. This means the voltage drop across the 2.5-k resistor is the same as the 2.5-V reference voltage, and the current flowing through this resistor is (2.5 V/ 2.5 k), or 1 mA.

Since no current flows into the op-amp inverting terminal, the full 1 mA is divided between the field effect transmitter (FET) and the bipolar transistor. If we assume the FET gate current is zero, then this same 1 mA is flowing out of the junction of the 10-k resistor and the transistor emitter, into the load (R_L).

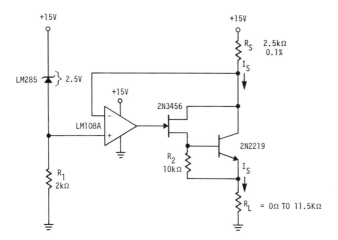

Figure 5-6 Precision 1 Milliamp Current Source

The current from this circuit is precise because it is controlled only by the reference voltage and the 2.5-*k* resistor. If we desire more current, then the 2.5-*k* resistor should be reduced.

Limits on the size of the load resistor (R_L) range from 0 Ω to 11.5 kΩ. The upper load resistor limit is determined by allowing a minimum 1-V drop across the transistors to maintain their correct operation.

In a similar fashion, we can generate a precision 1-mA current "sink" (current flows in rather than out) circuit as indicated in Fig. 5-7 by reversing the position of both the reference and the 2.5-*k* resistor.

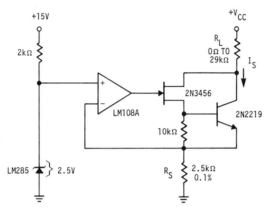

Figure 5-7 Precision 1 Milliamp Current Sink

As before, we have the 2.5 V across the 2.5-*k* resistor, and the sink current is 1 mA. The allowable range of load resistors is from 0 Ω to a value determined by the maximum collector/drain voltage of 30 V. Allowing, again, a minimum voltage across the transistors of 1 V, the maximum voltage across the transistors of 1 V, the maximum load resistor is (29 V/1 mA) or 29 kΩ.

The current source and sink circuits described have an output current capability limited only by the power dissipation of the 2N2219 output transistors. At 25°C, this transistor can dissipate 0.8 W. If we want an output current of 50 mA, then the maximum voltage drop across the transistor can be no greater than 0.8 W/50 mA, or 16 V.

When the required current is less than 1 mA, the circuits can be simplified by deletion of the output transistors as shown in Fig. 5-8.

Remember that the voltages at the input terminals of a closed loop op-amp must be the same, so the voltage drops across R_1 and the references are identical. We can obtain the required current by choosing R_1 to satisfy the relationship:

$$R_1 = \frac{2.5 \text{ V}}{I_S}$$

Notice that in the circuits of Fig. 5-8, the voltage across the load resistor R_L appears at the non-inverting input to the follower. As the load voltage increases, there is less voltage drop across the 10-K resistor and therefore less current change (ΔI)

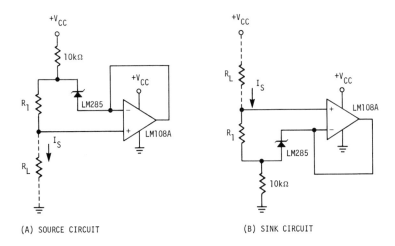

(A) SOURCE CIRCUIT   (B) SINK CIRCUIT

Figure 5-8 Low Current Source and Sink Circuits

through the reference. This current causes a slight voltage change (ΔV) across the reference of $\Delta V = \Delta I Z_{ZT}$, where Z_{ZT} is the dynamic (zener) impedance of the reference.

5-7 PRACTICAL EXAMPLES

EXAMPLE 5-1

We need a variable reference source to provide 0 to 10 V for calibration of the low voltage scales (1 V, 2.5 V, 10 V) of a VOM. The meter has a stated accuracy of 2% and the loading is 200 kΩ/V.

Accuracy of the calibrator should be at least 10 times more accurate than the device to be calibrated—in this case 0.2%.

Solution. The simple circuit shown below in Fig. 5-9 can be used for this calibration:

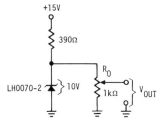

Figure 5-9 Calibrator Circuit

The LH0070-2 is a 10-V 0.05% reference, and R_O is a 10-turn Bourns Digital Knobpot with an output voltage accuracy of 0.1%. This gives a combined accuracy of 0.15%, which is within the required 0.2%.

We must, however, take meter loading into account. On the 1-V scale, the meter loading is 20 kΩ, and with 1-V output from the calibrator, the circuit is that of Fig. 5-10. The loading of the meter causes the output voltage (V_O) to be:

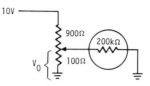

Figure 5-10   Equivalent Circuit with Meter Loading

$$V_O = \frac{10 \text{ V} \times R_{\text{EQ}}}{900 + R_{\text{EQ}}}$$

where:

$$R_{\text{EQ}} = \frac{100 \times 200 \text{ k}\Omega}{100 + 200 \text{ k}\Omega}$$

$$= 99.95 \ \Omega$$

Thus

$$V_O = \frac{10 \text{ V} \times 99.95}{999.95}$$

$$= 0.9996 \text{ V}$$

This means the output voltage is 0.4 mV low due to meter loading, which creates an error of 0.04%. Adding this to the 0.15% combined error from the reference and pot, we get 0.19%, which is just less than the required 0.2%. If we are just interested in checking the meter at the full scale points, we can use a fixed multiple output voltage divider rather than the potentiometer.

EXAMPLE 5-2

A deflection circuit requires a 1.0% linear repetitive sawtooth waveform that reaches a maximum value of +2 V in one ms. The fall time of the sawtooth waveform should be less than 1 μs.

Solution.   Let us use the circuit given in Fig. 5-8 A as a constant current source which charges a capacitor and generates the sawtooth waveform. The capacitor is discharged when it reaches the maximum value of 2 V by turning on Q_1. For the ramp to reach 2 V in one ms, the required current I_s is:

$$I_S = \frac{C \, \Delta V}{\Delta t}$$

$$= \frac{10^{-8} \text{F} \times 2\text{V}}{10^{-3}\text{s}}$$

$$= 20 \ \mu\text{A}$$

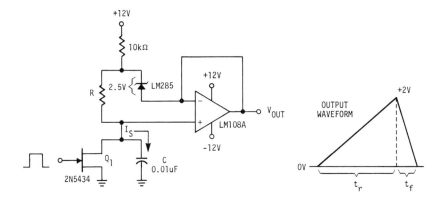

Figure 5-11   Precision Sawtooth Generator

This current is established by the references voltage drop across the resistor R. The value of R is then:

$$R = \frac{2.5 \text{ V}}{20 \text{ } \mu\text{A}}$$

$$= 125 \text{ k}\Omega$$

Linearity of the ramp voltage is a function of how constant the current remains during the capacitor charging time. Initially the capacitor voltage is zero and the output of the op-amp "follows" this value and is also zero. This means there is (12-2.5) V or 9.5 V across the 10-k resistor which causes 9.5 V/10 k = 9.5 mA to flow through the LM285 reference. When the capacitor charges to $+2$ V the voltage across the 10-k resistor is now 7.5 V and the reference current is now 7.5 mA. What we need to know is how much does the reference voltage change with a change of 2 mA in current through the device? The maximum dynamic resistance for the LM285 is 0.6 Ω— so the voltage change is (2 mA $\times$ 0.6 Ω) or 1.2 mV. This is a 0.048% change in the current I_s. The capacitor voltage, then, stays well within the required 1.0% linearity.

The last thing we need to determine is whether the required 1-μs discharge time can be achieved. Turning on Q discharges the capacitor through the FET's maximum on resistance of 20 Ω, creating a time constant of 0.2 μs. In 5 time constants, (1 μs) the capacitor voltage drops to 0.7% of 2 V or 14 mV. This creates an initial voltage on the capacitor when charging starts. But the voltage is still within 1% of the desired value.

5-8 CHAPTER 5—SUMMARY

- A voltage reference provides a constant voltage for power supplies and calibration circuits.

- Comparing a voltage reference with a zener diode, the reference has a lower dynamic resistance and has better temperature capability.
- Initial tolerances on voltage references range from 0.05% to 5%, and dynamic resistance is typically less than 1 Ω.
- A voltage reference can be used to increase the regulation of an IC voltage regulator.
- Temperature stability is so good for an LM199-type reference that it can replace a standard cell.
- Current sources provide a current that is independent of the value of the load resistor.
- A precision current source can be made from a voltage reference and an op-amp.
- Current flows out of a current source and current flows into a current sink.

5-9 CHAPTER 5—EXERCISE PROBLEMS

1. Determine the change in voltage across an LM329 voltage reference if the current through the device changes from 2 mA to 14 mA. Assume a zener impedance of 1.6 Ω.
2. Determine the maximum voltage change across an LM329B voltage reference caused by the temperature changing from 25°C to 65°C. Assume an initial voltage of 6.9 V.
3. State two advantages of a voltage reference over a zener diode.
4. Explain how the internal circuitry of a voltage reference lowers the "zener" impedance.
5. If the input voltage (V_{CC}) changes from 15 V to 18 V, find the voltage reference voltage change ΔV (see Fig. 5–12).
6. Select a 6.95-V reference from Table 5-1 that will not vary by more than 0.01% over a temperature range from −30°C to +60°C and has an initial tolerance of 2% or less.
7. Show a circuit that will increase the percent regulation capability of a LM317 regulator by a factor of 5. What is the minimum output voltage?
8. For the standard cell circuit of Fig. 5-4, find the change in output voltage if the input supply voltage changes by 1%.

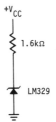

Figure 5-12   Circuit for Problem 5-5

9. Assume the supply voltage in the circuits in Fig. 5-8 is +15 V and $R_1 = 100$ K. Find the change in the current source if the load resistor changes from 10 kΩ to 200 kΩ.

10. How temperature-stable should the 100-K pot be which is used in the standard cell circuit in Fig. 5-4?

11. What effect does the initial accuracy of the LM329 reference, used in the expanded scale circuit (Fig. 5-5), have on the measurement accuracy?

12. How much error (%) does the voltage offset of the LM108A (reference Table 1-1) create in the precision current source of Fig. 5-6?

5-10 CHAPTER 5—LABORATORY EXPERIMENTS

Experiment 5A

Purpose: To verify the key parameters of a LM329 voltage reference.

Requirements: Connect up the circuit as shown below in Fig. 5-13.

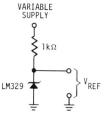

VARIABLE
SUPPLY

1kΩ

LM329

V_{REF}

Figure 5-13   LM329 Test Circuit

Measure:

1. Reference voltage at a current of 8 mA

2. Dynamic impedance with a current change from 0.6 mA to 15 mA. Compare with specific values in the data sheets in Appendix E.

Experiment 5B

Purpose: To show how the regulation of a LM317T regulator can be improved by adding a voltage reference.

Requirements: Use an input voltage to the LM317T regulator of +30 V and set up the regulator to provide a maximum output voltage of 25 V (see Fig. 3-10). Connect a load resistor of 100 Ω across the output.

Measure:

1. Load regulation with 25 V output (measure output voltage with and without load)
2. Add the voltage reference as shown in Fig. 5-3 and again measure the load regulation at 25 V.

6

Voltage Comparators

OBJECTIVES

Upon completion of this chapter you should understand the following concepts:
1. Purpose of a voltage comparator
2. Advantages of a voltage comparator over an operational amplifier
3. Zero crossing detector
4. Schmitt trigger circuit
5. Level detector circuit
6. Window comparator

6-1 INTRODUCTION

A voltage comparator is used to compare a varying input signal against a reference voltage. The comparator, like an op-amp, has two input terminals (inverting and non-inverting) and one output terminal. When the varying input voltage changes polarity relative to the reference voltage (becomes greater or less than the reference), the output of the comparator changes state. This can be clarified by the circuit shown in Fig. 6-1.

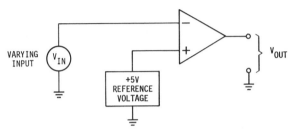

Figure 6-1   Voltage Comparator

Following the convention of an open loop operational amplifier, when the varying input becomes slightly (less than a millivolt) more positive than the +5-V reference, the output will swing from the high state to the low state (more positive to less positive). As soon as the varying input becomes less than the reference, the output will swing back from the low state to the high state.

6-2 VOLTAGE COMPARATORS VERSUS OPERATIONAL AMPLIFIERS

The circuit of Fig. 6-1 can use a conventional operational amplifier. However, voltage comparator integrated circuits are specifically made for this application with fast output switching capability, which allows the comparator to be used as an interface device with digital logic circuits without creating excessive delay in circuit operation. The delay results from the requirement to reach a particular voltage level at the logic gate before it can operate ($+2$ V for transistor logic—TTL). With a slow-changing output from the comparator, an appreciable time can elapse from when the comparator changes state to when the logic is triggered. This creates what is called a *propagation delay*. If we were using a 741 op-amp with its 0.5 V/μs slew rate, it would take 4 μs to reach the 2-V TTL trip point. A propagation delay of this magnitude could be intolerable in a given application.

Another advantage of the comparator over the conventional (741) type op-amp is a lower offset current. This means the comparator will trigger closer to the desired reference voltage.

6-3 ZERO CROSSING DETECTOR

The zero crossing detector is a special case of the comparator with a grounded reference. It can be used to sense the zero voltage crossings of a sine wave. At the zero crossing point, the output has a well-defined transition for both the positive and negative input slopes. The fast rise time of the comparator output can quickly trigger digital circuits, which allows measurement of the frequency or phase of the sine wave. Figure 6-2 illustrates a typical zero crossing detector circuit.

Input voltage offset can cause a slight error in the crossing point, and if this is critical, an offset pot should be used.

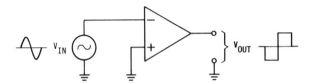

Figure 6-2   Zero Crossing Detector

6-4 LEVEL SENSING CIRCUIT

A level sensing circuit determines when a particular predetermined voltage level is present at the input. In the circuit shown in Fig. 6-3, the output changes from a high level to a low level when the input signal exceeds the reference voltage. If the input terminals are reversed, with the reference on the inverting terminal and the input on the non-inverting terminal, the output will go from low to high when the input exceeds the reference.

The input voltage could be from a liquid level sensor in a water tank, with the voltage from the sensor indicating the level in the tank. When the input voltage exceeds the reference, the change in the output signal from the comparator shuts off

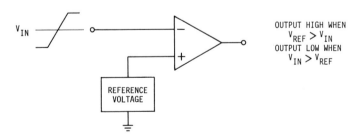

Figure 6-3   Level Sensing Circuit

power to the motor that pumps water into the tank. Thus, by adjusting the reference voltage, we can change the water level in the tank.

6-5 VOLTAGE COMPARATOR AS A SCHMITT TRIGGER

If noise is present on the input signal to a comparator, the output will "jitter" when the input signal is close to the reference value. This can be a problem if the output transitions are used to measure the frequency of the input signal. A circuit that eliminates this problem is the Schmitt trigger circuit shown in Fig. 6-4.

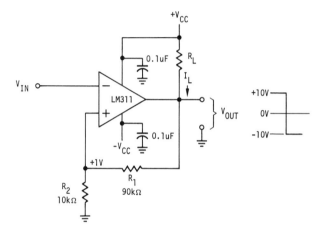

Figure 6-4   Schmitt Trigger

Notice that this is a comparator circuit with positive feedback from the output to the non-inverting input. The output will normally be high and with the voltage divider shown, the non-inverting input is biased to $+1$ V. When the input exceeds $+1$ V, the output will switch to -10 V and the non-inverting input will go to -1 V. Now in order for the output to return to $+10$ V, the input signal has to drop below -1 V. Thus, there is a 2-V difference between the trip points. This is called the circuit "hysteresis." To show how this increases noise immunity, consider the noise waveform in Fig. 6-5.

If a zero crossing detector is used, an output transition occurs each time the waveform passed through zero (eight times in the waveform shown in Fig. 6-5); but with the Schmitt trigger circuit, once the input waveform exceeded the $+1$-V upper trip point, a noise spike has to be 2 V in order to reach the lower -1-V trip point. The amount of hysteresis is varied by changing the ratio of the voltage divider resistors. If the anticipated noise is low (millivolts), the divider ratio is increased to provide a lower offset on the non-inverting input terminal. Note that the resistor R_L becomes part of the voltage divider when the output is high—so this resistor should be much smaller than R_1.

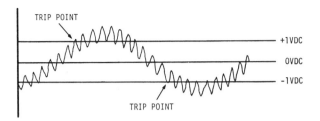

Figure 6-5 Noise Waveform

6-6 WINDOW COMPARATOR

Sometimes we need to sense when a voltage is between two limits—or when it is above an upper limit or below a lower limit. An example is a voltage which represents a gauging operation where the part must fall within two limits to be acceptable. A circuit that performs this function is called a "window comparator"; a version is shown in Fig. 6-6.

The output circuits use a light emitting diode (LED) to indicate when the input voltage is above the upper limit (V_H) or below the lower limit (V_L). The outputs of the comparators have open collectors that draw currents through the LEDs when the outputs go low. We could have used a relay or other device for the output indication, because the actual comparison is independent of the output circuit. Let us now look at the detailed circuit operation. When the input voltage exceeds the upper limit, the U_1 output (V_{01}) goes low to ground potential, which turns on the "high" LED. With this high input condition, the U_2 output (V_{02}) is at $+12$ V, and the "low" LED is turned off. If the input voltage is now lowered so that it is between the upper and lower limits, both U_1 and U_2 outputs will be high and no current will flow through either LED. The final situation is with the input voltage below the lower limit. This causes the U_1

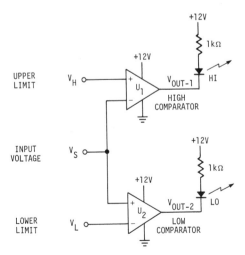

Figure 6-6 Window Comparator

output to be at $+12$ V and the U_2 output to be at ground potential, which forward biases the "low" LED.

The supply voltage used for the LEDs can assume any value providing it stays within the maximum input range of the comparator and the series resistor is changed to give the correct LED current.

A window comparator circuit which uses a single LED to indicate when the input is within a given range can be constructed by adding a transistor as shown in Fig. 6-7.

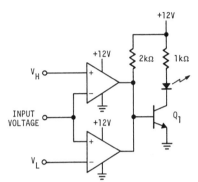

Figure 6-7   Single LED Window Comparator

With the input voltage within the upper and lower limits, both comparators' output transistors are turned off. This allows current to flow into the base of Q_1, turning on the LED.

If the input voltage is outside of the limits, one output transistor turns on and the base of Q_1 drops to about 0.2 V. Since this voltage is not great enough to bias Q_1, the LED is turned off.

6-7 A TYPICAL VOLTAGE COMPARATOR

An actual integrated circuit comparator is the LM311. This device has a high input impedance with a maximum input bias current of 275 nA and a maximum input voltage offset of 7.5 mV. The voltage on the input leads can be the same as the maximum supply voltages (± 15 V), which accommodates a large input voltage range. Both the input and the output circuits can be isolated from system ground, and the output circuits can drive loads connected to ground, to the positive supply, or to the negative supply. The partial input/output schematic of Fig. 6-8 illustrates these features.

Notice that the output (Pin 7) is uncommitted (open collector) and we have the option of connecting the GND (Pin 1) to system ground or to the negative supply. We can, therefore, connect the output load resistor in any of the three configurations of Fig. 6-9.

In Fig. 6-9(a) we have an output voltage swing from $+V$ to $-V$ or ground. The output voltage swing in (b) is from ground to $-V$ and in (c) is from ground to $-V$ or

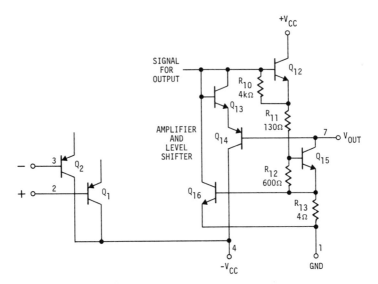

Figure 6-8 LM311 Input/Output Circuit

from +V to GND. We further have the option of connecting the load resistor (R_L) to a positive supply different from +V, providing we don't exceed 40 V more positive than −V. This gives us tremendous flexibility with our output signal—which is good because comparators interface with many different types of circuits.

If we don't connect Pin 1 to system ground, then the inputs are isolated from this connection. This can eliminate noise from appearing at the input via a common ground lead.

The typical switching time for the LM311 is 200 ns. If faster speed is required, the LM710 with a switching time of 40 ns could be used. However, the faster comparators require care in circuit layout because of the tendency to oscillate. It is a good idea to use 0.1-μF bypass capacitors on the chip supply pins.

A much used comparator is the LM339, which consists of four comparators on a single chip. This device has a rather slow switching time of 1.3 μs, but is a good

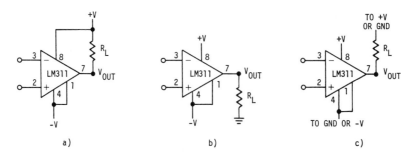

Figure 6-9 Load Resistor Configurations

choice in many low frequency applications. Like the LM311, the LM339 has open collector outputs.

6-8 VOLTAGE COMPARATOR APPLICATIONS

A bar display can use light emitting diodes (LED) to indicate the level of an input voltage. The actual display resembles a thermometer, with all lights mounted vertically (or horizontally) and the number of LEDs illuminated corresponding to the level of the input voltage. Figure 6-10 shows a circuit for a gasoline level indicator.

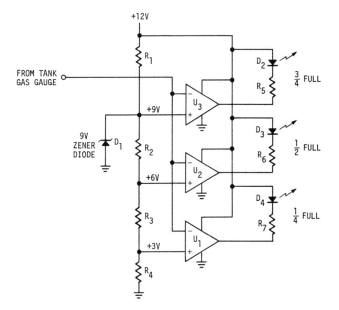

Figure 6-10   LED Gas Gauge

The circuit assumes that $+12$ V on the fuel gauge corresponds to a full tank and 0 V to an empty tank (not the case with an actual fuel gauge). If the tank is greater than three-quarters full, the voltage is greater than 9 V and all comparators have a "low state" on the output, which will cause all LEDs to illuminate. When the tank drops below three-quarters full, the top LED will extinguish. Below a quarter of a tank, all LEDs extinguish.

In an actual situation, one comparator circuit could be used for each gallon of fuel, i.e., a 10-gallon tank would use 10 circuits and the LEDs could be marked in gallons. For this application, the LM3914 dot/bar display device, which contains 10 comparators, could be used.

Another application for the voltage comparator is in a line voltage monitor circuit, which is used to disconnect a piece of equipment from the AC line if the line voltage falls outside of a predetermined range of values. Figure 6-11 shows such a

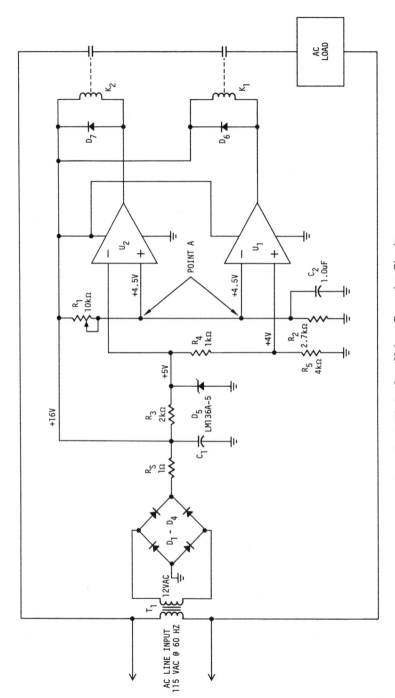

Figure 6-11 Under-Over Voltage Protection Circuit

circuit, which operates if the line voltage differs by more than 10% from the nominal value.

In operation, the LM136A-5 is a 5-V reference with a tolerance of $\pm 1\%$, which provides the upper limit voltage at the inverting input or U_2 and the lower limit of 4 V via the voltage dividers R_4 and R_5 to the non-inverting input of U_1. The voltage divider R_1 and R_2 is set by adjusting R_1 to provide 4.5 V at point A with the nominal line voltage. Variations in the AC line will cause point A to change. If the line voltage increases above 10%, the voltage at point A will exceed 5 V, and comparator U_2 output will switch high deactivating relay K_2, which will open the circuit to the load. With the line voltage lower than 10%, comparator U_1 will switch and relay K_1 will open the load circuit.

The relays used (K_1 and K_2) should have a minimum pull in voltage of about 14 V (low line) and a coil current of less than 20 mA. All resistors except R_5 and R_3 should be metal film, with R_4 and R_5 1% or better tolerance. C_2 is selected to reduce the ripple or voltage transients by a factor of 100 from the voltage appearing across C_1. In turn, C_1 should be large enough to allow less than 1-V ripple at the junction of R_5 and R_3.

A final application is a crowbar circuit, which protects against an overvoltage condition causing damage to circuit components. The term *crowbar* refers to a protection approach which consists of throwing a short circuit across the power supply terminals when an overvoltage occurs. Such a circuit, using a comparator, voltage reference, and a silicon controlled rectifier (SCR), is shown in Fig. 6-12.

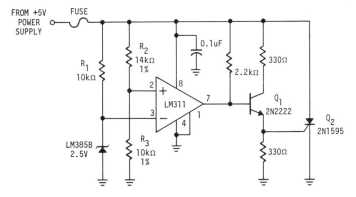

Figure 6-12   TTL Power Supply Crowbar Circuit

The circuit is designed to prevent more than 6 V from appearing on the transistor transistor logic (TTL) chips with a failure of power supply regulator. Maximum power supply voltage for the TTL is 7 V—so we are allowing a 1-V safety margin.

In operation, the LM385B provides 2.5 V $\pm 1.5\%$ at the inverting terminal of the comparator. With the power supply at +5 V, R_2 and R_3 establishes +2 V at the comparator non-inverting terminal. This voltage makes the comparator's output low, preventing forward biasing of the base Q_1. Thus, Q_1 and Q_2 are off. If the supply voltage increases to just greater than $+6$ V, the voltage at the non-inverting terminal exceeds the reference voltage and turns off the comparator output transistor. This allows the base of Q_1 to rise, turning on this transistor. When the voltage on the emitter of Q_1 reaches the gate trip point of Q_2 (maximum 3 V), the SCR fires. The SCR becomes a short across the power supply, causing the fuse to blow. Rating of the fuse is determined by the normal maximum circuit current flow. A fuse size twice this current should suffice—but the fuse should be a fast blow type.

Considering the tolerance of the reference voltage and R_2 and R_3, the firing point of the circuit can be off by 3.5% or 0.21 V. This means the circuit trip point could be as low as 5.79 V or as high as 6.21 V. If the power supply regulator is 5 V $\pm4\%$ (LM323 or LM309), the output could be as high as 5.2 V, which gives a margin of $(5.79 - 5.2)$ V or 0.59 V.

6-9 CHAPTER 6—SUMMARY

- A comparator compares two inputs, and the comparator output changes state when one input exceeds the other input.
- An op-amp can be used as a comparator if speed of response is not a problem.
- The output of a zero crossing detector changes state as the input signal passes through zero volts.
- Noise triggering of a comparator is avoided by using a Schmitt trigger circuit.
- A window comparator senses when the input is *within* a given voltage range.
- IC comparators usually have an open collector output stage, which allows for interface with various circuit devices, i.e., relays, lamps, etc.

6-10 CHAPTER 6—EXERCISE PROBLEMS

1. Give two advantages of a comparator IC over an op-amp when used in a comparator application.
2. What is the main reason for using a Schmitt trigger circuit?
3. Give an advantage and a disadvantage for the open collector output used on a comparator.
4. What is a "window" comparator?
5. Indicate the output circuit connections in order to achieve a 0-to-2-mA current swing through a 15-kΩ resistor. Power supply voltages of ±15 V are available.
6. Show a comparator circuit, the output of which goes high when the input voltage exceeds $+3$ V.

7. Determine all circuit values for a Schmitt trigger circuit with a total hysteresis of 100 mV using a LM311 comparator operating off ±9-V supplies. Load to be 20 mA.

8. Design a window comparator circuit using a LM311 which activates a relay when the input voltage is between 2 and 3 V. Use ±12-V supplies and specify all component values.

9. Determine the value of the capacitor C_1 for the circuit in Fig. 6-11 if the relays draw 15 mA each, the comparators are LM311, and the ripple at point C is to be less than 1%.

10. Redesign the circuit in Fig. 6-12 for a +15-V supply. Specify all component values and use a 5-V reference. Nominal trip voltage 16V.

6-11 CHAPTER 6—LABORATORY EXPERIMENTS

Experiment 6A

Purpose: To demonstrate the voltage comparator in a Schmitt trigger circuit configuration.

Requirements: Connect an LM311 comparator as a Schmitt trigger circuit with total hysteresis of 200 mV. Use power supply voltages of ±15 V. Couple in a 5-V peak-to-peak sine wave signal on the inverting input (see Fig. 6-4).

Measure:

1. Response time to reach +2 V (TTL "one" level)
2. Output fall time to reach 0.8 V (TTL "zero" level)
3. Hysteresis (put input signal on horizontal input to scope and output signal on vertical input)

Experiment 6B

Purpose: To demonstrate the operation of a window comparator circuit.

Requirement: Use the circuit determined in Exercise Problem 8.

Measure: The two trip points and compare with the predicted values.

7

Electronic Timers

OBJECTIVES

Upon completion of this chapter, you should know the following:
1. The basic operation of a 555 timer
2. The basic operation of a timer as an oscillator
3. Output circuit considerations
4. Triggering requirements
5. Limits for the maximum time internal
6. Advantages of the 3905 timer over the 555

7-1 INTRODUCTION

There are many applications that require a delay from the occurrence of one event until the start of another, e.g., the various cycles (rinse, wash, spin, etc.) of a modern washing machine, a traffic signal, etc. Electronic timers can provide this delay with times ranging from microseconds to hours.

In other applications, electronic timers can provide a time "window" during which a certain event is allowed to occur. This could be part of a burglar alarm sequence which allows a person to leave a building for a short period of time after the alarm is set. The timer would disable the alarm for a period of approximately 30 sec.

Timers also can be connected so the timing cycle continuously repeats. This is the astable or oscillator mode of operation.

Applications for timers are almost as varied as those for op-amps. This chapter should give the reader a basic understanding of these devices sufficient to follow most timer configurations—and, if necessary, to improve upon existing designs.

7-2 A BASIC ELECTRONIC TIMER

One of the more common types, and the most basic electronic timer, is the LM555. Figure 7-1 shows a functional block diagram of this timer in a "one-shot" circuit configuration.

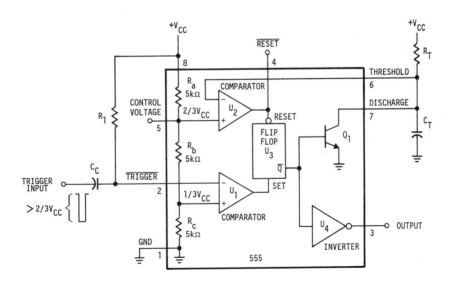

Figure 7-1   555 Basic Timer Circuit

7-3 TIMER CIRCUIT OPERATION

The overall circuit operation starts when a negative trigger pulse is applied to Pin 2 in Fig. 7-1, causing a positive pulse to occur at the output, which stays high for a period determined by the values of R_T and C_T.

Let us now look at the detailed operation of the timer circuit. If a negative pulse of sufficient amplitude is applied to the "trigger input," it will drive Pin 2 from V_{CC} to less than the internal voltage divider voltage of 1/3 V_{CC}, causing comparator U_1 output to go high. This "sets" the flip-flop, causing the $\bar{Q}$ output to go low, which turns off Q_1 and also causes the timer output to go high ($+V_{CC}$). With Q_1 turned off, timing capacitor C_T is allowed to charge via the external timing resistor R_T toward V_{CC}. When the voltage of C_T exceeds 2/3 V_{CC}, comparator U_2 output goes low and resets the flip-flop. This causes the flip-flop $\bar{Q}$ output to go high, dropping the timer output low (OV) and turning on Q_1. The turn-on of Q_1 abruptly discharges C_T and holds the voltage across it at the Q_1 saturation voltage. This completes the cycle, which causes the output to stay in the high state for the amount of time it took to charge the capacitor to 2/3 V_{cc}. So, by varying the time constant $R_T C_T$, we can vary the time interval that the timer output is held in the high state. Let us determine the actual relationship between R_T, C_T, and the total time, t_p, the output is high (timer interval).

The voltage on a capacitor (e_c) during charging is given by:

$$e_c = V_{cc}(1 - \varepsilon^{-t/RC})$$

But the time interval ends when $e_c = 2/3\ V_{cc}$.

Thus, at the completion of the timing cycle:

$$\frac{2}{3}\ V_{cc} = V_{cc}(1 - \varepsilon^{-t/RC})$$

and

$$\varepsilon^{-t/RC} = \frac{1}{3}$$

Taking the natural logarithm of both sides:

$$-t_p/RC = \ln\frac{1}{3}$$

Solving for t_p the output pulse width:

$$t_p = 1.1\ RC \qquad\qquad (7\text{-}1)$$

We can vary the output pulse, then, by simply changing either R or C.

If we connect an external voltage at the "control voltage" terminal (Pin 5), we can change the divider voltages. This changes the timer cycle time and also the required trigger voltage. Raising the control voltage increases the voltage to the non-inverting input of U_2 to above the 2/3 V_{cc} value, which means that C_T has to charge longer to reach the new voltage—so the timer interval is increased. The non-inverting input to U_1 will also be raised, which means the required negative input trigger level is reduced.

Connecting the center tap of a pot to Pin 5 allows the timing period to be accurately set to compensate for tolerance variations of R_T and C_T. One end of the pot should be connected to V_{cc} and the other to ground. A multiturn 1 kΩ wire-wound pot would be a good choice for this application.

The "on" time for the timer of 1.1 $R_T C_T$ could lead us to believe that we can get any time delay we want just by increasing the value of R_T and/or C_T. But there are practical limitations. The combined internal resistance at terminals 6 and 7 is about 20 MΩ. This resistance sets up a voltage divider with R_T as shown in Fig. 7-2.

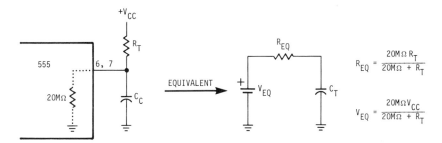

Figure 7-2   Voltage Divider Action Caused by Internal Leakage Resistance

The equivalent circuit shows that the internal resistance lowers both the effective supply voltage and the timing resistor. If, for example, we used an R_T of 20 MΩ, the effective supply voltage for charging the capacitor would halve—which means the capacitor could never charge to 2/3 V_{cc} and the timer would never shut off. Lower values of R_T could allow the capacitor to reach 2/3 V_{cc}, but the timing accuracy is still affected until we reach about 1 MΩ. So a conservative approach is to not exceed 1 MΩ for R_T.

Increasing the size of the capacitor beyond 4 μF takes us into electrolytic types, which have lower leakage resistance and looser tolerance. The leakage resistance of the capacitor, again, sets up a voltage divider with R_T. These considerations give a practical maximum timing interval for the 555 of about 15 minutes.

The trigger input signal should not be present when the timer turns off, otherwise the next cycle will begin automatically. A rule of thumb is to have the trigger pulse width less than one tenth of the timer interval. The trigger pulse should be negative and have an amplitude greater than 2/3 V_{CC}. Manual triggering can be accomplished by momentarily grounding Pin 2 via a push-button switch.

7-4 ELIMINATION OF INITIAL TIMING CYCLE

When power is first applied to the timer, a timing cycle could begin without an input trigger. We can see why this can happen with reference to Fig. 7-1. Comparator U_1 inverting input is connected to V_{cc} and the non-inverting is tied to 1/3 V_{cc}—so the "set input" to the flip-flop U_3 will be low. At the instant just after power is applied, the voltage across C_T will be zero (capacitor has had no time to charge), which holds the inverting input to U_2 low. The non-inverting input to U_2 will be at 2/3 V_{cc}, causing the reset input to the flip-flop to be high. With these input states to the flip-flop, its $\bar{Q}$ output can be high or low (indeterminate). If it is high, Q_1 will turn on and C_T will be shorted out. However, with the flip-flop $\bar{Q}$ output low, a timing cycle will begin. This can be undesirable in many applications because a momentary interruption of power caused, for example, by a lightning strike could cause an alarm to sound or equipment to start operating.

A solution to this problem is to connect the timing capacitor C_T as shown in Figure 7-3.

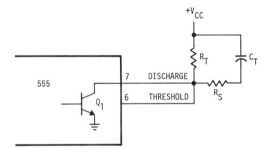

Figure 7-3 Circuit to Prevent Initial Timing Cycle

The capacitor has been connected in parallel with R_T rather than in series, but the timer operation is essentially the same. Now, when power is first applied, the inverting input to U_2 is at V_{cc}, causing the reset input to the flip-flop to be low. The flip-flop output $\bar{Q}$ will be forced high, turning on Q_1 and grounding pins 6 and 7, and there is no false timing cycle.

The actual circuit operation is as follows. With Q_1 on, C_T charges quickly through the low resistance of R_S (selected to limit the initial surge current to less than 100 mA) to V_{cc}. An input trigger pulse initiates the timing cycle by turning off Q_1. The voltage (V_{cc}) across C_T reduces as the capacitor discharges through R_S and R_T until the capacitor voltage is just less than 1/3 V_{cc}. At this time, the voltage from Pins 6 and 7 to ground is a little greater than 2/3 V_{cc}, and the inverting input to U_2 being more positive than the non-inverting input causes the comparator output to go low. This resets the flip-flop and turns on Q_1, which completes the timing cycle. If we neglect the resistance of R_S (much smaller than R_T), the timing cycle is still $1.1R_T C_T$.

7-5 ELECTRONIC TIMER AS AN OSCILLATOR

The monostable or "one-shot" operation described generates an output pulse for a time determined by R_T and C_T each time the timer is triggered by a negative pulse. When Q_1 is turned on at the completion of the timing cycle, a negative pulse is generated across capacitor C_T. If this pulse is coupled to the trigger input, the timer will recycle continuously. This is referred to as astable or free-running operation.

The 555 timer can be operated as an oscillator with the addition of a resistor R_1 between Pin 7 and $+V_{CC}$, a resistor R_2 between Pin 7 and Pin 6, and the timing capacitor C_T between Pin 6 and ground. A shorting jumper is connected between Pins 2 and 6 to enable astable operation. Figure 7-4 shows the LM555 in an oscillator circuit configuration.

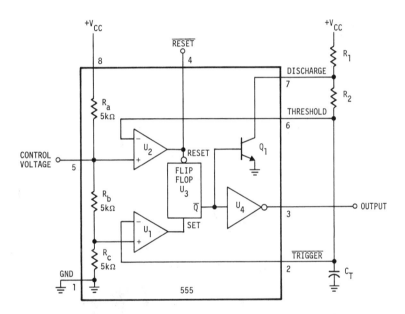

Figure 7-4   LM555 Oscillator Circuit

In operation, capacitor C_T charges to 2/3 V_{CC}, which causes the comparator U_2 output to change state. This resets the flip-flop U_3, which turns on Q_1 and effectively grounds Pin 7 and one end of resistor R_2. Capacitor C_T now discharges to ground through R_2. When the capacitor voltage drops to 1/3 V_{CC}, comparator U_1 output changes state, since the capacitor is connected directly to the U_1 input. Flip-flop U_3 is set and turns off Q_1. Capacitor C_T starts to charge via R_1 and R_2 toward V_{cc}, and the cycle repeats. The charge time is:

$$t_1 = 0.693(R_1 + R_2)C_T \qquad (7\text{-}1)$$

The discharge time is: $t_2 = 0.693R_2C_T$ (7-2)

The total period is: $T = t_1 + t_2$ (7-3)

$$= 0.693(R_1 + 2R_2)C_T$$

The frequency of oscillation is: $f = \dfrac{1}{T}$

$$= \frac{1.44}{(R_1 + 2R_2)C_T}$$ (7-4)

Notice that the timer "on" and "off" time and frequency are independent of power supply variations. The waveforms generated in the free-running mode of operation are shown in Fig. 7-5.

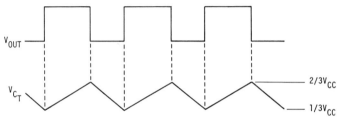

Figure 7-5   Oscillator Waveforms

Varying the voltage at the control voltage terminal will cause the oscillator frequency to change, which allows for frequency modulation (FM) of the output waveform.

If we make R_2 much greater than R_1, the output approaches a symmetrical square wave (equal "on" and "off" time). We can get a square wave by setting $R_1 = R_2$ and by connecting a diode in series with R_1 and also one in series with R_2 as shown in Fig. 7-6. This connection causes the charge and discharge time constants to be the same since during charging, current flows through R_1 and D_1, and when the capacitor is discharging, the flow is through R_2 and D_2.

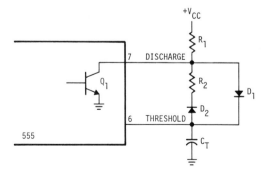

Figure 7-6   Timing Circuit for Square Wave Output

7-6 555 OUTPUT CIRCUIT

The internal output circuit of the 555 timer is shown in Fig. 7-7. It is a "totem pole" type output capable of sourcing or sinking 200 mA. In operation, Q_{28} is turned on during the timing cycle and Q_{26} is off. At the completion of the timing cycle, Q_{26} turns on and Q_{28} turns off. This type of connection charges or discharges any stray capacitance at the output quickly through the low "on" resistance of the transistors, and thus gives sharp rise and fall times at the output (typically 100 ns).

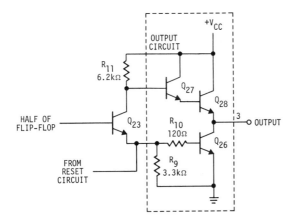

Figure 7-7 555 Timer Internal Output Circuitry

During the timing interval, the base of transistor Q_{23} is switched low, cutting off transistors Q_{23} and Q_{26}. With Q_{23} off, the collector of Q_{23} goes high, turning on transistors Q_{27} and Q_{28}, which are a Darlington pair. The output is pulled high to V_{CC}, minus the two base-to-emitter drops (0.7 V), and the drop across the 6.2-kΩ resistor (R_{11}). With a 15-V supply and a load of 200 mA, the output voltage is about 12.5 V. For the load to draw current during the timing cycle, the load should be connected from output to ground. This is the source connection.

When the timing cycle is complete, Q_{23} and Q_{26} are turned on and Q_{27} and Q_{28} are turned off. A load connected from the output to the supply voltage would now provide current into the collector of Q_{26}—which would sink this current. With a 200-mA load, the collector voltage of Q_{26} would be at about 2.5 V. Again, with a 15-V supply, this would leave 12.5 V across the load.

In summary, loads connected to ground are conducting current during the timing cycle, while those connected to V_{CC} are not.

7-7 555 RESET CIRCUIT

The reset input can be used to prevent the timing from starting or to terminate the timing cycle before the capacitor reaches the 2/3 V_{CC} value. This is accomplished by lowering the voltage on the reset Pin 4 to approximately 0.5 V. With reference to

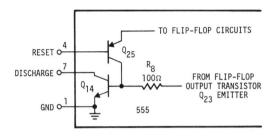

Figure 7-8 Reset Circuit

Fig. 7-8, transistor Q_{25} turns on when a low voltage is applied to the reset Pin 4. This will supply current to the base of transistor Q_{14} (Q_1 previously), turning on this transistor and discharging the external timing capacitor. The flip-flop is also reset by the current flow through the Q_{25} emitter. Thus, when the reset signal is removed, the timing cycle does not restart. However, if the trigger pulse is present when the reset signal is applied, the external timing capacitor will discharge, but the flip-flop will not be reset. This allows the timing capacitor to start charging again when the reset signal is removed.

7-8 ONE-SHOT TIMER APPLICATION

It is desired to have a battery saver circuit for a portable digital voltmeter (DVM) which will automatically shut off the power after one min of display. A possible circuit is shown in Fig. 7-9.

The meter "on" timer circuit is activated by momentarily depressing the push button to apply power to the 555. Since capacitor C_1 cannot charge instantaneously, the trigger input Pin 2 is held low; this causes the timer cycle to initiate. The output Pin 3 goes high, turning on transistor Q_1 and effectively grounding its collector. With Q_1 collector close to ground, the circuit for the DVM is completed and the meter turns

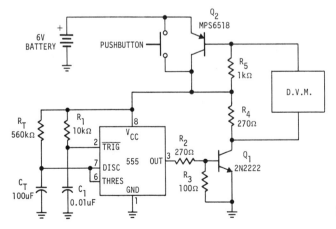

Figure 7-9 Battery Saver Circuit

on. Lowering the collector of Q_1 also turns on transistor Q_2, which shorts out the push-button switch and provides power to the circuit when the push button is released.

At the completion of the timing cycle (determined by R_T and C_T), the timer output goes low and Q_1 is turned off. The rising collector on this transistor turns off Q_2 and power is removed from the circuit.

Thus, current is only drawn from the battery during the timing cycle, and the problem of battery drain, if the meter power switch is left on, is eliminated. All of the components shown in Fig. 7-7 are mounted within the meter case.

7-9 OSCILLATOR APPLICATION

An example of an oscillator using a 555 timer is the power saver emergency flasher shown in Fig. 7-10.

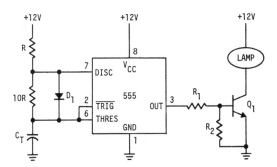

Figure 7-10   Power Saver Emergency Flasher

The circuit components are selected so that the lamp is turned on for 0.1 s and is off for 1 s. A conventional flasher is on and off for the same amount of time, resulting in greater battery drain.

You will recall that the output of the oscillator is high when the timing capacitor C_T is charging, and low when it is discharging. Since the capacitor is charged through resistor R and diode D_1 (diode shorts out resistor $10R$), and discharges through $10R$ (diode reversed biased), we get the required shorter on and longer off time—remembering that transistor Q_1 turns on when the output is high.

7-10 PRECISION TIMER

Another type of timer that has somewhat different capabilities than the 555 is the LM3905 precision timer. This timer offers greater versatility with better temperature stability (typically 0.003%/°C) than the 555 timer (typically 0.005%/°C). The LM3905 precision timer can operate with an unregulated power supply and is less sensitive to power supply and trigger voltage changes during the timing interval. Also, it has a logic capability for changing the sense of the output, i.e., high or low during the timing interval, and has an open collector and emitter on the output transistor. Because of lower internal leakage currents (2.5 nA versus 350 nA for the 555), the LM3905 is capable of providing constant timing periods from microseconds to hours. A functional block diagram of this timer is shown in Fig. 7-11.

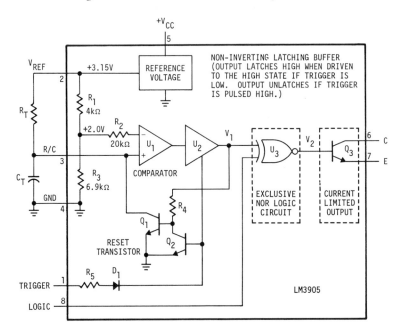

Figure 7-11   LM 3905 Functional Block Diagram

When power is initially applied, the reset transistor Q_1 is turned on, timing capacitor C_T is discharged, and latching buffer U_2 output V_1 goes high. Circuit operation is initiated when a *positive* pulse is applied to the trigger input Pin 1, which starts the timing cycle. When the trigger input pulse exceeds $+1.6$ V, it unlatches buffer U_2, causing its output to go low and turn off Q_1. Timing capacitor C_T now starts to charge from V_{REF} via timing resistor R_T. Resistors R_1 and R_3 form a voltage divider which establishes a $+2$-V input threshold level for comparator U_1. When the voltage on C_T exceeds the U_1 2-V input threshold level, U_1 toggles, latching U_2, which returns V_1 to its initial high state. With V_1 high, it turns on Q_1, which discharges C_T and ends the timing cycle. The timing cycle is determined simply by $R_T C_T$ without the 1.1 factor as required for the LM555 timer.

If the voltage on the trigger input is kept high when the timing cycle ends, transistor Q_2—which now has a collector supply voltage (since V_1 is high)—is turned on. This insures that the base voltage of Q_1 is low, thus maintaining Q_1 off. With Q_1 off, the timing capacitor will continue to charge toward the reference voltage V_{REF} ($+3.15$ V). During this period, V_1 will stay high, but buffer U_2 will not be latched. This means the output pulse is independent of the width of the input trigger pulse.

The logic input Pin 8 controls the sense of the output signal. If the logic input is high or open, the output transistor Q_3 will be turned off during the timing cycle. (The Exclusive-NOR U_3 has a low output when both inputs are different and a high output when both inputs are the same.) With the logic input grounded or low, the output transistor is turned on during the timing cycle. The sense of the output also can be changed depending on whether the output is taken from the collector or emitter of Q_3.

Voltages and currents for the output transistor must be consistent with the maximum device dissipation of 500 mW at 25°C. Output currents can be about 50 mA while the maximum collector voltage is 40 V. Since the collector is open, the collector supply voltage can be different from the voltage $(+V_{CC})$ used for the timing circuit.

As with the 555 timer, connecting C_T in parallel with R_T prevents the timer from cycling when power is applied.

7-11 EXAMPLE PROBLEMS

EXAMPLE 7-1

A burglar alarm system is required, consisting of entry switches and an alarm siren. If a switch is opened, the alarm should sound for 2 min and then shut off to take care of false-alarm situations. It is also required that upon activation of a "disable" switch the alarm system not be active for a period of 45 s. This allows entry or exit of authorized persons.

Solution.   The required 2-min and 45-s time intervals can be achieved with monostable operation of a timer. Let us develop the circuit using 555 timers as shown in Fig. 7-12.

The normal circuit operation is controlled by the alarm switches. If one switch opens, indicating that access is being attempted, the alarm timer (U_2) is triggered because the top end of R_{10} goes to ground, causing Pin 2 of U_2 to go low. The output of U_2 goes high, turning on Q_2, which provides a ground connection at point B for the alarm circuit. The alarm consists of two square wave generators—U_3, which oscillates around 1 kHz, and U_4, which is set for about 2 Hz. Note that the frequency of U_3 is caused to shift (which creates the siren effect) by connecting the bottom of its voltage divider (R_{14} and R_{15}) to the U_2 timing capacitor (C_6). This causes the voltage at the non-inverting input of U_3 to change, which in turn means that the voltage to which C_5 charges, moves, and the frequency of U_1 shifts. The output of U_3 is connected to the Darlington circuit $(Q_3$ and $Q_4)$, which provides the current gain to drive the speaker. Use of a negative supply voltage for the op-amps is avoided by use of the voltage divider R_{11} and R_{12}, which effectively provides ±6 V.

If the disable push button is depressed, U_1 is triggered and its output goes high, turning on Q_1. This brings the U_2 reset line to ground, which disables U_2. Now, building entry or exit is possible for the 45-second duration of U_1. Activation of the alarm with power turn-on is avoided by connecting the timing capacitor for U_2 to the positive supply rather than to ground.

Alarm switches normally used are magnetic reed switches. A small magnet is mounted on the door (or window), which causes the leaves of the switch to come together

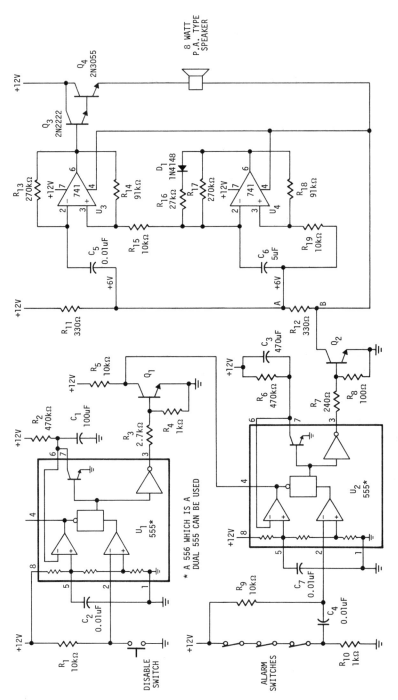

Figure 7-12 Burglar Alarm Circuit

and close the circuit. If the door is opened, the magnet moves away from the switch, causing the switch to open.

The power source for the alarm should be maintained even if the main AC power is turned off. This can be accomplished by floating a 12-V "Gel Cell"-type battery as shown in Fig. 7-13.

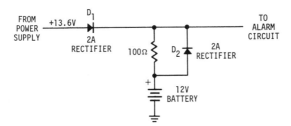

Figure 7-13 Standby Power Source

Notice that the power supply output voltage has been increased from +12 V to 13.6 V. This keeps the battery on trickle charge with 10 mA flowing through the 100-Ω resistor. If the input power is lost, rectifier D_1 reverse biases and D_2 is forward biased. Now the battery supplies the power for the alarm.

EXAMPLE 7-2

We wish to derive a 60_{Hz} TTL clock signal from a 1_{MHz} crystal oscillator. The accuracy of the clock signal should be the same as the oscillator over a 20°C range.

Solution.   The required frequency reduction is $1_{MHz}/60_{Hz}$, or 16,666 recurring to 1. We could use digital dividing circuits, but another approach is to use two LM322 timers (3905 timer with an adjustment terminal). Consider the circuit shown in Fig. 7-14.

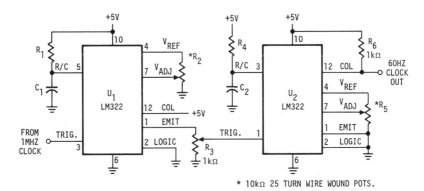

Figure 7-14 Frequency Reduction Circuit

To achieve the frequency reduction we will have each timer count down by a factor of 129 (the square root of 16,666). Since the 1_{MHz} clock has a period of 1 μs, U_1 will be set up for a time just less than 129 μs. Choosing values of 0.01 μF for C_1 and 13 kΩ for

R_1 will put us in the right timing range. By adjusting R_2 we can get the required time interval.

When the 129th clock pulse is received, U_1 will retrigger and the emitter of the output transistor (Pin 1) will go high, triggering U_2. This timer is set for a time interval of just less than the required 16,666 μs (within 129 μs). Using a 1 μF for C_2 and a 16-kΩ resistor for R_4 will put us in the ballpark. Fine adjustment will be accomplished by adjusting R_5.

The parts chosen must be consistent with an accuracy of one part in 129 or 0.77% over the required temperature change of 20°C. This amounts to a temperature coefficient of 38% parts per million per °C. The LM 322 has a coefficient of 30 parts, and using a ceramic capacitor for C_1, C_2 and a metal film resistor for R_1, R_4 will provide the required stability.

7-12 CHAPTER 7—SUMMARY

- A timer can cause circuit activation (or deactivation) for a predetermined amount of time.
- Timers can be connected in a continuous or oscillator mode.
- The most common timer is the 555 device with a timing equation of $1.1R_T C_T$.
- Triggering of the 555 requires a negative pulse of amplitude greater than 2/3 V_{CC}.
- The limit to how long the timer cycle can be depends on the internal leakage resistance of the timer and the leakage resistance of the timing capacitor (C_T).
- Maximum time period for the 555 is about 15 min, and the maximum value of the timing resistor (R_T) should be no greater than 1 MΩ.
- The control voltage terminal allows for "fine tuning" of the timer period or for frequency modulation when the timer is connected as an oscillator.
- Use of a "totem pole" output for the 555 provides fast rise and fall times.
- The reset terminal can be used to inhibit the start of a timing cycle or shorten an in-process timing period.
- The 3905-type timer has the following advantages over the 555:
  1. better temperature stability
  2. timing cycle independent of trigger
  3. less sensitivity to power supply changes
  4. uncommitted output transistor
  5. sense of output can be changed by a logic level input
  6. less leakage, which means longer timer intervals

7-13 CHAPTER 7—EXERCISE PROBLEMS

1. Determine the required input trigger waveform for a 555 timer operating from a +15-V power supply. Trigger pin is connected to the supply through a 10-kΩ resistor. Timer is set for 10 ms.

2. Show the complete circuit for a 555 timer which will turn on a 6-V, 50-mA lamp for 50 s. Circuit to be activated with a push-button switch and a +12-V supply is available.

3. Determine the output voltage swing for a 555 timer with a source load of 150 Ω and a power supply voltage of +15 V.

4. Design a 5-kHz oscillator circuit using a 555 timer. Output signal across 1-kΩ load resistor is to be high 60% of the time. Power supply voltage is +15 V. Use 10 nF for C_T.

5. Design a 20-s battery saver using a 555 timer for a digital voltmeter operating off +6 V with a current drain of 200 mA. Circuit is to be activated by the momentary depression of a push-button switch. Determine all component values.

6. With reference to Fig. 7-10, determine all component values for an emergency flasher with 0.1 s "on" time and 1 s "off" time. Use $C_T = 1\ \mu F$.

7. Show a circuit using an LM3905 timer to turn on a +2-V, 20-mA LED for 1 min. Also show the circuit change to have the LED normally on and turned off for 30 sec. Use a +12-V power supply and $C_T = 10\ \mu F$.

8. Design a zero dissipation timer using a LM3905 to turn on a relay for 30 sec; timer to be activated by a push-button switch. Relay operates on +6 V at 100 mA.

9. Redesign the circuit in Fig. 7-12 using a 3905 timer rather than a 555.

10. Generate a new circuit and specify all component values for a frequency divider circuit using a 555 timer rather than a LM322 (reference Fig. 7-14).

11. Determine the value of a resistor R_2 connected to ground which will provide a voltage at Pin 5 of a 555 timer such that the timer interval is RC rather than 1.1 RC.

12. Why is a "totem pole" output used on a 555 timer?

7-14 CHAPTER 7—LABORATORY EXPERIMENTS

Experiment 7A

Purpose: To demonstrate the use of a 555 as a "one-shot" timer (monostable).

Requirements: Connect up a 555 timer as a 1-ms "one shot." Use a +12-V power supply and trigger from a pulse generator. Select a resistor for a source or sink load. Sketch waveforms for timing capacitor, trigger input and output on same time axis—note DC levels. Increase the frequency of the pulse generator so that the pulse period is shorter than the timer period. This will cause frequency division, i.e., the pulse frequency will be greater than the frequency out of the timer.

Measure:

1. Output pulse width
2. Output voltage swing
3. Output rise time
4. Output fall time
5. Voltage level required to trigger circuit

Compare with specified values.

Observe: Frequency division.

Experiment 7B

Purpose: To demonstrate the use of a 555 as a free-running oscillator (astable).

Requirements: Connect up a 555 timer as a 10-kHz oscillator with output voltage high 60% of the time period. Use power supply of 15 V. Select a load resistor.

Measure:

1. Frequency
2. Duty cycle
3. Output rise time
4. Output fall time
5. Capacitor voltage

Compare with computed or specified values.

Experiment 7C

Purpose: To demonstrate the use of a LM3905 timer as a "one-shot" timer.

Requirements: Connect up the LM3905 timer to illuminate a light emitting diode (LED) for five minutes after the circuit is triggered. After completion, connect up a new second circuit which extinguishes the LED for two minutes after circuit is triggered. (*Caution—Do not* connect the logic pin to a supply voltage that is greater than +5 VDC.)

Measure: Time intervals and compare with predicted values.

8

Voltage-Controlled Oscillators

OBJECTIVES

Upon completion of this chapter, you should know the following concepts:
1. Purpose of a voltage-controlled oscillator
2. Basic block diagram
3. Modulation circuits
4. Frequency modulator
5. Device applications

8-1 INTRODUCTION

One of the fundamental requirements of frequency modulation (FM) is to vary an oscillator frequency with an input voltage (modulating voltage). Initially, this was accomplished by oscillators designed with reactance tubes—a device whose capacitance can be varied by changes in its input bias voltage. When the reactance tube input bias voltage (modulating voltage) would change, so would the oscillator output frequency, hence, FM. Today, the reactance tube has been replaced by the varactor diode for high frequency applications, but at lower frequencies the integrated circuit-type voltage-controlled oscillator (VCO) is used. This chapter is devoted to the description of this device.

8-2 A TYPICAL VOLTAGE-CONTROLLED OSCILLATOR INTEGRATED CIRCUIT

One of the more common types of integrated circuit VCO devices is the 566. This device has a maximum operating frequency of 1 MHz with a 10-to-1 range of frequency variation with a change in modulating input voltage. Figure 8-1 illustrates a basic block diagram of the 566 VCO.

8-3 VOLTAGE-CONTROLLED OSCILLATOR OPERATION

Referring to Fig. 8-1, the current source/sink provides a constant charging or discharging current to timing capacitor C_T. The amount of current is controlled by the

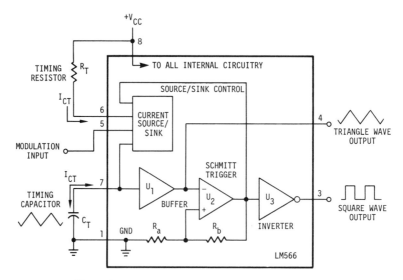

Figure 8-1 Voltage Controlled Oscillator Block Diagram

timing resistor R_T. Increasing the resistance of R_T decreases the capacitor charging/discharging current. Control of the charging current is also possible by changing the voltage at the modulating input. The voltage at Pin 6 is internally maintained at the same voltage as Pin 5. Thus, if the modulating voltage at Pin 5 is increased, the voltage at Pin 6 increases, resulting in less voltage across R_T and, therefore, less charging current.

The voltage developed on capacitor C_T is applied to the Schmitt trigger circuit U_2 via the buffer amplifier U_1. The output voltage swing on the Schmitt trigger is approximately V_{CC} to $0.5V_{CC}$. Resistors R_a and R_b form a positive feedback loop from the output of U_2 to its non-inverting input. With equal dividing resistors R_a and R_b, the non-inverting input swing is from $0.5V_{CC}$ to $0.25V_{CC}$. If the voltage on the timing capacitor C_T exceeds $0.5V_{CC}$ during charging, it will cause the Schmitt trigger output to go low ($0.5V_{CC}$). A low level on the output of U_2 causes the current source to change to a sink (discharging C_T). When C_T discharges to $0.25V_{CC}$, the output of the U_2 will swing high (V_{CC}), causing the current sink to return to a source (charging C_T). Since the source and sink currents are equal, it takes the same amount of time to charge C_T as it does to discharge this capacitor. This results in a triangular voltage waveform on C_T which is available as a buffered output at Pin 4. A square wave appears at the output of the Schmitt trigger and is inverted by inverter U_3 for a second output at Pin 3. If the current from the source/sink is increased, the charge/discharge time for the capacitor is reduced and the output frequency is increased (shorter period).

8-4 DETERMINING VCO OUTPUT FREQUENCY

The output frequency can then be changed by three methods:

1. Changing the value of C_T,
2. Changing the value of R_T,
3. Changing the voltage at the modulating input terminal.

We can determine the actual frequency of oscillation from the time it takes to charge and discharge the capacitor. The basic equation for charging a capacitor from a constant current source is:

$$i = C \frac{dV}{dt} \tag{8-1}$$

$$\frac{dV}{dt} = \frac{i}{C} \tag{8-2}$$

where dV is the voltage change on the capacitor during the time change dt.

The total voltage on the capacitor changes from $0.25V_{CC}$ to $0.5V_{CC}$.

Thus
$$dV = 0.5V_{CC} - 0.25V_{CC} \tag{8-3}$$
$$= 0.25V_{CC}$$

and
$$\frac{0.25V_{CC}}{dt} = \frac{i}{C_T} \tag{8-4}$$

therefore
$$dt = \frac{0.25\ V_{CC}\ C_T}{i} \tag{8-5}$$

The triangular waveform on the capacitor has a period $T = 2dt$ (equal charging and discharging time). The frequency of oscillation is:

$$f = \frac{1}{T} \tag{8-6}$$

$$f = \frac{1}{2dt} \tag{8-7}$$

Substituting dt from Eq. 8-5, the frequency of oscillation is:

$$f = \frac{i}{0.5\ V_{CC}\ C_T} \tag{8-8}$$

but
$$i = \frac{V_{CC} - V_5}{R_T} \tag{8-9}$$

Where V_5 is the voltage at Pin 5, then:

$$f = \frac{2(V_{CC} - V_5)}{C_T\,R_T\,V_{CC}} \tag{8-10}$$

For best operation, the resistance of R_T should be within 2 to 20 kΩ.

In actual operation, R_T and C_T are selected for the correct center operating frequency and the modulating input voltage is varied to give a shift in the output frequency (frequency modulation). The range of allowable variation of the modulating input signal is from $0.75V_{CC}$ to V_{CC}, which yields an output frequency variation of about 10 to 1. With no modulation input signal, the voltage at Pin 5 is biased at $7/8\ V_{CC}$. This allows us to simplify Eq. 8-10 to give the unmodulated frequency f_o:

$$f_o = \frac{2(V_{CC} - \frac{7}{8}V_{CC})}{C_T\,R_T\,V_{CC}}$$

$$f_o = \frac{1}{4\ C_T\,R_T} \tag{8-11}$$

If we want to determine what input modulation voltage (ΔV) is required to give a given output frequency deviation, (Δf), we can assume the original frequency is f_o and the new frequency f_1, then:

$$\Delta f = f_1 - f_o$$

$$= \frac{2(V_{CC} - V_5 + \Delta V)}{C_T R_T V_{CC}} - \frac{2(V_{CC} - V_5)}{C_T R_T V_{CC}}$$

$$= \frac{2\Delta V}{C_T R_T V_{CC}} \tag{8-12}$$

Solving for ΔV:

$$\Delta V = \frac{\Delta f \, C_T R_T V_{CC}}{2} \tag{8-13}$$

Substituting $R_T C_T$ from Eq. 8-11:

$$\Delta V = \frac{\Delta f \, Vcc}{8 f_o}$$

8-5 MODULATION INPUT CIRCUIT

A typical modulation input circuit is shown in Fig. 8-2.

The two resistors, R_1 and R_2, are used to set up the required midpoint operating bias of $0.875 V_{CC}$. The actual modulating input signal is coupled through coupling capacitor C_C.

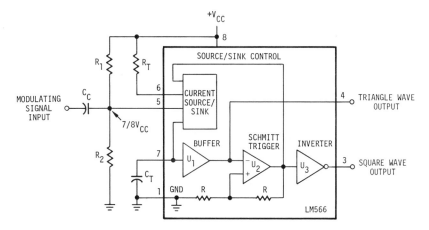

Figure 8-2 Modulation Input Circuit

To minimize the voltage drop across the coupling capacitor, it should be selected to provide a reactance of no greater than one tenth of the load resistance at the lowest modulating frequency. The input load in the circuit is R_1 in parallel with R_2, assuming zero reactance between V_{CC} and ground.

8-6 A TYPICAL FREQUENCY MODULATOR

To further illustrate the VCO operation, we will use the frequency modulator circuit shown in Fig. 8-3 as an actual numerical example. Assume the desired center frequency f_a of the circuit is 10 kHz and the modulating input signal has a frequency (f_M) of 100 Hz with an amplitude of 200 mV peak to peak. The applied V_{CC} is +15 VDC. We would like to find the circuit component values and the frequency deviation.

The DC level at the modulating input is:

$$V_5 = \frac{V_{CC} R_2}{R_1 + R_2} \tag{8-14}$$

$$= \frac{15 \text{ V} \times 82 \text{ k}}{82 \text{ k} + 12 \text{ k}}$$

$$= 13.1 \text{ V}$$

This is approximately 7/8 V_{CC}.

The coupling capacitor can be determined by making the capacitor reactance equal to one tenth of the parallel combination of R_1 and R_2 or:

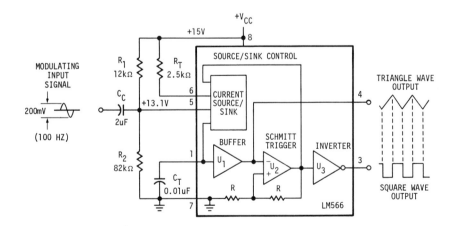

Figure 8-3   Frequency Modulator

$$X_c = \frac{R_1 \times R_2}{10(R_1 + R_2)} \tag{8-15}$$

$$= \frac{(12\ k \times 82\ k)}{10(12\ k + 82\ k)}$$

$$= 1\ \text{k}\Omega$$

also
$$C_c = \frac{1}{2\pi f_M X_c} \tag{8-16}$$

$$= \frac{1}{2\pi \times 100\ \text{Hz} \times 1k}$$

$$= 1.59\ \mu\text{F}$$

The closest standard capacitor size is 2 μF.

With the required VCO center frequency of 10 kHz, we can select 0.01 μF as a reasonable value for C_T and solve for R_T using Eq. 8-10.

$$R_T = \frac{2(V_{CC} - V_5)}{f_c\ C_T\ V_{CC}} \tag{8-17}$$

$$= \frac{2(15\ \text{V} - 13.1\ \text{V})}{10\ \text{kHz} \times 10^{-8}\text{F} \times 15\ \text{V}}$$

$$\sim 2.5\ \text{k}\Omega$$

We can determine the maximum amount of deviation of the carrier by using Eq. 8-12.

$$\Delta f = \frac{2\Delta V}{C_T R_T V_{CC}}$$

$$= \frac{2 \times 0.02\ \text{V}}{10^{-8} \times 2.5 \times 10^3 \times 15\ \text{V}}$$

$$\sim 100\ \text{Hz}$$

Thus, the maximum excursions of the carrier will be 9.95 kHz and 10.05 kHz, corresponding to the positive and negative peaks of the input sine wave. The output can be taken from Pin 3 or Pin 4, depending on whether it is desired to have a frequency modulated square wave or triangular wave.

If the modulation output signal is digital, the carrier will assume a frequency corresponding to a one and another frequency corresponding to a zero; thus, an abrupt change occurs in the carrier frequency. This method of digital modulation is known as frequency shift keying (FSK). Remember that when a signal passes through a capacitor, there is no DC component passed, and the waveform will center itself so there is equal signal "area" above and below the waveform. Thus, as the code changes, the one and zero voltage values will change; i.e., if the number of ones in the code increases, the signal will shift down. This means that the actual voltage for a one and zero will change, which will cause a change in the one/zero frequency. Therefore, AC coupling should not be used for digital input signals.

A circuit which will provide the necessary DC coupling is shown in Fig. 8-4.

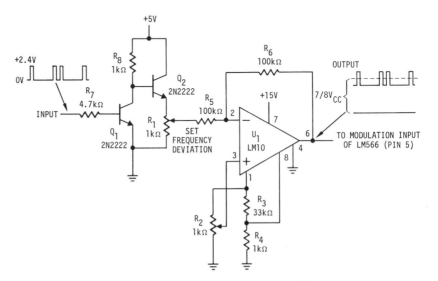

Figure 8-4 Digital Modulation Circuit for VCO

It will be recalled that the LM10 op-amp contains a 200-mV reference, a reference amp, and an op-amp. In this application, the reference amplifier is set up for a gain of 34 $(R_3/R_4 + 1)$, which provides a voltage at Pin 1 of (34×200) mV or 6.8 V. The gain of the op-amp is two $(R_6/R_5 + 1)$ for the input voltage taken from the wiper of R_2. By adjusting R_2, we can get the desired $7/8V_{CC}$ voltage at the op-amp output—which is connected to the modulation input of the VCO. Assuming the digital input is at TTL levels, a one level is greater than 2.4 V and a zero level is less than 0.4 V. The one level turns on Q_1 and causes Q_2 to turn off, making the pot voltage zero, while the zero level turns off Q_1 and the emitter voltage of Q_2 rises to 4.3 V (the supply minus the 0.7-V base-emitter drop).

Let us apply this digital modulation circuit to generate the Kansas City Standard, which is an FSK system, using 1200 Hz for a zero and 2400 Hz for a one.

Increasing the voltage at the modulation input of the 566 causes the output frequency to go down. This is contrary to what we want for the K.C. Standard. Q_1 serves as a logic inverter, giving us the correct output sense. If we use a +12-V supply for the VCO, the DC voltage at Pin 5 should be 10.5 V $(7/8V_{CC})$. We can adjust R_2

to set this value. With a one input from the logic, the voltage at Pin 5 should be at its minimum value (highest frequency). Let us use 10.5 V for this value. We should now select the values of R_T and C_T for 2400 Hz. Selecting a 0.01-μF capacitor for C_T, we can determine R_T from Eq. 8-11.

Adjusting R_2 will take care of the tolerances of R_T and C_T and get exactly 2400 Hz.

The voltage change on Pin 5 to give 1200 Hz can be determined from Eq. 8-13; recognizing that $f = 1200$ Hz.

$$\Delta V = \frac{1200 \times 10^{-8} \times 10^4 \times 12}{2}$$

$$= 0.72 \text{ V}$$

By setting R_1 with a zero logic input, we can achieve this value.

8-7 CHAPTER 8—SUMMARY

- A voltage-controlled oscillator enables us to convert voltage variations into frequency variations.
- VCOs can be used to provide frequency modulated output signals and also single frequency tones.
- The most common VCO is the 566, which has a maximum operating frequency of 1 MHz.
- A sine wave and a square wave output are provided by the 566.
- When digital signals are used to modulate a VCO, the output shifts between two discrete frequencies. This is called frequency shift keying (FSK).
- Because the average value of a digital data stream changes with time, DC coupling must be used when the VCO is modulated with this type of signal.
- To establish the center frequency of the VCO, the modulation input to the 566 must be DC biased to $7/8 V_{CC}$.
- The kHz/V factor for a VCO indicates how much frequency deviation of the output occurs with a given input voltage change.

8-8 CHAPTER 8—EXERCISE PROBLEMS

1. What will happen to the output frequency of the 566 VCO if the internal resistor R_a is greater than R_b (reference Fig. 8-1).
2. Find the square wave output voltage swing from a 566 VCO if the supply is 12 V.
3. Show a circuit which will allow manual switching of an output tone from 5 kHz to 10 kHz. Use a 566 VCO and a +12-V supply.
4. Design a VCO with a nominal frequency of 5 kHz and show the AC coupled input modulation circuit for 50-Hz modulation signal. Specify all parts and use +12 V for $+V_{CC}$. Use a value of 0.01 μF for C_T.
5. For the Exercise Problem 4, determine the input modulation swing to give an output frequency increase of 100 Hz.
6. Determine the kHz/V factor for a LM566 at a center frequency of 5 kHz and a center frequency of 10 kHz with a $V_{CC} = 12$ V.
7. Find the excursions of the center frequency if the modulation signal has a plus and minus 0.2-V level; the center frequency is 8 kHz and the supply is 15 V.
8. Show the circuit for a sweep frequency generator with a linear output sweep from 2 kHz to 20 kHz. Sweep rate is to be 60 Hz—consider a LM555 with a constant current source charging the timing capacitor C_T for the modulating signal. Supply voltage is to be +12 V.

8-9 CHAPTER 8—LABORATORY EXPERIMENT

Purpose: To demonstrate the relationship between the input voltage and output frequency for a LM566 voltage-controlled oscillator and also to show how frequency modulation can be achieved.

Requirements:

1. Connect up a LM566 using a +12-V supply and set the nominal frequency at 5 kHz. Apply a DC voltage to the modulating input and determine the output frequency change versus input voltage change over the device operating range.
2. With the modulation input set for a DC value of 7/8 V_{CC}, couple in a 1-V peak-to-peak 50 Hz sine wave and sketch the FM modulation of the triangular wave.

Measure:

1. For Requirement 1, measure frequency change as a function of input voltage (kHz/V).
2. For Requirement 2, determine frequency excursions (minimum and maximum values).

Compare results with the computed values.

9

Phase Locked Loops

OBJECTIVES

At the completion of this chapter you will know the following phase locked loop concepts:
1. Basic operation
2. Lock-in range
3. Capture range
4. FM demodulation
5. Synthesizer operation

9-1 INTRODUCTION

A phase locked loop (PLL) is a circuit which contains a voltage-controlled oscillator (VCO) that is made to oscillate at the same frequency as an incoming signal. When the VCO frequency is the same as the incoming signal, it is said to be *locked in*. The frequency of the VCO is controlled by an error signal that is generated by a comparison of the incoming signal with the VCO output frequency. Comparison of the signals is accomplished by a phase comparator within the PLL.

Applications for the PLL are quite varied and include demodulation of frequency modulated (FM) signals, frequency multiplication and division, signal regeneration, and generation of different frequencies from a fixed single source— frequency synthesis.

9-2 PHASE LOCKED LOOP OPERATION

Phase locked loop operation can be best explained by looking at the LM565, which is a typical phase locked loop device and is shown in block diagram form in Fig. 9-1.

With reference to Fig. 9-1, an input signal is applied to the phase comparator (Pins 2 and 3) and is compared with the signal from the internal voltage-controlled oscillator (VCO). If a frequency or, therefore, a phase difference occurs between the incoming and VCO signals, an error signal is generated at the output of the phase comparator. This error signal is amplified and filtered at the output by a low pass filter consisting of resistor R_a (internal to the PLL) and capacitor C_2 and fed back to the VCO as a control signal. The filtered signal is also used as the output signal when the

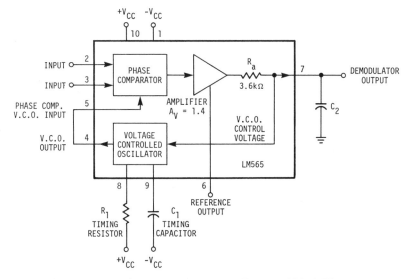

Figure 9-1   LM565 Phase Locked Loop Functional Block Diagram

PLL is operated as an FM demodulator. If the input frequency and VCO output frequency differ, the feedback control signal shifts the VCO frequency until it matches the incoming frequency. When a match occurs, the VCO is in the locked condition. If the incoming signal frequency is varied, the VCO output frequency will follow, provided that the incoming signal frequency stays within a hold-in (lock-in) range of the PLL. A phase difference of 90° exists if the incoming signal is at the same frequency as the VCO idle frequency.

If the incoming signal frequency is below the VCO idle frequency, the lock-in phase ranges from 90° to 0°; with the incoming signal frequency above, the phase is from 90° to 180°. When in lock, the output voltage from the phase detector follows the cosine of the phase difference between the incoming signal and the VCO frequency. If the input signal and the VCO idle frequency are the same, the output of the phase detector is zero (cos 90° = 0). When the incoming signal frequency is above the VCO idle frequency, the phase detector output ranges from 0 to −0.7 V, and when the incoming signal frequency is below, it ranges from 0 to +0.7 V. If the incoming signal is a high level, a sine wave will become a square wave and the phase detector output response will become linear rather than cosine. The phase conversion gain K_D of the phase detector is:

$$K_D = \frac{1.4 \text{ V}}{\pi \text{ rad}} \tag{9-1}$$

Since 90° = $\pi/2$ rad, the output voltage from the phase detector, as stated above, varies between ±0.7 V. The amplifier following the phase detector has a gain of 1.4, so the output voltage from the amplifier is (0.7 V × 1.4), or ±0.98 V.

If the incoming signal frequency is moved from the VCO frequency so that the phase shift falls outside the 0° and 180° limits, then the phase detector output is reduced (see Fig. 9-2). Since it requires more output than is available from the phase detector to shift the VCO frequency to the incoming frequency, the circuit falls out of lock.

In order to determine the actual lock-in range of the PLL, we need to consider all the circuit constants. K_D was given above as 1.4 V/πrad. The amplifier gain is 1.4

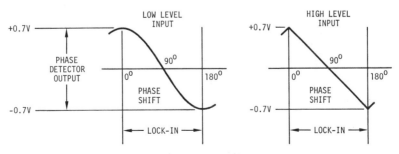

Figure 9-2 Phase Detector Characteristics

and the VCO conversion factor K_O is given as:

$$K_o = \frac{50 f_o \text{ rad}}{V_{CC} \text{ V/s}} \qquad (9\text{-}2)$$

where V_{CC} is the total supply voltage and f_O is the free-running (idle) frequency of the VCO. The actual frequency lock-in (f_L) range can be determined as follows:

$$2 \pi f_L = K_O K_D A_V \theta_d \qquad (9\text{-}3)$$

where θ_d is the maximum phase shift within the lock-in range. Re-equating, f_L is:

$$f_L = \frac{K_O K_D A_V \, \pi/2}{2 \pi} = \frac{K_O K_D A_V}{4} = \frac{50 f_O}{V_{CC}} \times \frac{1.4}{\pi} \times \frac{1.4}{4} \qquad (9\text{-}4)$$

$$f_L = \pm \frac{7.8}{V_{CC}} f_O$$

This equation shows that the higher the center frequency, the greater the lock-in range, but raising the supply voltage decreases the lock-in range.

The lock-in range can also be reduced by lowering the amplifier gain. This is done by connecting a resistor between the demodulated output (Pin 7) and the reference output (Pin 6). Using resistors from $30\ K$ down to $0\ \Omega$ reduces the lock-in range from $\pm 60\%$ down to $\pm 20\%$ of f_o.

The idle frequency f_o of the VCO is adjusted by means of timing resistor R_1 and timing capacitor C_1 and is determined by:

$$f_o = 0.25/R_1 C_1 \qquad (9\text{-}5)$$

For best operation, R_1 should be within the range of 2 to 20 kΩ, with 4 kΩ being the optimum value.

If the incoming signal is lower than the VCO frequency, and is then brought closer, capture will occur. Increasing the frequency of the input signal still further will cause the VCO to follow the signal up to the upper lock-in frequency. The VCO will then drop back to its idle frequency (f_o) determined by R_1 and C_1 and will no longer track the incoming signal.

Looking further at the phase comparator output, the phase comparator can be considered a multiplier for low level inputs which multiplies the input signal ($E_s \sin 2\pi f_s t$) by the VCO signal ($E_o \sin 2\pi f_o t + \theta_o$). Thus, the phase comparator output is:

$$e_{\text{peak}} = K E_s I_o (\sin 2\pi f_s t)(\sin 2\pi f_o + \theta_o) \qquad (9\text{-}6)$$

where K is the comparator gain (or attenuation constant) and θ_o the phase shift between the incoming and VCO signals when in the locked condition. Using trigonometric expansion:

$$e_{\text{peak}} = K E_s E_o \cos(2\pi f_s t - 2\pi f_o t - \theta_o) - \cos(2\pi f_s t + 2\pi f_o t + \theta_o) \qquad (9\text{-}7)$$

At lock, $f_s = f_o$. Thus, the difference term $(2\pi f_s - 2\pi f_o)$ is zero and

$$KE_sE_o\cos(2\pi f_s - 2\pi f_o - \theta_o) = KE_sE_o\cos(-\theta_o) \qquad (9\text{-}8)$$

We are left with

$$e_{\text{peak}}KE_sE_o\cos(-\theta_o) - KE_sE_o\cos[2\pi(2f_o) + \theta_o] \qquad (9\text{-}9)$$

This yields a double frequency term $KE_SE_O \cos [2\pi(2f_o) + \theta_o]$ and a DC term $KE_sE_o \cos (\theta_o)$, which varies as a function of phase (θ_o) between the two signals. This last term is the useful output term from the phase comparator. Looking at this term, we find it is zero when the phase shift is 90°. Thus, if the incoming signal and VCO are at the same frequency, the phase shift is 90° (cos 90° = 0) since an error signal is not required to shift the VCO to the signal frequency.

The low pass filter (R_a and C_2) is used to limit the bandwidth of the output to pass the wanted signal (FM demodulation) and reduce the double frequency term ($2f_o$) to an insignificant value. The 3-dB down frequency of the filter ($R_a = X_{C2}$) is chosen somewhat greater than the wanted demodulated signal.

The nominal DC level of the demodulator output (Pin 7) is seven eighths of the total supply voltage. The reference output (Pin 6) is also at this voltage and enables a differential output to be used, i.e., Pin 6 and 7, connected to the inverting and non-inverting inputs of an op-amp. Since the 7/8 V_{CC} is common mode, only the wanted signal will appear at the output of the op-amp.

If the incoming signal frequency is far removed (outside of the lock-in ranges) and is then brought closer, a point is reached where the VCO frequency is "captured" by the incoming signal frequency. This occurs when the error voltage developed across the low pass filter formed by R_A and C_2 is sufficient to move the VCO to the incoming frequency.

Before the VCO has been captured by the incoming signal, the error voltage is a sine wave with a frequency that is the difference between the VCO and the input signal. This sine wave becomes lower in frequency as the signal frequency gets closer to the VCO. With a lower frequency error signal, the error voltage across the low pass filter capacitor C_2 increases. The VCO frequency swings back and forth with greater excursions at the slower rate as the signal frequency comes closer. When the error voltage is sufficient to cause the VCO to reach the incoming signal frequency, capture occurs. Once captured, the VCO will follow the incoming frequency over the lock-in range.

Remembering the circuit operation just described, we will develop an equation for the capture frequencies.

On the slope portion of a low pass filter the output voltage (V_{OUT}) can be approximated by:

$$V_{OUT} = V_{IN}X_C/R_a$$

The maximum output from the amplifier was determined previously to be 0.98 V. This is the input voltage to the filter, so:

$$V_{\text{OUT}} = \frac{0.98 \text{ V}}{3.6 \text{K} \times 2\pi f_c C_2}$$

with f_C the difference frequency between the VCO and the incoming frequency. At the point of capture, the following relationship holds:

$$\frac{f_c}{f_L} = \frac{V_{\text{OUT}}}{0.98 \text{ V}}$$

And substituting for V_{OUT} and solving for f_C:

$$f_c = \pm \sqrt{\frac{f_L}{2\pi R_a C_2}} \tag{9-10}$$

The plus and minus signs on the capture and hold frequencies indicate deviation of both sides of the VCO center frequency. It should be realized that the capture frequency must always be equal to the lock-in frequency (without C_2) or less than the lock-in frequency.

9-3 HIGHER FREQUENCY OPERATION

A PLL that can operate up to 50 MHz is the NE564 shown in block diagram form in Fig. 9-3.

Differences in the block diagram of this device and that of the 565 are that the VCO is fed directly from the phase comparator, and the addition of a DC restorer and a Schmitt trigger circuit for the digital signal output.

The DC restorer is effectively an integrator that can provide an output which is the average of the input signal, or it can also be used as a low pass filter. Coupling the average level into one side and the demodulated digital signal into the other side of the Schmitt trigger makes the output independent of changes in the DC level caused by center frequency drift.

A Schmitt trigger is used to reduce the effect of noise causing false transitions of the digital output signal. The hysteresis of the circuit can be changed by varying the voltage at Pin 15.

The design equations for the 564 are as follows:

$$\text{VCO frequency } f_o = \frac{1}{2.5 \times 10^3 \, C_1} \tag{9-11}$$

where C_1 is the capacitor between Pins 12 and 13.

The analog output voltage (V_0) is determined by the equation

$$V_o = \frac{(f_s - f_o)}{-1.8 \text{ MHz}} \tag{9-12}$$

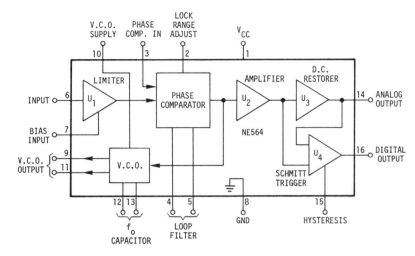

Figure 9-3   Block Diagram NE564

Lock-in range is fixed at about $\pm 35\%$ of f_O and can be reduced by lowering the current into terminal 2. The capture frequency is typically $\pm 15\%$ of f_O.

An FM demodulator using the NE564 is shown in Fig. 9-4.

The 80 pF capacitor sets the frequency of the VCO at the incoming carrier frequency of 5 MHz. Appearing at the analog output is the demodulated signal.

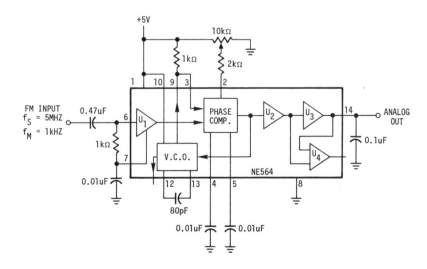

Figure 9-4   5 MHz F.M. Demodulator Circuit

9-4 FREQUENCY DEMODULATOR EXAMPLE

In order to more fully illustrate PLL operation, an example of a frequency demodulator is given.

Let us assume that we wish to FM demodulate a 100-kHz carrier modulated with a 1 kHz sine wave which deviates the carrier frequency ±10 kHz.

First, we choose a 565 PLL and set up the VCO frequency to 100 kHz. If we pick a 1 nF capacitor for C_1, then R_1 from Eq. 9-5 is:

$$R_1 = \frac{0.25}{C_1 f_o} \qquad\qquad (9\text{-}13)$$

$$= \frac{0.25}{10^{-9}\,\text{F} \times 10^5\,\text{Hz}}$$

$$= 2500\,\Omega$$

In order to determine the output voltage we will get with a 10 kHz deviation of the carrier, we first need to determine the lock-in frequency f_L by using Eq. 9-4. (Assume total supply voltage $V_{CC} = +12$ V.)

$$f_L = \frac{7.8\,f_o}{V_{CC}}$$

$$= \frac{7.8 \times 100\,\text{kHz}}{12\,\text{V}}$$

$$= 65\,\text{kHz}$$

If the carrier is deviated by 65 kHz, we will get 0.7 V out of the phase detector (Eq. 9-1) and (0.7 V × 1.4) or 0.98 V out of the amplifier. But we are deviating the carrier by 10 kHz, not 65 kHz, so the voltage out (V_{OUT}) of the amplifier is:

$$V_{OUT} = \frac{0.98\,\text{V}}{65\,\text{kHz}} \times 10\,\text{kHz}$$

$$= 150\,\text{mV}$$

The amplifier output voltage is the FM demodulated output we desire and, in this case, is a 1-kHz sine wave with a peak-to-peak amplitude of 2 × 150 mV = 300 mV.

Also appearing at the amplifier output is the unwanted double frequency component (Eq. 9-9) of peak amplitude 0.98 V. The low pass filter capacitor (C_2) value must be picked to reduce this component below the wanted signal. If we pick the 3-dB down frequency of the filter at 2 kHz (twice the modulation frequency), then C_2 is:

$$C_2 = \frac{1}{2\pi \times 3.6 \text{ K} \times 2 \text{ kHz}}$$

$$= 0.022 \ \mu\text{F}$$

Using this capacitor, the response is down about 40 dB at 200 kHz (-20 dB per decade roll-off). This means that the response at 200 kHz is a factor of 100 less than the response at 1 kHz. Thus, with the filter circuit, the signal amplitude remains at 300 mV but the double frequency term is reduced to 0.98 V/100 = 9.8 mV. Since we have determined C_2, the capture frequency is established and is, from Eq. 9-10:

$$f_c = \pm \sqrt{\frac{65 \text{ kHz}}{2\pi \times 3.6 \text{ K} \times 2.2 \times 10^{-8} \text{ F}}}$$

$$= \pm 11.4 \text{ kHz}$$

This means that if the signal carrier frequency is within 11.4 kHz of the PLL, VCO frequency, capture will occur. The complete circuit schematic is shown in Fig. 9-5.

The circuit has been set up using a single $+12$-V supply, which requires that Pin 2 be biased to $+6$ V. A nominal DC voltage of 10.5 V will appear at Pin 7.

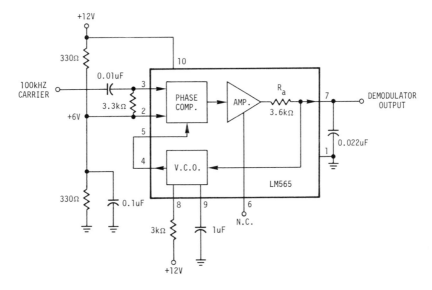

Figure 9-5. 100KHz F.M. Demodulator Circuit

9-5 FREQUENCY SYNTHESIS

This operation involves generating various frequencies from a fixed source—typically, a crystal oscillator. The new frequencies have the same long-term frequency stability as the crystal, but without the need for a different crystal for each frequency. A phase locked loop can be used for this application.

If we desire a frequency which is a multiple of the crystal frequency, we can use the PLL circuit shown in Fig. 9-6.

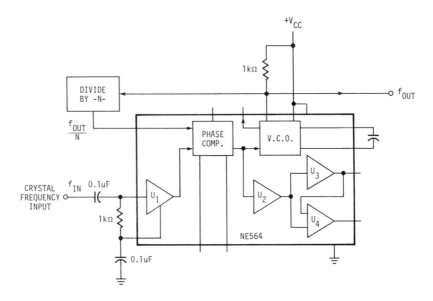

Figure 9-6   Frequency Multiplier

Notice that the VCO output is not fed directly to the phase detector, but passes through a digital divide by N counter first, which reduces the VCO output frequency by the countdown factor N. Since the closed loop ensures that the two frequencies into the phase detector are the same, the VCO output is N times the crystal frequency.

The output signals from the synthesizer are not always simple multiples of the crystal frequency. A heterodyne conversion system shown in Fig. 9-7 allows for nonharmonically related outputs.

Added components from the previous circuit are the second crystal frequency f_2 and the mixer. A mixer is a non-linear device that generates the sum and difference of two input frequencies as well as harmonics. In this application, all output frequencies from the mixer are filtered out except for the difference frequency $f_O - f_2$ (f_O is greater than f_2). Since the input frequencies to the phase detector must be equal, f_1 is:

$$f_1 = \frac{f_{\text{OUT}} - f_2}{N} \tag{9-14}$$

then f_{OUT} is

$$f_{\text{OUT}} = Nf_1 + f_2 \tag{9-15}$$

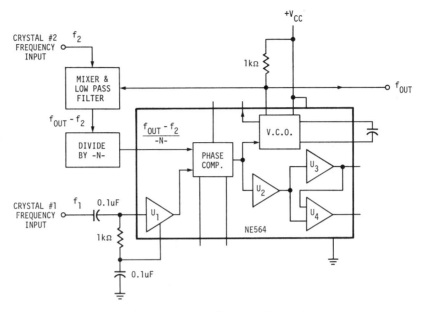

Figure 9-7 Heterodyne Frequency Conversion

Selecting suitable choices for N, f_1, and f_2, we can generate the desired output frequency. In practice, the counter factor N is made variable so we can select several output frequencies.

As an example of a heterodyne synthesizer, let us assume we need to generate local oscillator frequencies for a 44-channel CB receiver with an IF frequency of 7.8 MHz. The CB channel range is from 26.965 MHz to 27.405 MHz in 10-kHz increments. Thus, the required local oscillator range is 34.765 MHz to 35.205 MHz (adding IF frequency).

With the required 10 kHz increments, f_1 should be at this frequency. Crystal controlled oscillators are at higher frequencies than 10 kHz—so let us assume we have a 1 MHz crystal available for the oscillator. We can then reduce this frequency by a factor of 100 by use of decade counters.

The frequency f_2 can be determined by assuming that the lowest value of N (corresponding to the lowest output frequency of 34.765 MHz) is 1, then, from Eq. 9-15, f_2 is:

$$f_2 = f_{OUT} - Nf_1$$
$$= 34.765 \text{ MHz} - 10 \text{ kHz}$$
$$= 34.755 \text{ MHz}$$

In order to obtain the full range of possible signals, N must be varied from 1 to 44.

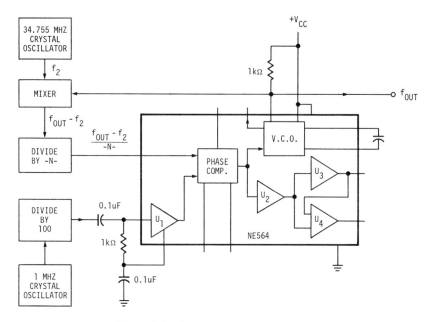

Figure 9-8 C.B. Frequency Synthesizer

A complete block diagram of the CB frequency synthesizer is shown in Fig. 9-8.

We can see that in order to vary the receiver frequency, the N factor in the divide by N counter must be varied. This is accomplished by using a digital programmable counter, i.e., 74192, MC4018, etc. These counters are programmed by changing the levels on four control lines, which allows selection of up to 16 combinations. Using two cascaded counters allows the 44 selections required for the example problem.

Integrated circuits exist such as the MM55108 PLL frequency synthesizer, which can perform most of the functions described above; however, it still requires the use of external mixers, crystal oscillators, and VCOs.

9-6 CHAPTER 9—SUMMARY

- A phase locked loop is used to demodulate FM signals or to generate new frequencies.
- Once it is captured by the incoming signal, a VCO in a phase lock loop has the same frequency as the signal.
- The phase error between the incoming signal and the VCO creates the voltage necessary to shift the VCO frequency so that it is identical to the incoming frequency.

- If the incoming signal is injected at the same frequency as the VCO, the phase shift for the 565 will be 90° and the error voltage will be zero.
- The lock-in range is the range of frequencies over which the VCO will follow the incoming signal.
- At the limits of the lock-in range the phase shift is either O° or 180°.
- The capture frequency is that frequency at which there is sufficient error voltage to move the VCO frequency to the same frequency as the incoming signal.
- Capture frequency is determined by the low pass filter capacitor C_2 and the lock-in range and can never be outside of the lock-in range.
- When used for FM demodulation, the output signal is the error voltage created by the VCO tracking the input carrier.
- By putting a frequency divider (N) between the VCO and the phase detector, we can scale up the frequency of the VCO by the factor N. This is used in frequency synthesis.

9-7 CHAPTER 9—EXERCISE PROBLEMS

1. Is there an error signal out of the phase detector when "in lock"?
2. Why does the capacitor C_2 in Fig. 9-1 affect the capture range?
3. What is the phase shift between the 565 internal VCO and the incoming signal just as the circuit drops out of lock at the low frequency end?
4. Calculate the analog output voltage from a NE564 if the idle (VCO) frequency is 20 MHz and the incoming signal is at 21 MHz.
5. Specify the components for a 565 PLL to operate at a frequency of 10 kHz. (Use a 10-nF capacitor for C_1.) If the supply is +12 V, determine the lock-in range and capture frequency.
6. With reference to Problem 5, determine the output signal voltage if the incoming signal carrier deviates by 1 kHz.
7. Determine the value of the PLL filter capacitor (C_2) if the maximum modulation frequency is 500 Hz. With a carrier frequency of 10 kHz, calculate the rejection of the double frequency term by the filter capacitor.
8. Compute the change in output voltage from a 565 if an incoming signal is applied which is 2 kHz from the VCO idle frequency and the lock-in range is ±5 kHz.
9. Show the complete circuit for an FM modulator and demodulator, using a 566 and a 565. Carrier frequency to be 10 kHz, modulation frequency to be 100 Hz with a peak amplitude of 200 mV. Determine the amplitude of the demodulated signal. Use a single +12-V supply.
10. Show a PLL circuit with a 100-kHz crystal oscillator which will provide an output frequency of 700 kHz.
11. A heterodyne synthesizer is required to provide 10 frequencies from 5 MHz to 9.5 MHz in 500-kHz increments. Show circuit and specify the crystal frequencies and components used.

12. Find the capture frequency for the circuit in Problem 5 if $C_2 = 47nF$.

9-8 CHAPTER 9—LABORATORY EXPERIMENTS

Experiment 9A

Purpose: To demonstrate the frequency lock capability of a phase locked loop (PLL).

Requirements: Connect up the LM565 PLL for a nominal operating frequency of 10 kHz. Select C_2 for a carrier modulation frequency of 500 Hz. Couple in a sine wave generator and note capture and lock-in frequency ranges by viewing both the incoming signal and the VCO signal.

Measure:

1. Lock-in range—compare with computed value.
2. Capture frequency—compare with computed value.
3. Double frequency rejection measured at C_2—compare with computed value.

Experiment 9B

Purpose: To demonstrate a VCO and PLL connected in a circuit to provide a complete frequency modulation and demodulation system.

Requirements: Connect up a 566 VCO and a 565 PLL for a nominal frequency of 100 kHz. Use +12 V for the power supply voltage. (Use a voltage divider to bring the PLL input up to +6 V.) Modulate VCO with a 100-Hz, 200-mV peak sine wave and verify correct output. Verify correct operation of the PLL and then connect VCO output to PLL input. Sketch input modulating signal to the VCO and the output demodulated signal from the PLL. (*Hint:* Attenuate output signal from the VCO to about 100 mV.)

Measure:

1. Frequency deviation of VCO—compare with spec values.
2. Amplitude of demodulated output—compare with computed value.
3. Double frequency amplitude—compare with computed value.

10

Frequency-to-Voltage Conversion

OBJECTIVES

This chapter will cover the following points relative to frequency-to-voltage conversion:

1. Basic tone decoder circuit operation
2. Determination of tone decoder center frequency and bandwidth
3. Tone decoder output circuitry
4. LM2907 frequency to voltage converter operation
5. Selection of LM2907 circuit components
6. Possible application

10-1 INTRODUCTION

Frequency-to-voltage conversion occurs when an input frequency creates an output DC voltage at one particular frequency, or in another form, the output DC voltage is proportional to the input frequency. The first type of conversion is typical of a tone decoder that is used to sense when a particular frequency is present. Examples are used in telephone touch-tone decoder, end-of-message tones on tapes, or in tone signaling systems, like the three of eight tones used in train speed control.

If the output voltage is proportional to the input frequency we have a linear frequency-to-voltage convertor. This device can be used as a tachometer to sense speed or to sense frequency drift.

Let us begin by looking at a tone decoder similar to the PLL described in the previous chapter.

10-2 TONE DECODER

A very common tone decoder is the 567. This decoder can operate from 0.01 Hz to 500 kHz with a controllable bandwidth up to 14%. The output is TTL compatible and has a current sink capability of 100 mA. Shown in Fig. 10-1 is the block diagram for this integrated circuit.

The device contains a phase locked loop which consists of the I (inphase) phase detector, amplifier U_1, and the VCO. Operation of this part of the circuit is as described for the 565 PLL. An incoming signal within the capture range of the PLL will cause an error signal out of the I phase detector that will shift the VCO so that its frequency is the same as that of the incoming signal. If the incoming signal frequency is at the free-running (idle) frequency of the VCO, no error signal will appear at the output of the phase detector. This frequency (f_O) is called the center frequency of the tone decoder.

The actual output from the circuit is generated from the Q (quadrature—90° out of phase) detector, which has its maximum output signal when the I phase detector has zero output—which is the center frequency. The Q output provides a voltage to the inverting side of comparator U_2 that is greater than V_{REF}, causing the open collector output transistor of U_2 to turn on. With the transistor on, U_2 sinks current from the load resistor R_L.

As the incoming signal moves away from the center frequency of the tone decoder, the output from the Q detector is reduced. When it drops below V_{REF}, the output of U_2 goes high. Bandwidth of the tone decoder is determined from the range of frequencies over which the output of U_2 is low.

The design formula for the center frequency (f_O) of the decoder is:

$$f_o = \frac{1.1}{R_T C_T} \tag{10-1}$$

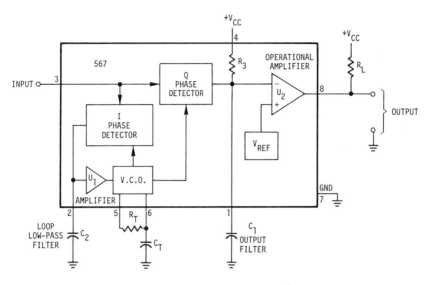

Figure 10-1   LM 567 Tone Decoder

where R_T and C_T are the timing components for the internal VCO. As with the 566, R_T should be between 2 and 20 kΩ.

If the input signal ranges from 20 mV to 200 mV, the bandwidth (B) in hertz is approximated by the equation:

$$B = 10.7 \sqrt{\frac{V_s f_o}{C_2}} \qquad (10\text{-}2)$$

where V_S is the rms value of the input signal in volts and C_2 (in microfarads) is the PLL low pass filter capacitor, which controls the capture range. Input signals above 200 mV give the same bandwidth as the 200-mV value. If a very narrow bandwidth is required, the value of C_2 could become too large. Connecting resistance from Pin 2 to ground reduces the amount of signal fed back to the VCO and narrows the bandwidth. With a 200-mV input signal, a 500-Ω resistor will give a bandwidth of 2% of f_o, while a 4-kΩ resistor will provide a bandwidth of 14%.

10-3 TONE DECODER EXAMPLE

We are now ready to look at an application of the tone decoder. Let us go through a basic example of detecting a 10-kHz tone with a required bandwidth of 500 Hz and a predicted input signal level of 100 mV.

Selecting 0.01 μF for C_T and rearranging Eq. 10-1 to solve for R_T:

$$R_T = \frac{1.1}{C_T f_o}$$

$$= \frac{1.1}{10^{-8} \times 10^{-4}}$$

$$= 11 \text{ k}\Omega$$

Solving for C_2 in Eq. 10-2:

$$C_2 = \frac{(10.7)^2 \, V_s f_o}{(B)^2}$$

$$= \frac{(10.7)^2 (0.1 \text{ V})(10 \text{ kHz})}{(500 \text{ Hz})^2}$$

$$\sim 0.47 \, \mu\text{F}$$

It is suggested that for noise and transient suppression, a time corresponding to 10 periods of the input frequency be allowed before the output is sampled. This can be controlled by making the output filter capacitor C_1 have a value:

$$C_1 = 260 \, \mu\text{F}/f_O$$

With our value of 10 kHz for f_O

$$C_1 = 0.026 \, \mu\text{F} \ (\text{use } 0.027)$$

The complete circuit, with the component values we have determined, is given in Fig. 10-2.

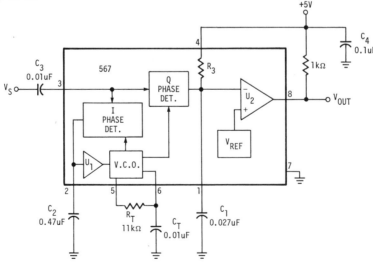

Figure 10-2   10KHz Tone Detector Circuit

The minimum value for the input coupling capacitor (C_3) is determined by allowing a maximum 10% drop across the capacitor at 10 kHz. Since the input resistance of the 567 is 20 kΩ, C_3 reactance must be less than 2 kΩ.

$$C_3 = \frac{1}{2 \times 10^3 \times 10^4 \times 2\pi}$$

$$\sim 8 \text{ nF (use 10 nF)}$$

Because of the fast switching of the output, a power supply bypass capacitor (C_4) is suggested located close to the 567 terminals.

10-4 MULTIPLE-TONE DECODER SYSTEMS

Some systems require that more than one tone be decoded before an output is provided. This can be accomplished by using several decoders and by "anding" the outputs. Consider a system where four tones are used and the presence of any two tones will cause a particular output to go high. To implement this circuit, the tone decoder components are determined for the required frequencies and the six outputs could be connected as shown in Fig. 10-3.

Nor gates provide the necessary outputs. If A and B tone decoders are turned on, their outputs go low; with zero on both inputs to NOR gate U_1, its output goes high. Notice that no other gate has the required zero inputs—so no other outputs will be high. The 1 k pull-up resistors are used to hold the inputs to the gates high when the decoder output transistors are turned off.

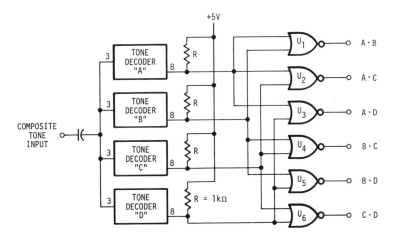

Figure 10-3   Two of Four Tone Decoder

Thus far we have looked at a frequency-to-voltage convertor that provides an output at one particular frequency. We will now study the other type of frequency-to-voltage conversion, where the output is proportional to frequency.

10-5 *LINEAR FREQUENCY-TO-VOLTAGE CONVERSION*

In certain applications, such as in motor controllers, speed control systems, tachometers, etc., it is required that an output voltage vary in proportion to change in the input frequency. A device that will perform this function is the LM2907 frequency-to-voltage converter integrated circuit shown in block diagram form in Fig. 10-4.

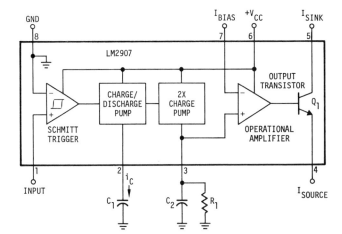

Figure 10-4   Frequency-to-Voltage Converter

Referring to Fig. 10-4, the input signal is applied through Pin 1 to the non-inverting input of a Schmitt trigger circuit, which has a ± 15-mV hysteresis for noise immunity. If the input signal goes negative, the output of the Schmitt trigger goes low, causing the charge/discharge pump to charge capacitor C_1 with a charging current (I_1) of approximately 180 μA. This results in an increase in the voltage across C_1 from a quiescent $1/4$ V_C to $3/4$ V_C. On the positive cycle of the input waveform, C_1 discharges with the same current flow from $3/4$ V_C to $1/4$ V_C. Thus, C_1 is alternately charged and discharged when the negative and positive transitions of the input signal occur. Each time C_1 is charged or discharged, capacitor C_2 is charged by the 2X (2 times) charge pump with the same amount of charging current that entered C_1. Therefore, C_2 is charged twice for each cycle. The charge on C_1 is:

$$q_1 = C_1 \left(\frac{3}{4} V_c - \frac{1}{4} V_c \right)$$

$$q_1 = \frac{C_1 V_c}{2} \tag{10-3}$$

Over one input cycle C_2 receives twice this charge, or:

$$q_2 \text{ (charge)} = C_1 V_C \tag{10-4}$$

The resistor R_1 connected across C_2 causes a discharge over one cycle of:

$$q_2(\text{discharge}) = \frac{V_3 T}{R_1} \tag{10-5}$$

where V_3 is the average voltage on Pin 3 and T is the period of the input signal.
In equilibrium

$$q_2(\text{charge}) = q_2(\text{discharge})$$

and from Eqs. 10-4 and 10-5:

$$C_1 V_c = \frac{V_3 T}{R_1}$$

Substituting $f = 1/T$ and solving for V_3:

$$V_3 = V_C R_1 C_1 f \tag{10-6}$$

(*Note:* f is the input frequency.)
Equation 10-6 shows that the DC voltage at Pin 3 is proportional to the input frequency, i.e., if the frequency doubles, V_3 will also double.
The maximum input frequency is limited by the time it takes to charge C_1 by $V_C/2$, or:

$$t = \frac{C_1 V_c}{2 I_1} \tag{10-7}$$

with I_1 the current from the constant current source.
A complete charge/discharge cycle for the capacitor C_1 takes $2t$; therefore, the maximum frequency equation is:

$$f_{\max} = \frac{1}{2t} = \frac{I_1}{C_1 V_c} \tag{10-8}$$

Equation 10-8 applies to symmetrical inputs, i.e., sine waves and square waves. If pulse inputs are used, the minimum pulse width is given by Eq. 10-7.
It should be apparent from these relationships that the maximum operating frequencies are reached with symmetrical input waveforms (equal positive and negative excursions).
Since C_2 charges from the constant current source and discharges through R_1, a ripple voltage will appear across this capacitor. In order to determine the value of the

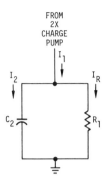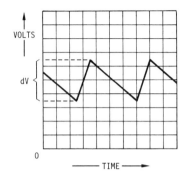

Figure 10-5   Charging Circuit For Capacitor C_2

ripple, we must consider the current flow during the charge time, as indicated in Fig. 10-5.

From the circuit in Fig. 10-5:

$$I_1 = I_2 + I_R \tag{10-9}$$

Recognizing that $I = dq/dt$ and using Eq. 10-3:

$$\frac{C_1 V_c}{2\ dt} = \frac{C_2 dV}{dt} + \frac{V_3}{R_1}$$

$$dV = \frac{C_1 V_c}{2\ C_2} - \frac{V_c R_1 C_1 f\ dt}{C_2 R_1}$$

Using the relationship $dq = Idt$ for Eq. 10-3:

$$dt = \frac{C_1 V_c}{2\ I_1}$$

$$dV = \frac{C_1 V_c}{2\ C_2} - \frac{(V_c C_1)^2 f}{2\ C_2 I_1}$$

Therefore, the ripple voltage is:

$$dV = \frac{C_1 V_c}{2\ C_2} \left(1 - \frac{V_c C_1 f}{I_1} \right) \tag{10-10}$$

We need now to consider practical values for C_1, R_1, and C_2. The capacitor C_1 also provides internal compensation for the charge/discharge pump and, therefore, is required to have a minimum value of 100 pF. If we increase C_1, the output voltage V_3 increases for a given frequency (Eq. 10-6). Increasing C_1 also increases the time to charge the capacitor to the required $V_C/2$ (I_1 fixed) and therefore lowers the maximum operating frequency.

There are two things to consider with R_1. The first is that the output impedance of the charge pump is 10 MΩ, therefore, R_1 should be much smaller than this value (less than 100 kΩ) in order for the charge pump to remain a constant current source. Secondly, as R_1 decreases in value, the output voltage V_3 for a given input frequency also decreases.

Increasing the value of C_2 decreases the output ripple but also slows the response of V_3 (slew rate) to fast input frequency changes. This is illustrated in the following equation for the rate of change of output voltage V_3:

$$\frac{d\,V_3}{dt} = \frac{I_1}{C_2} \qquad (10\text{-}11)$$

Referring back to Fig. 10-4, the output voltage V_3 from the charge pump is connected to the non-inverting input of the operational amplifier. This operational amplifier has a gain of 200,000 and a unity gain slew rate of 0.2 V/μs. The output transistor can source or sink 50 mA with either a collector or emitter load. The conventional closed loop connections would be from the transistor emitter (Pin 4) to the operational amplifier inverting input (Pin 7). By adding gain resistors to the op-amp, the voltage across R_1 can be boosted to a higher voltage at the output.

10-6 A TYPICAL FREQUENCY-TO-VOLTAGE CONVERTER APPLICATION

A typical application for the LM2907 frequency-to-voltage converter is a tachometer output circuit as shown in Fig. 10-6.

We shall assume that the required full-scale reading of the meter is 10,000 RPM and the minimum input corresponds to 500 RPM. Also, the meter movement is 100 μA full scale with an internal resistance of 50 Ω. The ripple on the meter should be kept below 3% and a 12-V power supply is available. With these assumptions we need to determine values for C_1, R_2, R_1, and C_2.

The maximum value of C_1 is controlled by Eq. 10-8. We can solve for C_1 knowing that $V_{CC} = 12$ V, $I_1 = 180$ μA, and the maximum speed of 10,000 RPM with an eight-tooth rotor corresponds to $8 \times 10^4/60$ or 1,333 Hz.

$$C_1 = \frac{I_1}{f_{max}\,V_{CC}}$$

$$= \frac{1.8 \times 10^{-4}\text{ A}}{1.333 \times 10^3_{Hz} \times 12\text{ V}}$$

$$= 11.3 \text{ nF (use 10 nF)}$$

Before we can specify R_2, we need to select the maximum output voltage. With a 12-V supply we can allow 2 V minimum drop across the output transistor (Q_1). The drop across the meter is 0.1 mA $\times$ 50 or 5 mV, so we can develop a maximum output

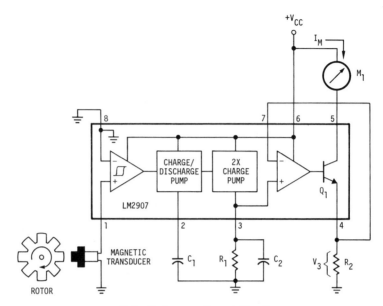

Figure 10-6   Tachometer Output Circuit

voltage (V_M) of 10 V across R_2. With the maximum meter current of 0.1 mA flowing through this resistor, the value is (10V/0.1 mA) or 100 kΩ.

For the determination of R_1 we need to solve for this resistor in Eq. 10-6.

$$R_1 = \frac{V_M}{V_c\, C_1 f}$$

$$= \frac{10\text{ V}}{12\text{ V} \times 10^{-8}\text{F} \times 1.333 \times 10^3\text{ Hz}}$$

$$= 62.5\text{ k}\Omega\text{ (use 62 k}\Omega)$$

The last component to determine is the filter capacitor (C_2). Three percent ripple amounts to (10 V × 0.03) or 0.3 V for dV. The worst-case ripple will occur at the lowest frequency (500 Hz). Rearranging Eq. 10-10 to solve for C_2:

$$C_2 = \frac{C_1 V_c}{2\, dV}\left(1 - \frac{V_c C_1 f}{I_1}\right)$$

$$= \frac{10^{-8}\text{F} \times 12\text{ V}}{0.6\text{ V}}\left(1 - \frac{12\text{ V} \times 10^{-8}\text{F} \times 5 \times 10^2\text{ Hz}}{1.8 \times 10^{-4}\text{ A}}\right)$$

$$= 48\text{ nF (use 50 nF)}$$

With the value of C_2 known, we can determine the maximum rate of change of the input frequency from Eq. 10-11.

$$\frac{dV_3}{dt} = \frac{1.8 \times 10^{-4}}{5 \times 10^{-8}}$$

$$= 3600 \text{ V/s}$$

This corresponds to a maximum allowable input frequency change of 480/kHz/s, which is much greater than is probable from a mechanical-type input.

To compensate for device and component tolerance, it is suggested that R_2 be a variable resistor. Using a calibrated-input frequency source set for the maximum RPM, R_2 can be adjusted for a full-scale reading on the meter.

Another application for the LM2907 is motor speed control—where a tach on a motor shaft feeds a signal into the frequency to voltage-converter circuit and the LM2907 collector output is used to control the motor current (via a power Darlington).

Since the output voltage is a function of C_1 (Eq. 10-6), we can fix the input frequency, and by connecting an unknown capacitor in place of C_1, use the circuit for capacitance measurement. To change range, R_1 can be switched to different values.

10-7 CHAPTER 10—SUMMARY

- Frequency-to-voltage conversion occurs when an output voltage indicates an input signal presence or the output voltage is proportional to the input frequency.
- The tone decoder described provides an output "low" when the input frequency is within the bandwidth of the decoder.
- A 567 tone decoder consists of a phase locked loop and a quadrature phase detector.
- The bandwidth of the 567 can be reduced by increasing the value of the loop low pass filter capacitor C_2. It can also be reduced if the loop gain is decreased by connecting a resistor from Pin 2 to ground.
- An output voltage proportional to input frequency is provided by the LM2907 frequency-to-voltage converter.
- The output voltage of the LM2907 increases with input frequency, the supply voltage, and the value of R_1 and C_1.
- The minimum value of C_1 is the 100 pF required for frequency compensation, but the maximum operating frequency is reduced as C_1 increases.

- If R_1 is reduced, the output is decreased, and if R_1 is greater than 100 kΩ, the charge pump becomes less of a constant current source.
- Increasing the value of the filter capacitor (C_2) reduces the ripple, but also reduces the ability of the LM2907 to follow fast input frequency changes.

10-8 CHAPTER 10—EXERCISE PROBLEMS

1. Find the value of the filter capacitor (C_2) for a 567 tone decoder if the desired bandwidth is 300 Hz, the input voltage 100 mV, and the center frequency 5 kHz.

2. Determine the approximate value of a resistor to connect from Pin 2 of a 567 to ground in order to reduce the bandwidth to 8% with a 200 mV input signal.

3. Compute the value of the output capacitor (C_1) for a 567 if the center frequency is 5 kHz.

4. Determine the minimum input pulse width to a 2907 convertor if the power supply is 10 V and C_1 is 10 nF.

5. Find the maximum slew rate in V/μs at the output of a 2907 frequency-to-voltage converter if the value of C_2 is 100 nF.

6. Show a circuit that uses a 2907 to measure capacitance.

7. State the differences between a 567 tone decoder and a 565 PLL.

8. Determine all circuit components for a circuit that will detect the presence of a 50-mV 2-kHz signal which can vary from 1900 Hz to 2100 Hz. Use an LED to indicate signal decoding, a C_T value of 47 nF, and a power supply of +5 V.

9. Show the output circuitry for a three-of-five tone decoder.

10. Determine the output voltage (V_3) for an LM2907 with an input sine wave frequency of 5 kHz if the supply voltage is 12 V, R is 1 kΩ, and $C_1 = 0.001$ μF.

11. Find the maximum operating frequency for the circuit in Problem 10.

12. Design a tachometer circuit for an automobile using an LM2907. A 1-mA meter with internal resistance of 100 Ω is available for display. The maximum RPM is 5,000 (minimum 500 RPM) for a six-cylinder engine, and the ripple on the meter should be kept below 5%. (Remember, there are three sparks per RPM.) Power supply of +12 V to be used, and let $R_1 = 30$ kΩ.

10-9 CHAPTER 10—LABORATORY EXPERIMENTS

Experiment 10A

Purpose: To demonstrate the operation of a 567 tone decoder.

Requirements: Construct the tone decoder described in Exercise Problem 8.

Measure:

1. The frequency range over which the LED is on.
2. The bandwidth for input levels of 20, 100, 150, and 200 mV.

Compare the results with the computed values.

Experiment 10B

Purpose: To build a tachometer using a LM2907 frequency-to-voltage converter.

Requirements: Construct the tachometer circuit of Exercise Problem 12 using a 12-V power supply. Use a 100-Ω resistor to simulate the meter resistance and a pulse generator to simulate the tach.

Measure:

1. Output voltage across the 100-Ω resistor for five different input frequencies.
2. The ripple across the 100-Ω resistor.

Compare the results with the computed or specified values.

11

Analog-to-Digital Converter

OBJECTIVES

At the conclusion of this chapter, you should know the following concepts about analog-to-digital converters (A/D):
1. Definition of an analog signal
2. Why analog signals are converted to digital
3. How a successive approximation A/D operates
4. How a flash A/D operates
5. How a dual slope A/D operates
6. How a tracking A/D operates

11-1 INTRODUCTION

An analog signal is one that can assume any voltage value, e.g., sine waves, speech waveforms, etc. In contrast, a digital binary signal can assume only two values called "1" and "0"; for transistor transistor logic (TTL) the 1 state is approximately +5 V and the 0 state is 0 V.

If we wish to transfer analog signals from one point to another with minimum interference from noise, we need to convert the analog signal to a digital signal (pulse code modulation). The conversion process consists of taking sample points on the analog waveforms (see Fig. 11-1) and converting each point into a digital word. The more points we take, the closer the sampled waveform follows the original waveform.

In the example shown in the Fig. 11-1, the analog signal has a maximum value of 3.1 V. Let us assume that after processing and transmission, we would like the analog signal to be restored to within 0.1 V of its original value. We can determine the required resolution by dividing the 3.1 V into 0.1-V increments. With a signal range of 0 to 3.1 V, we have 32 steps of 0.1 V, i.e., 0 V, 0.1 V, etc., which gives a required resolution of 1 part in 32.

To represent these 32 steps as a digital word requires five bits. (With five bits we can get 32 combinations of ones and zeros.) Zero volts can be represented by the binary-coded decimal count of 00000, 0.1 V by 00001, and so on. At each particular sample time, a digital word consisting of five bits is generated for the particular analog value.

Analog-to-digital (A/D) converters are integrated circuit devices that perform the described conversion process. There are four basic types of A/D converters:

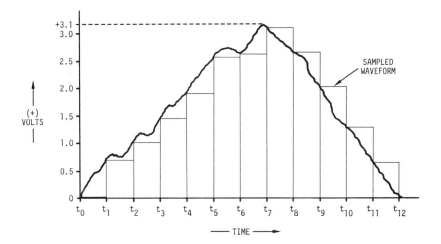

Figure 11-1 Sampling an Analog Waveform

1. Successive approximation
2. Flash
3. Dual slope
4. Tracking

The successive approximation A/D convertor is used where reasonably fast conversion is required. The flash-type A/D convertor is used for fast-changing analog input signals. Dual slope conversion is slow, but offers excellent rejection of power supply noise. Finally, the tracking convertor is faster than the dual slope, but slower than the successive approximation or flash types.

11-2 SUCCESSIVE APPROXIMATION A/D CONVERTER

A typical successive approximation device is the 8-bit A/D convertor. This device can resolve the input signal into 256 increments (2^8) using 8 bits. A basic block diagram of this type of A/D converter is shown in Fig. 11-2. In operation, the analog input voltage level is compared with an internal reference voltage divider. When a match is achieved between the input and a tap on the divider, a digital word is generated.

Now for the detailed operation. The analog signal is applied to the analog input, and the conversion process is initiated by the START CONVERSION signal along with the CLOCK signal. With +5.12 V applied across the 256-resistor network, voltages are available via the analog switches for comparison with the analog input signal.

The conversion process starts with closure of the center switch on the resistor ladder network. This applies +2.56 V (half of +5.12 V) to the inverting input of the comparator. If we assume the analog input signal is +3.5 V, then the comparator output will be in the high state. With this condition, the switch to the center of the resistor network is opened and the switch midway between the top and center is closed. This samples a voltage of +3.84 V (+5.12 V $\times$ 0.75), which is greater than the analog input and the comparator output will go low. A switch is now closed between 0.75 and 0.5 of the resistive divider, yielding a voltage of +3.2 V, and the comparator output goes high.

The successive approximation process continues until the sampled voltage matches the input. This occurs for this example when the voltage at the top of the 175th resistor is sampled. At this point, the voltage from the divider matches the input and is converted into an eight-bit digital word via the selection and control logic. If a match is not possible—i.e., the input voltage falls between two voltage divider taps—the comparator cycles back and forth. The control logic recognizes this condition and converts the highest tap value to a digital word. A "low" output appears on the $\overline{\text{INTR}}$ terminal, indicating the process is complete. If the TRI-STATE ENABLE line is made high by a "low" on the $\overline{\text{READ}}$ line, the eight-bit word will appear on the eight output lines from the eight-bit latch.

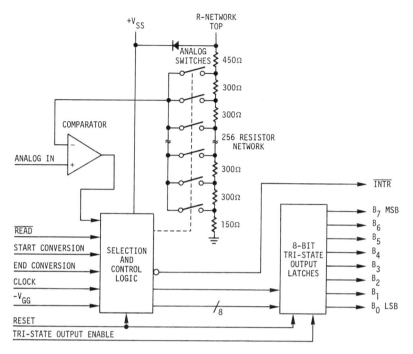

Figure 11-2 Successive Approximation Converter Basic Block Diagram

To determine what would appear on the output with the $+3.5$-V input used for the example above, we need to look at the binary representation of the point where the match between the analog input and the resistive divider occurred. It will be recalled that the match was at the top of the 175th resistor. Thus, we need the binary representation of 175. The significance of each bit is indicated as follows:

Bit No.:	B_7	B_6	B_5	B_4	B_3	B_2	B_1	B_0
	2^7	2^6	2^5	2^4	2^3	2^2	2^1	2^0
	128	64	32	16	8	4	2	1

The most significant bit is $2^7 = 128$ and the least is $2^0 = 1$. The number 175 is greater than 128, therefore the 2^7 bit $= 1$. If we subtract 128 from 175, we are left with 47, and since 2^6 (64) is greater than 47, the 2^6 bit is zero. The 2^5 (32) is less than 47, thus the 2^5 bit $= 1$, and subtracting 32 from 47 yields 15. The 2^4 bit (16) is greater than 15, so this bit is zero. The 2^3 bit (8) is a 1 and the difference (7) is greater than the 2^2 bit (4), so this bit is also a 1. Finally, the difference (3) is greater than the 2^1 bit (2). Thus, it is a 1 and 2^0 (1) is equal to the difference (1) and this bit is also a 1.

The complete representation of 175 is then 10101111, and this appears at the output of the eight-bit tri-state latch.

11-3 A PRACTICAL SUCCESSIVE APPROXIMATION A/D CONVERTER

An actual successive approximation A/D converter is the ADC 0804 shown in block diagram form in Fig. 11-3.

This converter is an eight-bit device and is microprocessor compatible. It operates off of a nominal +5-V power supply and has a total unadjusted error of ± 1 bit. The external clock can be as high as 1.46 MHz and is divided down by a factor of 8 to provide the internal clock.

The converter is initialized by having the chip select ($\overline{CS}$) and the write ($\overline{WR}$) lines go low (the bar over the abbreviation indicates active low). This sets the start flip-flop U_2 and the resulting high on the Q output resets the shift register U_4. It also sets the interrupt flip-flop U_5 and inputs a 1 to flip-flop 1 (U_3). The clock transfers the 1 to the output of this flip-flop, which is connected to the D input of the eight-bit shift register U_4 and also the AND gate U_6. When the clock signal goes high, and either $\overline{CS}$ or $\overline{WR}$ goes high, the gate output resets the start flip-flop. This removes the reset signal from the eight-bit counter, and the 1 at the input to this counter from flip-flop 1 can now be shifted through the counter to start the conversion process.

The counter stage outputs cause switches on the ladder network to close, and the comparisons with the input signal are made until the best match is obtained. After 8 clock pulses (64 external clock pulses), the 1 at the input to the counter will have reached the counter output. This provides a 1 to an input of gate G_2 (U_7), which, along with the high on the $\overline{Q}$ output of latch 1 (U_8), causes the gate U_7 output to go high. A high on this transfer line causes the digital word, corresponding to the analog input, to be transferred from the ladder network to the tri-state latch.

The 1 at the D input to latch 1 (U_8) is shifted by the next clock pulse to the output of the latch. This causes the $\overline{Q}$ output of the latch to go low, setting the interrupt flip-flop, which in turn makes the $\overline{INTR}$ line go low. A low on the $\overline{INTR}$ line signals the microprocessor that data is available.

When the microprocessor is ready to take the data, it causes the $\overline{CS}$ and $\overline{RD}$ (read) lines to go low, which enables the tri-state latch and the digital word to appear on the data bus. With $\overline{CS}$ and $\overline{RD}$ low, the interrupt flip-flop is reset and the $\overline{INTR}$ line goes high.

If both the $\overline{CS}$ and $\overline{WR}$ lines go low while a conversion is in process, the conversion stops; and when either $\overline{CS}$ or $\overline{WR}$ go high, a new conversion starts.

The converter can be operated in a free-running mode by tying the $\overline{CS}$ line low and connecting the $\overline{INTR}$ output to the $\overline{WR}$ terminal. A chip test circuit connected in this fashion is shown in Fig. 11-4 (see p. 220).

The power supply voltage used is +5.12 V. Input voltages on any other terminal of the chip should not exceed this value or be less than 0 V, or the device will be damaged (consider spikes riding on any input signal). The differential input ($V_{IN}^{(+)}$

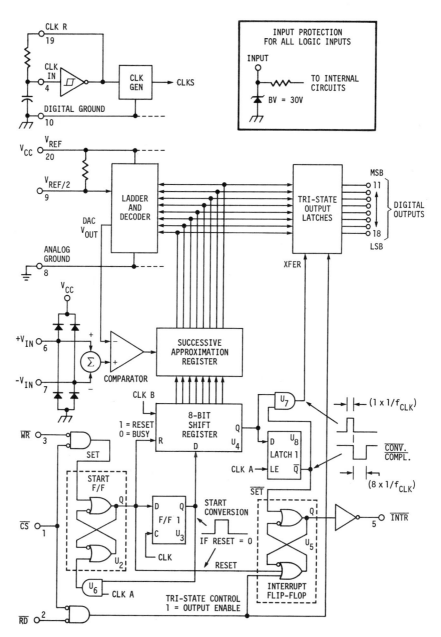

Figure 11-3 Block Diagram of ADC 0804

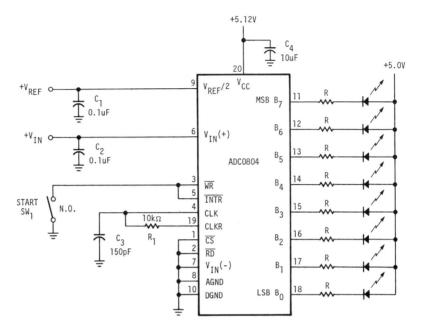

Figure 11-4   Free Running Analog-Digital Converter

and $V_{IN}^{(-)}$) can be used to reduce common mode noise, or the $V_{IN}^{(-)}$ input can be used to subtract a voltage from the analog input, e.g., removal of an offset voltage. In the circuit shown, the $V_{IN}^{(-)}$ terminal is grounded, which provides an unbalanced input.

Additional flexibility is provided by allowing the ladder reference voltage to be changed. Let us assume that the analog input voltage variation is from 0 V to 1 V. If we use a 5.120-V reference, then our input varies only over about one fifth of the available range, so our resolution is reduced by a factor of 5. But, by connecting an external reference voltage of 0.5 V to the $V_{REF}/2$ terminal (pin 9), we force the full-scale reference to be 1 V and recover our resolution. If we don't connect a voltage to the $V_{REF}/2$ terminal, then one half of the supply voltage (2.560 V) will appear at this terminal.

The clock signal can be provided externally (by connecting to the CLK terminal) or internally by connecting a timing resistor R_1 and capacitor C_3 as shown in Fig. 11-4. The equation for the clock frequency is:

$$f_{CLK} \sim \frac{1}{1.1\, R_T\, C_T} \tag{11-1}$$

With the components shown in Fig. 11-4, the clock frequency is about 600 kHz. The time for the complete conversion process is about 72 clock pulses (64 plus up to 8 start pulses), or 120 μs. Referring again to Fig. 11-4, the analog input signal is connected to $V_{IN}^{(+)}$, and the write (WR) line is momentarily grounded to start the process. The timing diagram of Fig. 11-5 shows the signals on the various lines for the conversion sequence.

At the completion of the conversion, the interrupt ($\overline{INTR}$) line will go low. If the read ($\overline{RD}$) line now goes low, the tri-state output latches will be enabled and data will appear on the output line.

The digital output LED display can be decoded by dividing the eight bits into two hex characters—the four most significant and four least significant bits. For example, if the readout was 10101111, this is "AF" in hex, and we can convert to decimal by $(10 \times 16) + (15) = 175$. The least significant bit is 5.12 V/256 = 20 mV. Thus, the analog input voltage must have been 175×20 mV = 3.5 V.

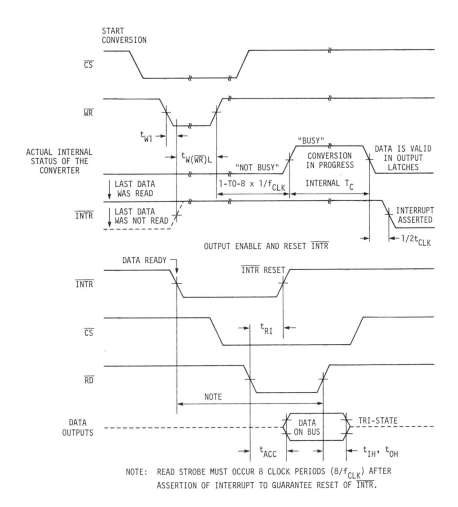

Figure 11-5 Timing Diagram

11-4 FLASH CONVERTER

When we convert a sine wave to a digital representation, we have to decide how many samples to take over the complete sine wave cycle. The more samples we take, the closer the resultant data approaches a sine wave. This would be apparent if we converted the digital data back to analog. Few points taken over the sine wave would appear as a very noticeable stepped waveform. Let us assume for comparison purposes that 20 samples will give a sufficient representation of the sine wave. (For data transfer, Nyquist[1] states that two samples are sufficient.) As the frequency of the sine wave increases, the samples move closer together and the time available to convert from analog to digital (conversion time) decreases. With the ADC0804 example, we determined that the conversion time was 120 μs. Multiplying this time by the required number of samples yields a total time over one cycle of 2.4 ms, which means that the maximum frequency we can convert with 20 samples per cycle is 1/2.4 ms or 416 Hz. Conversion time, then, limits maximum input frequency.

Flash-type A/D converters have conversion times considerably shorter than the successive approximation types. The CA3308 C-MOS type eight-bit flash, for example, has a conversion time of 66.7 ns, which raises the sine wave input frequency, with 20 samples, to 750 kHz. We shall see from the circuit description how this faster conversion is achieved.

A functional block diagram for the CA3308 is shown in Fig. 11-6.

As with the ADC0809, a 256 resistor divider is used—but instead of sampling the divider voltages in a step-by-step sequence, all samples are made at the same time.

A phase 1 (ϕ_1) and a phase 2 (ϕ_2) clock are used with a 180° phase difference between the two signals. The conversion starts with the ϕ_1 clock high, which closes the switches connecting the divided reference voltages to the amplifier (comparator) input capacitors and also shorts the output of each amplifier to its input. This is called the *auto balance* phase and serves to balance the amplifiers at their trips points of half the power supply voltage ($V_{DD}/2$). The input capacitors charge up during this phase to the difference voltage between the voltage taken off the reference network V_{RN} and $V_{DD}/2$ (see Fig. 11-7 for a single path).

It will be realized that those input capacitors at the top of the reference divider will charge positive on the left side of the capacitor, while those at the bottom of the reference divider will be charged negative. The amount of voltage across the capacitor is a function of how much the divided reference at a particular tap point differs from $V_{DD}/2$.

The balancing of the second set of amplifiers, which also occurs during the auto balance phase, removes any tracking problems with the first set of amplifiers.

At the start of the sample phase, the ϕ_1 clock goes low and the ϕ_2 clock high. So, all the switches closed by the ϕ_1 clock are opened, and the ϕ_2 controlled switch

[1]Nyquist, H., "Certain Factors Affecting Telegraph Speed" (1924) and "Certain Topics in Telegraph Transmission Theory" (1928), Transactions A.I.E.E.

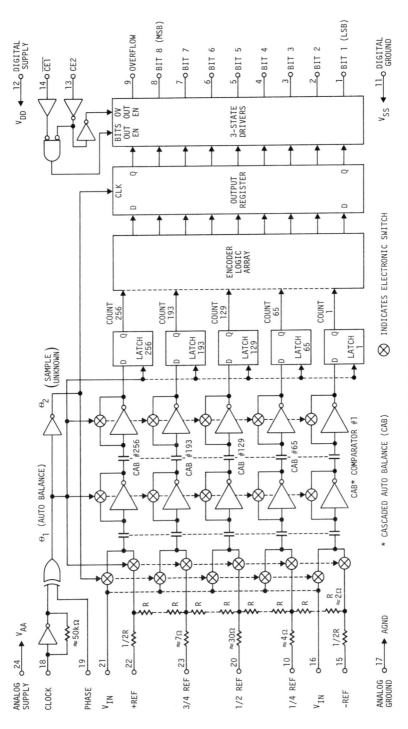

Figure 11-6 Block Diagram For CA3308

223

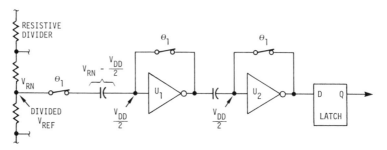

Figure 11-7   Auto Balance for 1 of 256 Paths

connecting the input capacitor to the analog input is closed. The input capacitor can be thought of as a battery charged to the difference voltage $(V_{RN} - V_{DD/2})$. If we also consider the input voltage (V_s) as a battery, then the voltage appearing at the high impedance input to the amplifier is the algebraic sum of these two voltages. Figure 11-8 shows the equivalent circuit. (Notice that C_1 can have either polarity.)

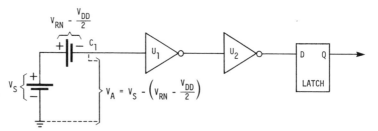

Figure 11-8   Sample Phase Input Equivalent Circuit

As shown in the figure, the voltage at the input to the amplifier (V_A) is:

$$V_A = V_S - (V_{RN} - V_{DD}/2) \qquad (11\text{-}2)$$

This voltage can be positive or negative or even zero (unlikely). If it is positive, a one will enter the latch—if negative, a zero.

During the next auto balance cycle, the 256 bits of information in the data latches are converted into a nine-bit data word (eight bits plus overflow). The rising edge of the ϕ_2 clock starts the next sample phase and shifts the data word into the output registers. Tri-state buffers are used at the output of these registers and are enabled by two chip select lines—$\overline{CEI}$ when it is low or CE2 when it is high. With CE2 high, the overflow bit is also enabled.

The sequence described above is with the phase input (Pin 19) to the "exclusive or" circuit in the "high" state. Holding this line low will reverse the ϕ_1 and ϕ_2 operations. A timing diagram with the phase input high is shown in Fig. 11-9.

The actual conversion time is from the start of a sample period to when data is shifted into the output register. We can see from Fig. 11-9 that this takes one complete clock period, and so with a 15-MHz clock the conversion time is 1/15 MHz

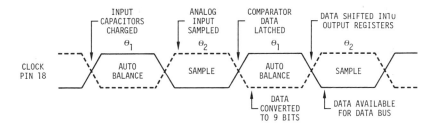

Figure 11-9   Timing Diagram for the CA3308

or 66.7 ns.

A complete circuit using the CA3308 is shown in Fig. 11-10.

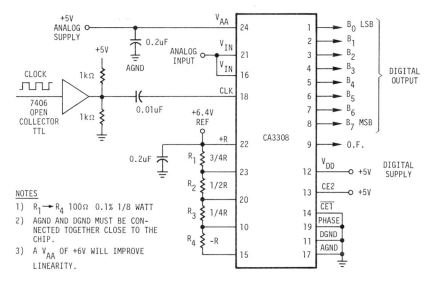

Figure 11-10   Typical Circuit for the CA3308

Adding the external precision resistors R_1 through R_4 to the reference divider increases the overall divider accuracy.

A TTL buffer (7406) is used to isolate the clock input signal. The two 1 kΩ resistors on the output of the buffer provide a clock signal of about 2.5 V peak to peak.

11-5 DUAL SLOPE CONVERTER

As mentioned previously, the dual slope A/D converter has good noise immunity, but is restricted to measuring slow varying or DC inputs. This is accomplished by allowing the analog signal to be input typically for a time equal to the period of the AC line frequency (16.67 ms). Line noise riding on the positive and negative

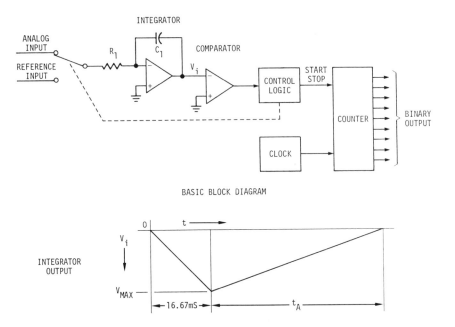

BASIC BLOCK DIAGRAM

Figure 11-11   Dual Slope Analog-Digital Converter

alternations of the sine wave will tend to cancel out.

A simplified block diagram for the dual slope converter is shown in Fig. 11-11.

If a positive analog input (V_a) is applied to the integrator input, the output will fall in accordance with the equation:

$$V_{\mathrm{OUT}} = \frac{-V_A\, t}{R_T\, C_T} \tag{11-3}$$

At the end of the sample period, the integrator output voltage will be:

$$V_{\mathrm{MAX}} = \frac{-V_A\,(16.67\text{ ms})}{R_T\, C_T} \tag{11-4}$$

With the completion of the sample period, the counter is started and the input switched to the reference (negative voltage if the analog input is positive). The negative input to the integrator causes the output to ramp positive from the minus V_{MAX} value. When the output voltage just crosses zero, the comparator output changes state and, via the control logic, stops the counter. The number N stored in the counter is proportional to the time (t_a) the integrator takes to ramp back to 0 V. So:

$$V_{\mathrm{MAX}} = V_{\mathrm{REF}}\,\frac{t_a}{R_T\, C_T} \tag{11-5}$$

and, if we equate Eqs. 11-4 and 11-5, we get:

$$\frac{V_A \,(16.67 \text{ ms})}{R_T \, C_T} = \frac{V_{\text{REF}} \, t_a}{R_T \, C_T} \tag{11-6}$$

Solving for t_a:
$$t_a = \frac{V_A \,(16.67 \text{ ms})}{V_{\text{REF}}} \tag{11-7}$$

Finally, since $16.67 \; ms/V_{\text{REF}}$ is constant and N is proportional to t_a:

$$N = KV_A \tag{11-8}$$

If we determine the constant K factor, we are able to predict the value of N for a given analog input voltage. The output count is a binary representation of the analog input.

The dual slope convertor is often used in digital voltmeters, and in this application the counter output is converted to a seven-segment readout display.

11-6 EXAMPLE OF A DUAL SLOPE CONVERTER

A typical dual slope converter is the ICL7126. This device is 3½ digit CMOS and the chip includes seven-segment decoders, LCD drivers, voltage reference, and a clock. The full-scale input voltage can be as high as 2 V. A low supply current of 100 μA at 9 V means the device can be battery powered.

The block diagram for the analog section is shown in Fig. 11-12.

Operation of the device is divided into three phases:

1. Auto-zero (A/Z)
2. Signal integrate (INT)
3. Reference integrate or de-integrate (DE)

Switches on the block diagram are indicated with a crossed circle; the phase abbreviations are alongside. Each of the sets of switches are closed during the particular phase, i.e., all A/Z switches close during the auto-zero phase. We will now look at what happens during each phase.

Several things occur during the auto-zero phase. Analog inputs (HI and LO) are disconnected from the chip and internally shorted to analog common (32). The reference capacitor C_{REF} is charged to the reference voltage. Finally, the output of the comparator is shorted to the inverting input of the integrator. This allows capacitor C_{AZ} to charge to a voltage created by the total voltage offset of the buffer, integrator, and comparator.

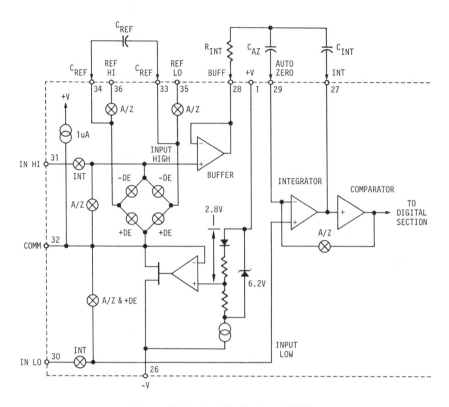

Figure 11-12 Analog Section of 7126

The signal integrate phase connects the analog input to the integrator for a time corresponding to 1000 clock pulses. At the end of this phase, a voltage is developed on the integrating capacitor (C_{INT}) which is proportional to the input voltage. The capacitor C_{AZ} subtracts out any offset error from this voltage.

During the de-integrate phase, the reference voltage replaces the analog voltage as an integrator input. The reference voltage stored on capacitor C_{REF} is used rather than the reference source directly to allow for ease of polarity reversal. If the analog input is positive, then the reference input must be negative in order to discharge the integrating capacitor (and vice versa). This is controlled by the reversing configuration of the DE($+$) and DE($-$) switches. The sense of the comparator output tells the logic which set of switches to close. It can take a period of time from 0 to 2000 clock pulses for the integrating capacitor to discharge, with the time being proportional to the analog input (V_A). The actual digital reading displayed is:

$$\text{Digital Count } N = 1000(V_A/V_{REF}) \tag{11-9}$$

With a maximum count of 2000 and a full scale V_A of 2 V, we can see from this equation that V_{REF} must be 1 V. If we desire a full-scale value less than 2 V, we simply reduce V_{REF}.

The digital section of the converter is shown in Fig. 11-13 (see p. 230).

Carrying on the description from the analog section, the comparator output indicates to the logic control the polarity of the analog input and also when the integrating capacitor voltage is zero (comparator output changes state) at the end of the reference input phase.

The clock signal is derived from an external oscillator connected to Pin 40, a crystal between Pins 39 and 40, or an RC oscillator using Pins 38, 39, and 40 as shown in Fig. 11-13. Division of this oscillator by four is performed to obtain the internal clock signal. It takes 4000 counts of the internal clock (or 16,000 oscillator cycles) for a complete conversion cycle. This consists of:

1. signal integrate one fourth of the total 1000 counts
2. reference de-integrate—0 to 2000 counts
3. auto-zero, the remainder—1000 to 3000 counts

If we want 3 conversion cycles per second, we need an oscillator frequency of (16,000 Hz $\times$ 3) or 48 kHz. It is recommended that for maximum rejection of 60-Hz pickup, the signal integrate time period should be a multiple (n) of 16.666 ms (the period of 60 Hz), which means that the oscillator frequency should be:

$$f_{OSC} = 2.4 \text{ MHz/n} \qquad (11\text{-}10)$$

The control logic allows the clock signal to be counted during the de-integrating phase. At the end of this phase, the accumulated count is stored in the latch. The stored count is converted into a seven-segment code and applied to the individual liquid crystal segments via the drivers. To prolong the life of the crystal display, an AC voltage must be developed between the segment and the crystal backplane. This is accomplished by applying the divided down (by 200) clock signal to the backplane.

With the oscillator frequency at 48 kHz, the frequency of the signal applied to the back plane is 48 kHz/4 $\times$ 200, or 60 Hz.

If the segments are driven at the same amplitude and frequency and are in phase with the back-plane signal, the segment will be off. When the signals are out of phase the segment will be on (light).

A meter application of the 7126 is shown in Fig. 11-14 (see p. 231).

With the reference set for 100 mV, the full-scale reading is 200 mV. Looking at the components used in the circuit, the 1-MΩ resistor and the 0.01-μF capacitor between the analog input lines serves as a noise filter. An integrating resistor of 180 kΩ is recommended for the full-scale reading of 200 mV. With 2 V full scale, the resistor would have to be increased to 1.8 MΩ to maintain the same charging current.

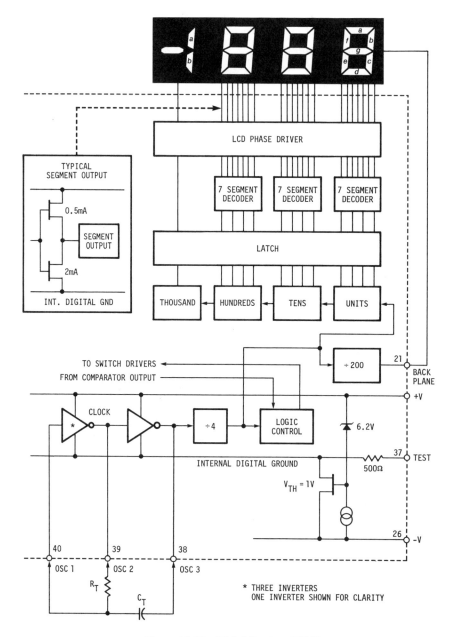

Figure 11-13 Digital Section of 7126

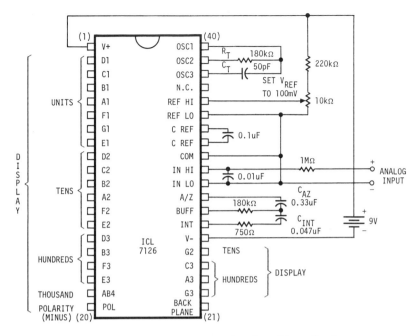

Figure 11-14   Portable 200 mV Full Scale Meter Circuit

For 3 readings per second (48-kHz oscillator) the integrating capacitor (C_{INT}) is 0.047 μF. If we reduced the readings to 1 per second (16-kHz oscillator) C_{INT} would have to be increased by a factor of 3 to 0.15 μF. The 750-ohm resistor placed in series with C_{INT} compensates for comparator delay.

A 48-kHz clock (f_C) is established from the equation:

$$f_C \sim 45/R_T C_T \qquad (11\text{-}11)$$

With a recommended C_T of 50 pF, the R_T value of 180 kΩ is obtained.

The auto-zero capacitor (C_{AZ}) should have a value of 0.33 μF for minimum noise with a full-scale input of 200 mV. For a 2-V full-scale operation, a value of 0.033 μF is recommended to speed recovery from an input overload. Finally, a value of 0.1 to 1.0 μF is considered sufficient for the reference capacitor.

11-7 TRACKING CONVERTER

This converter is faster than the dual slope and uses a technique whereby an internal signal is caused to track the analog input. The device operation will be discussed with reference to Fig. 11-15.

The digital-to-analog convertor (DAC) block takes the binary (digital) output from the counter and converts it to an analog voltage (described in the Chapter 12).

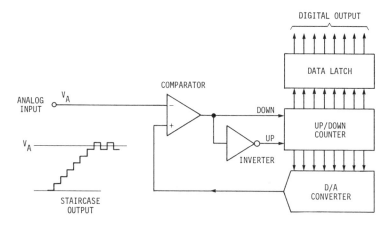

Figure 11-15   Tracking Analog-Digital Converter

In operation, we assume a positive analog input voltage to the comparator inverting input terminal and that the output of the DAC, applied to the comparator, is initially zero. With these conditions, the comparator output will be low, causing the "up" line to the counter to be high.

The counter will start counting up and the DAC output will increase in a staircase fashion, each voltage step corresponding to a particular binary count. When the staircase voltage exceeds the analog input, the comparator output will go high and cause the counter to decrement. The lower number in the counter means the staircase will drop back one step. This will cause the comparator output to go low and the counter to be incremented. The staircase will step up and the cycle will repeat.

Circuits within the convertor sense the up and down transitions of the staircase and, recognizing that the closest match exists between the analog input and the D/A convertor output, shifts the number in the counter to the data latch.

We will see in the next chapter that the tracking converter can be implemented with a DAC (digital-to-analog converter) and a microprocessor.

11-8 CHAPTER 11—SUMMARY

● An analog-to-digital converter changes points on a continuously varying analog waveform into two-level digital data.

● The four basic types of A/D, listed in order of shortest conversion time, are:

1. flash
2. successive approximation
3. tracking
4. dual slope

- An eight-bit flash converter uses a 256 resistor reference voltage divider to establish the voltage on 256 capacitors. These voltages are *simultaneously* compared with the analog input to establish the digital output.

- The successive approximation A/D convertor also uses a 256 resistor network, but comparisons with the analog input take eight steps.

- A tracking A/D consists of a comparator, an up-down counter, and a digital-to-analog converter (DAC). The counter and the DAC generate a voltage staircase to match the input voltage, and the count at match becomes the digital output.

- Although the slowest, the dual slope has the best noise immunity. The device consists of an integrator, comparator, and a counter. A higher-level input voltage results in a higher-level output from the integrater in a given time (16.67 ms). The count to bring the output back to zero with a fixed reference input is the digital output corresponding to the analog input.

11-9 CHAPTER 11—EXERCISE PROBLEMS

1. It is desired to convert an analog voltage to digital with a resolution of at least one part per thousand (0.1%). How many bit A/D convertor is required?

2. Determine the digital output code for an ADC0804 A/D with a +5.12-V reference and an input of 3 V. Find the conversion time if the clock is running at 800 kHz.

3. Determine the analog input voltage range to an ADC0804 if the output is 01101101 and the reference is +5.12 V.

4. Determine the reference voltage to be applied to Pin 9 of an ACD 0804 if the input signal has a maximum value of 2 V.

5. Find the conversion time for the four types of A/D converters with an input voltage of +3 V, full-scale readings of 5.12 V, clock at 300 kHz, and, for the dual slope, a reference of 2 V.

6. If the reference voltage for a CA3308 is 2.56 V, and the logic supply (V_{DD}) 5 V, find the voltage across the 68th input capacitor during the auto balance phase.

7. Calculate the conversion time for a CA3308 flash converter if the clock frequency is 10 mHz.

8. Determine the digital count for an ICL7126 dual slope converter at the completion of the de-integrate phase if the reference voltage is 150 mV and the input voltage 100 mV.

9. A digital voltmeter is required using an ICL7126 with a full-scale reading of 2 V, and 2 displays per second. Determine the oscillator frequency and all circuit components. Use a 9-V battery as a power source.

10. Derive Eq. 11-9 for the optimum oscillator frequencies to give maximum 60-Hz rejection with a dual slope converter.

11-10 CHAPTER 11—LABORATORY EXPERIMENTS

Experiment 11A

Purpose: To demonstrate conversion from an analog signal to a digital representation, using an A/D convertor.

Requirements: Connect the ADC0804 device as shown in Fig. 11-4. Apply five different DC inputs and verify output binary states in accordance with the predicted values.

Measure: Conversion time and compare with the specified value.

Experiment 11B

Purpose: To build a digital voltmeter with an LCD readout using a ICL7126 A/D converter.

Requirements: Design the circuit to meet the requirements specified in Exercise Problem 9.

Measure: The LCD output reading with five different analog input voltages of known value. Determine the error and compare with the specified ±1 LSB.

12

Digital-to-Analog Conversion

OBJECTIVES

At the completion of this chapter, you will know the following digital-to-analog converter concepts:
1. Reasons for conversion
2. Typical device configurations
3. Accuracy considerations
4. Reference adjustments

12-1 INTRODUCTION

Mathematical operations on analog signals are more easily performed today with the use of the digital computer. To enable digital processing, the analog signal is normally changed into a digital representation (analog-to-digital conversion), usually consisting of eight-bit digital words. After processing, it may be necessary to convert these words back into an analog signal for a particular application. A device that can perform this function is the digital-to-analog converter (DAC).

The digital-to-analog converter takes a digital word input and translates it into a specific output voltage. Thus, when the bits change in a digital word, a change occurs in the output voltage. For example, assume that the digital word consists of 4 bits, giving us 16 (2^4) possible 4-bit combinations. Let us further assume that the 16 combinations represent 0 to 15 V (in 1-V steps) with 0000 = 0 V and 1111 = 15 V (binary-to-decimal equivalence). Therefore, with an input word of 0100, the output of the DAC is 4 V. The word 1010 gives an output of 10 V and so on.

Integrated-circuit DACs typically operate by having the digital input word control the amount of current flowing through an output load resistor. Following the previous example, the current is increased in 15 equal increments as the input word varies from 0000 to 1111. If the current is 1 mA and the load resistor is 1 kΩ, then the output voltage ranges from 0 to 15 V in 1-V steps.

12-2 A TYPICAL DIGITAL-TO-ANALOG CONVERTER

One type of integrated circuit DAC is the DAC0800. This device converts digital inputs to analog outputs with a resolution of 8 bits or 256 (2^8) steps. Digital inputs to the device can be from TTL, CMOS, PMOS, etc. The full-scale output current (I_{FS}) for the device is typically 2 mA. A block diagram for the DAC0800 is shown in Fig. 12-1.

Referring to Fig. 12-1, the reference current I_{REF} is established from the current flowing around the inverting input of the operational amplifier U_1 into the collector of Q_0. This same current flows through the emitter resistor R of Q_0 to the negative supply voltage $-V_{CC}$ and sets up a voltage drop V_R that is equal to $I_{REF} \times R$. This same voltage drop appears from the emitter to $-V_{CC}$ for each transistor, Q_1 through Q_8. Since the emitter resistor of Q_1 has twice the resistance of Q_0 emitter resistor, the current through Q_1 is one half of I_{REF}. The other transistors have additional series resistors, and an analysis shows that the current in each transistor is one half that of the preceding stage, or:

$$I_{Q_0} = I_{REF}$$

$$I_{Q_1} = I_{REF}/2 \qquad (2^{-1})$$

$$I_{Q_2} = I_{REF}/4 \qquad (2^{-2})$$

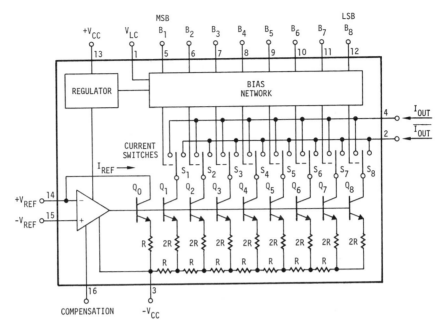

Figure 12-1 DAC0800 Block Diagram

$$I_{Q_3} = I_{REF}/8 \qquad (2^{-3})$$

$$I_{Q_4} = I_{REF}/16 \qquad (2^{-4})$$

$$I_{Q_5} = I_{REF}/32 \qquad (2^{-5})$$

$$I_{Q_6} = I_{REF}/64 \qquad (2^{-6})$$

$$I_{Q_7} = I_{REF}/128 \qquad (2^{-7})$$

$$I_{Q_8} = I_{REF}/256 \qquad (2^{-8})$$

When the current switches S_1 through S_8 are in the left position, all current flows from the I_{OUT} terminal Pin 4. The total current (full scale) with all switches in this position is the total of I_{Q_1} through I_{Q_8} or $255/256 \times I_{REF}$. This same current flows from the $\overline{I_{OUT}}$ terminal Pin 2 if all switches are in the right position. The position of each switch is controlled by the logic inputs on lines B_1 through B_8. A logic high (logic one) on any input line puts the associated switch in the left position. For example, if only the B_3 input line is high and all other digital inputs are low, then only the current (I_{Q_3}) from transistor Q_3 flows from the I_{OUT} terminal. Since all other digital inputs are low, with their associated switches to the right position, the current flowing from $\overline{I_{OUT}}$ is $(255/256 \times I_{REF}) - (I_{REF}/8)$.

12-3 A DIGITAL-TO-ANALOG CONVERTER CIRCUIT

An actual circuit using the DAC0800 is shown in Fig. 12-2.

The external 10-V reference is used in conjunction with the 4.5-K resistor to establish a 2-mA reference current. A reference current trim adjustment provided by the 50-kΩ potentiometer R_2 is used to get the full-scale current (I_{FS}). Full scale is adjusted by putting all digital inputs high (to logic reference) and adjusting R_2 until $I_{OUT} = I_{FS}$. The reference current I_{REF} can be measured by inserting an ammeter in series with the 4.5-kΩ resistor R_1 or by measuring the voltage drop across R_1 and then computing the current through it. Full-scale current I_{FS} is determined by:

$$I_{FS} = \frac{V_{R1}}{R_1} \times \frac{255}{256} \tag{12-1}$$

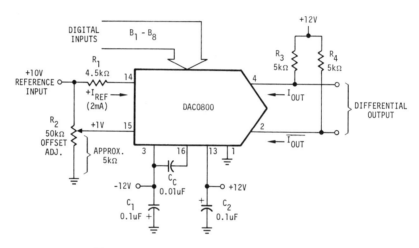

Figure 12-2   Digital-to-Analog Converter Circuit

When R_2 is properly adjusted, the voltage at the non-inverting input will be $+1.0$ VDC. Since the voltages at the op-amp inputs must be the same, V_{R1} is $(10\text{ V} - 1\text{ V})$ or 9 V, I_{REF} is 2 mA, and I_{FS} is 1.992 mA.

The maximum voltage drop across each 5-kΩ output load resistor R_3 and R_4 is $V_L = 1.992$ mA $\times$ 5 k$\Omega = 9.96$ V. The output voltage difference between the output terminals Pin 2 and Pin 4 varies from $+9.96$ V to -9.96 V, for a total excursion of 19.92 V, or twice the voltage drop across one output load resistor. If we wish to have a single-ended output, we can take the output voltage from Pins 2 or 4 to ground, but we must connect both load resistors into the circuit. It should be pointed out that the voltage drop across R_3 and R_4 is limited to a value that allows for about 4 V minimum between the negative supply $-V_{CC}$ and the output terminals (Pin 2 or Pin 4). The 4 V is necessary for correct operation of the internal circuitry of the D/A converter. If we had used 10 kΩ instead of 5 kΩ values for the load resistors, we would have a voltage drop of $V = I_{REF} \times R_L$ or 20 V. With $+10$ VDC supplying the load resistors and a -12 VDC supply voltage (a difference of 22 V), we would have an insufficient voltage (2 V) for operation of the device with these load resistors.

12-4 MICROPROCESSOR COMPATIBLE D/A

With many D/A applications in microprocessor-based systems, compatibility with the microprocessor is an important consideration. The DAC0830 CMOS type DAC is provided with control line inputs, which allow for simple integration with a microprocessor.

The block diagram for the DAC0830 is given in Fig. 12-3.

A key feature of this DAC is the double buffering of the input data which requires that the data passes through two independently controlled eight-bit latching registers. This allows for holding of the current DAC data in the DAC register and the next data word in the input register to allow fast updating of the DAC output on demand. It also allows several DACs in a system to have their output analog signals provided simultaneously via a common strobe signal.

The TTL level control lines shown on the block diagram provide the following functions.

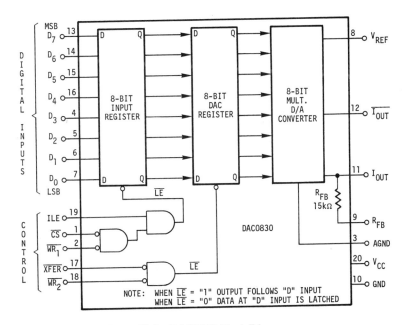

Figure 12-3   DAC0830 Block Diagram

1. Input latch enable (ILE) allows input data to latch into the input register with $\overline{CS}$ and $\overline{WR_1}$ low.

2. Chip Select ($\overline{CS}$) active low—provides selection of a particular DAC.

3. Write register 1 ($\overline{WR_1}$) active low—causes data to be written into the input register.

4. Write register 2($\overline{WR_2}$) active low—causes data to be written into the DAC register.

5. Transfer ($\overline{XFER}$) active low—allows the $\overline{WR_2}$ to pass. Can be used as a strobe signal.

These signals can originate at the microprocessor's control bus or can be generated from discrete logic decoding of address inputs (memory mapping).

The multiplying feature of this DAC can be used by applying the multiplying analog voltage to the V_{REF} terminal. An input range of ± 10 V is provided for this function. If straight digital-to-analog conversion is required, then V_{REF} is tied to a precision voltage source.

Output currents (I_{OUT} and $\overline{I_{OUT}}$) can either flow into or out of the device, depending on the polarity of the load resistor source, and have a maximum full-scale value of:

$$I_{FS} = \frac{V_{REF}}{15 \text{ K}} \times \frac{255}{256} \tag{12-2}$$

The current path equivalent circuit is shown in Fig. 12-4.

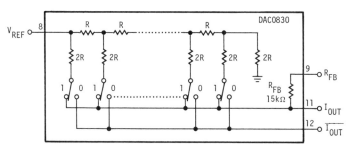

Figure 12-4   Internal Current Circuit

As with the DAC0800, the logic input controls the switch positions and the full-scale current (I_{FS}) flows through the I_{OUT} terminal if all switches are in the "one" position. The output current can be converted to voltage by connecting an op-amp to the output terminals as shown in Fig. 12-5 ($R_{FB} = 15$K).

In the circuit I_{OUT} will not flow into the high impedance terminals of the op-amp, but through the internal feedback resistor R_{FB}. Since the $\overline{I_{OUT}}$ terminal is grounded, the inverting terminal of the op-amp is at virtual ground and the output voltage (V_{OUT}) is:

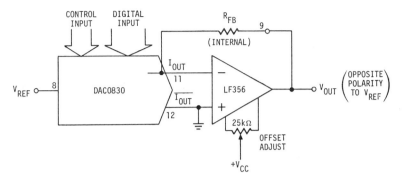

Figure 12-5 Circuit to Provide Voltage Output

$$V_{OUT} = -I_{OUT}\, R_{FB}$$

$$= -I_{OUT} \times 15\ K$$

The offset pot is adjusted with a digital input of all zeros ($I_{OUT} = 0$) for a $V_{OUT} = 0$. Full-scale adjustment can be made by having a 1- K pot in series with R_{FB} and adding a 1 MΩ resistor from the V_{OUT} terminal to the I_{OUT} terminal. With a digital input of all ones, the pot is adjusted until the full-scale output voltage (V_{FS}) is:

$$V_{FS} = -I_{FS}\, R_{FB}$$

Substituting I_{FS} from Eq. 12-2:

$$V_{FS} = \frac{-V_{REF} \times 255 \times 15\ K}{15\ K \times 256} \tag{12-3}$$

$$V_{FS} = -\frac{V_{REF} \times 255}{256}$$

The op-amp power supplies must be sufficient (typically 2 V greater than V_{FS}) to accommodate the full output swing. Since the op-amp inverts, a positive output voltage requires a negative reference voltage.

Another method for obtaining an output voltage rather than current, but without an op-amp, is to use the circuit of Fig. 12-6.

The circuit connections have been changed so that the reference voltage is inserted at the I_{OUT} terminal, and the output voltage (V_{OUT}) is taken from the V_{REF} terminal (8). This provides an output voltage which ranges from 0 V to 255 $V_{REF}/$ 256. Restrictions on this circuit are that V_{REF} be positive, and no greater than +5 V, and the supply voltage V_{CC} be at least 9 V more positive than V_{REF}. Loading of the output circuit should also be considered, since the output resistance can be as high as 20 kΩ.

By grounding the CS, WR$_1$, WR$_2$, and XFER terminals and tying the ILE terminal high, the DAC0830 can be operated in a flow-through configuration like the

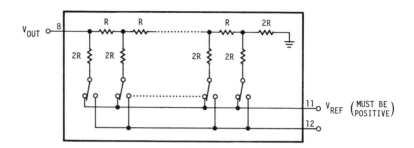

Figure 12-6 Voltage Mode Output

DAC0800. This can be used for a test circuit configuration as shown in Fig. 12-7.

Switches on the data lines allow manual selection of one or zero inputs. The output voltage (V_{OUT}) will be:

$$V_{OUT} = -\frac{V_{REF}\,N}{256}$$

$$= -\frac{10\,V \times N}{256}$$

where N is the decimal number corresponding to the binary input data. For eight-bit accuracy, the output voltage should differ from the predicted value by no more than $10\,V \times 0.004$, or $\pm\,20$ mV.

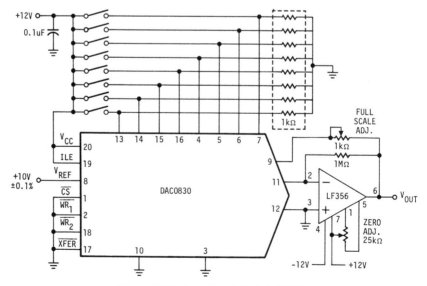

Figure 12-7 Test Circuit for DAC0830

To check the input control functions, a switch that allows connection to the positive supply or ground can be added to the $\overline{CS}$, $\overline{WR_1}$, $\overline{WR_2}$, $\overline{XFER}$, and ILE lines to allow either one or zero inputs. By operating the switches in a given sequence, data transfer through the registers and D/A section can be observed.

12-5 MICROPROCESSOR INTERFACING

Both the DAC0800 and the DAC0830 can be interfaced with a microprocessor. However, the DAC0800 requires a data latch and control (address) circuitry, since the digital word on the microprocessor data bus is only present for about a microsecond. A peripheral interface adapter (PIA) like the 6821 can be used to provide the latch and polling or interrupt functions. The PIA takes and stores the digital word appearing on the data bus which is destined for the DAC. It then provides the word to the DAC until it is updated by new data from the microprocessor. A block diagram for interfacing a DAC0800 with a 6800 microprocessor is shown in Fig. 12-8.

Gates U_1, U_2, and U_3 are used for address decoding of the memory-mapped address lines to enable selection of the PIA. The gate connections are set up to decode addresses 4R XS where R ranges from 0 to 7, S ranges from 0 to 3, and X can have any hex value. Address lines A_0 and A_1 are used for PIA register selection.

Table 12-1 is a 6800 machine language program that initializes the PIA and then takes data from each of 35 memory locations (0040 through 0062 hex) to generate a symmetrical waveform at the output of the DAC. Going from 0040 to 0061 generates the positive slope portion of the waveform, while 0062 to 0041 creates the negative slope portion.

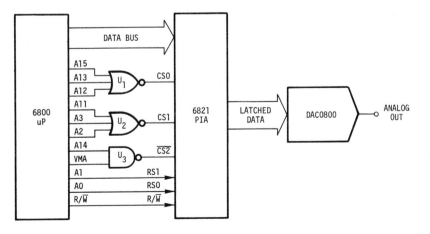

Figure 12-8   Interfacing a DAC0800 With a 6800 Microprocessor

Table 12-1 Function Generator Program

	0000	8600	
	0002	B7 4000	
	0005	8604	
	0007	B7 4001	
	000A	86 FF	Initialize PIA
	000C	B74002	
	000F	86 04	
	0011	B7 4003	

0014	CE 0040	LDX 0040	Load index REG
0017	A6 00	LDAX 00	Load AccA indexed
0019	B7 4002	STAA 4002	Output data to DAC
001C	BD 0036	JSR 0036	Go to delay sub
IF	08	INCX	Increment index REG
0020	8C 0062	CPX 0062	Compare with highest memory location
0023	26 F2	BNE F2	Branch back if not equal
0025	A6 00	LDAX 00	Load Acc indexed
0027	B7 4002	STAA 4002	Output data to DAC
002A	BD 0036	JSR 0036	Go to delay sub
002D	09	DECX	Decrement index REG
002E	8C 0041	CPX 0041	Compare with one greater than lowest memory location
0031	26 F2	BNE F2	Branch back if not equal
0033	09	DEC X	Decrement index REG
0034	20DE	BRA DE	Branch back to beginning
0036	C6 05	LDAB Y	Load 'B' AccA with Y
0038	01	NOP	
0039	01	NOP	
003A	5A	DEC B	
003B	C1 00	CMP 00	Time delay for output data
0031	26 F9	BNE F9	
003F	39	RTS	

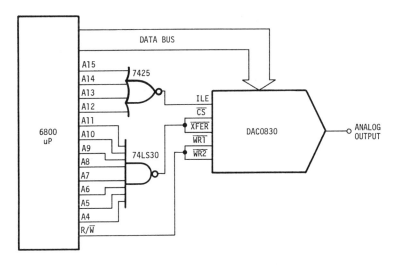

Figure 12-9   Interfacing DAC0830 With a 6800

Using the DAC0830 with a microprocessor is considerably simpler, as shown in the block diagram of Fig. 12-9.

The address decoding is more selective in this case—the hex address being 0FFR where R is any odd hex number. Writing data into the DAC causes the data to enter the first register and is latched into the second register when the $\overline{\text{WRI}}$ line goes high. Also, during the write time, previous data is transferred from the second register into the DAC and converted to an analog output.

The program for transferring data is similar to that given in Table 12-1, minus the PIA initialization.

In the previous chapter we saw the DAC used as part of a tracking analog to digital converter. By interfacing a DAC with a microprocessor, we can generate a tracking or a successive approximation type A/D convertor. For this application, the circuit can be connected as shown in Fig. 12-10.

The microprocessor provides an increasing count on the data lines, which is translated to a staircase output voltage from U_1. When the staircase voltage just exceeds the analog input, U_2 output goes low, which provides an interrupt signal to the microprocessor. This notifies the microprocessor that a match has been reached, and the data count corresponding to the analog input is stored in memory.

A faster conversion can be achieved by using a microprocessor with the successive approximation method. The microprocessor program is different, but the same circuit configuration as Fig. 12-10 can be used.

At the start of the successive approximation conversion, the microprocessor provides a word on the data lines with the most significant bit = 1 and all other bits = 0. This produces an output voltage corresponding to half the full-scale value. If the DAC output is greater than the analog input, an interrupt is sent to the

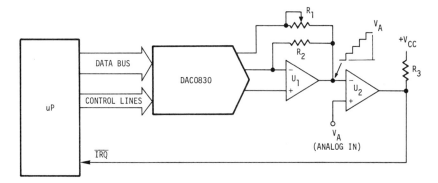

Figure 12-10   Microprocessor Controlled A/D Tracking Convertor

microprocessor and the MSB is set to zero. But if the DAC output is less than the analog input, the MSB is left at one. The microprocessor now makes the second most significant bit a one, and the comparator output determines whether it remains or is removed. The process continues until all eight bits have been tried and the resultant digital word corresponds to the analog input.

This program is obviously more complicated than the previous one, but a time-saving maximum of eight comparisons are made rather than as many as 256 for the tracking-type converter.

12-6 CHAPTER 12—SUMMARY

- Integrated circuit digital-to-analog convertors translate a digital input signal into a current output.

- Basically, a DAC consists of a ladder network with digital control of the scaled current through the branches, the output current being the sum of the branch currents.

- The number of bits for the DAC defines the resolution, e.g., a 10-bit device can provide 1024 different output levels.

- Accuracy of a DAC can be no greater than the accuracy of the voltage reference used.

- The analog output can be multiplied by a signal applied to the voltage reference terminal.

- DACs with a complementary current (I_{OUT}) output can provide a differential voltage swing of twice the output of a single-ended circuit.

- Since DACs are often used with microprocessors, compatability with these devices is important.

• The built-in logic and latch makes the DAC0830 easier to integrate with a microprocessor.

12-7 CHAPTER 12—EXERCISE PROBLEMS

1. Is the output from a DAC a voltage or a current?
2. If the reference current should be 10 times better than the A/D accuracy, what should be the percent reference accuracy for a 10-bit A/D?
3. How does the DAC 0830 differ from the DAC 0800?
4. Why is a PIA sometimes used with a DAC when interfacing a computer?
5. The reference current for DAC0800 D/A converter is 2 mA. If the input digital word is 11010101, find the output voltage across a 5-kΩ resistor.
6. The output voltage from a DAC0800 converter is 6.4 V. If the reference current is 1mA and the load resistor is 10 kΩ, find the input digital word.
7. Show the output circuit for a DAC0800 converter using 3-kΩ load resistors feeding into an operational amplifier with a gain of 2. Assume a 2mA reference current and determine the total output voltage excursions from the operational amplifier.
8. Design a circuit using a DAC0830 to provide a maximum output voltage of +9.96 V with a digital input of 11111111.
9. With reference to Fig. 12-7, calculate the percent variation in R_{FB} which can be accommodated by the 1-K pot.
10. The circuit in Fig. 12-11 is a DAC gain-controlled amplifier (volume control). Determine the gain for an input word of 10000000 and for 00000001 (*hint*—use circuit of Fig. 12-4 as the amplifier input resistor).

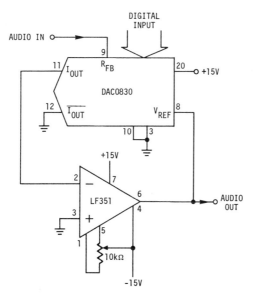

Figure 12-11 DAC Gain Control

12-8 CHAPTER 12—LABORATORY EXPERIMENTS

Experiment 12A

Purpose: To demonstrate the conversion of parallel digital input signals into a discrete analog output signal.

Requirements: Connect up the circuit as shown in Fig. 12-2. Use the $+10$ V for a logic one input and use ground for logic zero input. With all ones input, adjust R_2 for the required full-scale current.

Measure: Output current and compare with the predicted values for five different input logic bytes (eight-bit words).

Experiment 12B

Purpose: To demonstrate the latch capability and data conversion of a DAC0830.

Requirements: Set up the circuit as shown in Fig. 12-7, but connect the $\overline{\text{WR1}}$ and $\overline{\text{WR2}}$ input lines so they can be independently connected high or low.

Measure:

1. Starting with both $\overline{\text{WR1}}$ and $\overline{\text{WR2}}$ high, input an eight-bit digital word and momentarily zero $\overline{\text{WR1}}$. Now momentarily zero $\overline{\text{WR2}}$ and verify that the analog output corresponds to the digital input.
2. Repeat Step 1 for four different digital eight-bit words.

13

Data Acquisition

OBJECTIVES

Upon completion of this chapter you should know the following:
1. The block diagram of a data acquisition system
2. The purpose and operation of a sample and hold circuit
3. The purpose and operation of a multiplexer
4. The purpose and operation of a demultiplexer

13-1 INTRODUCTION

A data acquisition system can be used to transfer analog data from various remote input sources into a central computer system. Figure 13-1 shows a block diagram of this type of system.

Signals from the various transducers are amplified and fed into the multiplexer (mux); the mux looks at each input one at a time and provides a signal output which time shares the input signals. This composite signal is coupled to a sample and hold circuit, which holds the signal steady on a capacitor while A/D conversion takes place. If the output of the A/D is located a great distance from the computer, conversion to serial data and then to FSK can occur to improve the signal-to-noise ratio over the transmission line. The line driver serves to boost the signal level and impedance match the transmission line, while the line receiver restores any signal losses encountered during transmission. FSK demodulation is accomplished by a phase lock loop, and the serial digital output from the PLL is converted to parallel data for entry on the computer data lines. Combining or modification of the data occurs within the computer, and the DAC converts the parallel digital data back to an analog form. Finally the demultiplexer (de-mux) can process the time-shared data and restore it to individual analog signals for application to display or control devices.

Thus far we have discussed all items on the block except the sample and hold circuit, the multiplexers, and the digital serial-to-parallel and parallel-to-serial translators.

We will cover the sample and hold and the multiplexers in this chapter. The digital translators are not a linear device, so they will not be explained; for this application reference should be made to devices like the UART or the ACIA.

13-2 SAMPLE AND HOLD CIRCUIT

Sometimes it is required that a time-varying analog voltage be sampled for viewing, measuring, or signal processing. Normally, we can look at a voltage directly with a meter, but it can be difficult to hold the meter probe on a small point while operating and looking at the meter face. If we can sample and then hold the voltage, we could remove the probe from the circuit and then view the meter. We could also take a reading and hold and then display the difference between the first reading and a second reading, giving an expanded scale of the difference.

In other applications, we may wish to convert an analog voltage into a digital signal. It takes a certain amount of time for this conversion to occur (conversion time), and during this time the analog voltage must be kept steady, otherwise errors will occur in the digital signal.

The "sample and hold" function described can be performed with an integrated circuit. Figure 13-2 shows the functional block diagram of a LF398 monolithic sample and hold integrated circuit which uses an external hold capacitor.

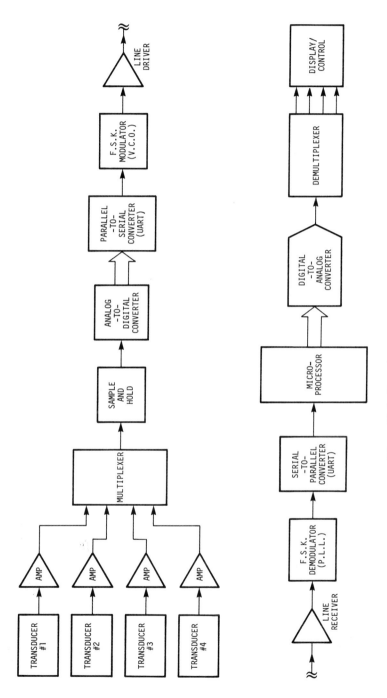

Figure 13-1 Data Acquisition Block Diagram

251

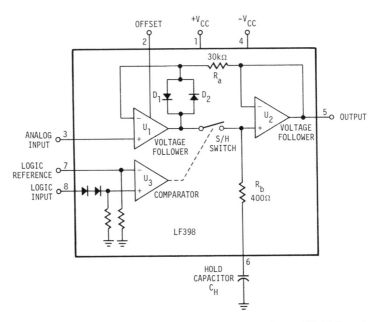

Figure 13-2   LF398 Functional Block Diagram with External Hold Capacitor

13-3 SAMPLE AND HOLD DEVICE OPERATION

Referring to Fig. 13-2, analog signals to be sampled are applied to the input terminal (Pin 3) and appear at the output of voltage follower U_1. Voltage follower U_1 has a high input impedance and unity gain. If the sample and hold (S/H) switch is closed (sample mode), the external hold capacitor charges to the value of the input voltage from a 400-Ω source. When the switch is opened (hold mode), the sample voltage is held on the capacitor and appears at the output of voltage follower U_2 for application to other circuits, e.g., metering circuits, analog-to-digital converter, etc. Voltage follower U_2 is used to prevent the capacitor from discharging through the output (Pin 5) to other circuits. The back-to-back diodes CR_1 and CR_2, which are connected to the output of U_1, allow the output of U_1 to overshoot by 0.7 V (diode forward voltage drop), speeding up the charging of the hold capacitor.

13-4 SAMPLE AND HOLD DEVICE CONTROL INPUTS

The input control circuit can be considered a comparator with two diodes in series with the non-inverting input terminal. When the signal applied to the logic input (Pin 8) exceeds +1.4 V (two diode drops), the comparator output goes high and closes the S/H switch. Thus, a logic 1 (high) on the logic input puts the sample and hold circuit in the sample mode. When the sample mode is to be terminated, a logic 0 (low) is then applied to the logic input, causing the comparator output to go low and open the S/H switch. If we desire to have the S/H switch close when a low logic input signal is applied, then the circuit in Fig. 13-3 can be used.

Notice that the logic input signal is applied to Pin 7 rather than to Pin 8, and that Pin 8 is biased up to +2.8 V, which puts the non-inverting input at +1.4 V. The comparator output is high (S/H switch closed) if a logic input signal of less than +1.4 V is applied to Pin 7. When the logic input signal is greater than +1.4 V, the comparator output voltage is low, thus opening the S/H switch (hold mode). The control logic is now reversed; a high input puts the circuit in the hold mode while a low input puts the circuit in the sample mode.

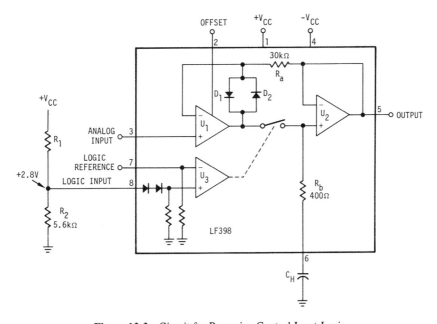

Figure 13-3   Circuit for Reversing Control Input Logic

13-5 *SAMPLE AND HOLD DEVICE TERMINOLOGY*

There are several important terms associated with the sample and hold circuit which describe its operational capabilities. These terms are listed below and are then further defined.

1. Acquisition time
2. Aperture time
3. Dynamic sampling error
4. Gain error

5. Hold settling time

6. Hold step

Acquisition time (t_A) is the time required to acquire a new analog input voltage with a change of 10 V from the previous sample. The voltage change is specified because the time it takes the input circuits to settle is a function of the amount of voltage change. The input signal must be "acquired" before the device switches to the hold mode.

Aperture time (t_{AP}) is the time during which the analog input voltage must not change after the hold command is given, otherwise the hold signal will be in error. It is determined from the time of the initiation of the hold command to the time it takes the internal circuits to disconnect the hold capacitor from the input signal.

Dynamic sampling (V_S) error is the change in millivolts that occurs across the sampling capacitor during the aperture time.

Gain error is the percentage error between the input signal and the output signal. The gain should be unity.

Hold settling time (t_{HS}) is the time required for the output to settle within 1 mV of the required output voltage after the initiation of the hold command.

Hold step (V_H) is the amount of voltage step that occurs in the output when switching from sample to hold with a steady input voltage. The step represents feed-through from the logic circuitry.

A graphical representation of these terms is given in Fig. 13-4. The final value should equal the input voltage.

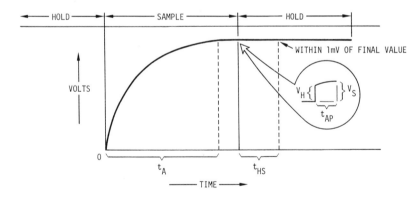

Figure 13-4   Sample and Hold Parameters

13-6 DETERMINING SAMPLE AND HOLD CIRCUIT PARAMETERS

In the sampling mode of operation, when the S/H switch is closed, the hold capacitor starts to charge to the new input voltage. The charging current for the capacitor is from a resistance source of typically 400 Ω. To charge the capacitor to within 1% of the input voltage, a minimum acquisition time is required which can be approximated from the equation:

$$t_A = 5\,RC \qquad (13\text{-}1)$$

where C is the capacitance in farads, R is the internal 400-Ω resistance and 5 stipulates 5 RC time constant to charge the capacitor. Let us assume the following:

$$C = 0.01\ \mu\text{F}$$

$$R = 400\ \Omega$$

and substituting into Eq. 13-1,

$$t_A = 5 \times 10^{-8}\text{F} \times 400\ \Omega$$

$$= 20\ \mu\text{s}$$

This indicates that the S/H switch should be closed for a minimum of 20 μs for the values given in order to charge the capacitor to within 1% of the new input value.

During the hold time the capacitor must retain its charge; however, some leakage will occur across the capacitor. Thus, the capacitor used should be a low-leakage type such as polypropylene or Teflon. (These capacitors also exhibit low sagback, i.e., drop in capacitance after a quick charge.) Also, leakage current from the sample and hold circuit will flow into the capacitor (maximum 200 pA at 25°C for the LF398).

For short sample times we require small capacitance, but for long hold time the capacitance should be large because voltage change on the capacitor is:

$$dV = \frac{i_C\, t_H}{C} \qquad (13\text{-}2)$$

where i_C is the current flowing out of the capacitor and t_H is the hold time. Figure 13-5 illustrates the equivalent circuit during the hold time.

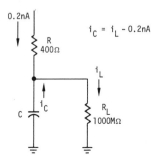

Figure 13-5 Sample Capacitor Discharge Circuit

To determine how much drop in voltage occurs across the capacitor during the hold time, assume the following:

$$V_C = 10 \text{ V}$$

$$t_H = 1 \text{ ms}$$

$$C = 0.01 \ \mu\text{F}$$

$$R_L = 1000 \text{ M}\Omega$$

then
$$i_L = \frac{V_C}{R_L} \qquad (13\text{-}3)$$

$$= \frac{10 \text{ V}}{1000 \text{ M}\Omega}$$

$$= 10 \text{ nA}$$

Therefore
$$i_C = i_L - i_{S/H\text{-leak.}} \qquad (13\text{-}4)$$

$$= 10 \text{ nA} - 0.2 \text{ nA}$$

$$= 9.8 \text{ nA}$$

Substituting into Eq. 13-2:

$$dV = \frac{9.8 \times 10^{-9}\text{A} \times 10^{-3} \text{ S}}{10^{-8}\text{F}}$$

$$= 0.98 \text{ mV}$$

Thus, the capacitor voltage changes by 0.98 mV over the hold time of 1 ms. This represents a change of one part in 10,000, which is better than 2^{13} or 13-bit accuracy. A good rule of thumb is to have the S/H accuracy three bits better than the required overall data acquisition accuracy. In a practical situation, the capacitor and circuit leakage resistance could be much less than 1000 MΩ; therefore, the drop in voltage would be greater.

Sample and hold times are determined from the system requirements. If an accuracy of 1% is tolerable (six bits), then five times constants can be used, but greater accuracy requires a longer sample time. Hold time is determined from the circuit following the S/H. If it is an A/D convertor, then the hold time must be at least as long as the A/D conversion time, but the longer the hold time the greater the voltage droop on the hold capacitor.

13-7 MULTIPLEXERS

Many times there are several analog signals—e.g., from transducers monitoring security, fire, air conditioning, etc.—in a building for input into a computer system. Since the computer can process only one signal at a time, a device is needed to perform the input selection process. Such a device is an analog multiplexer (a-mux).

Conversely, when several analog devices need to be controlled by a single computer, the computer can accomplish this through the use of an analog demultiplexer (a-demux), a device with a single input and several selectable outputs.

Multiplexer and demultiplexer devices operate in a similar manner but perform the complementary function of each other.

13-8 TYPICAL MULTIPLEXER

An analog multiplexer device can have 4, 8, or 16 inputs and 2, 3, or 4 input control lines to select which input line will be connected to the output. An example of an eight-channel multiplexer is the LF13508, which is shown in block diagram form in Fig. 13-6.

As the truth table indicates, the binary states of the three input control lines (A_0, A_1, A_2) determine which switch (S_1 thru S_8) is closed (on) and consequently which input is selected to be the output. For example, with all input control lines low (less than 0.7 V) and the enable input line high (greater than 2 V), then switch S_1 is closed. As long as the enable input is high, any one of the eight switches can be turned on. However, if the enable input line goes low, all switches open (off) and remain open regardless of the logic states on the input control lines.

An ideal switch has 0 Ω across its contacts when closed and thus, no voltage drop. Multiplexer solid state switches (FETS), however, have resistances in the range of 200 to 500 Ω. Therefore, when current flows, a small voltage drop occurs. When an external load resistor is connected across the output of the multiplexer, its

TRUTH TABLE

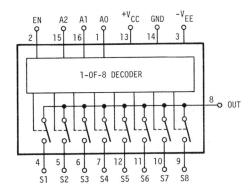

EN	A2	A1	A0	SWITCH CLOSED
H	L	L	L	S1
H	L	L	H	S2
H	L	H	L	S3
H	L	H	H	S4
H	H	L	L	S5
H	H	L	H	S6
H	H	H	L	S7
H	H	H	H	S8
L	X	X	X	NONE

H = HIGH   L = LOW
X = DON'T CARE

Figure 13-6   Eight Channel Multiplexer

resistance must be greater than the resistance value of the closed switch to reduce the voltage drop across the switch. To approach the condition of an ideal switch, the voltage drop across the multiplexer switch should be limited to one thousandth of the input voltage (10-bit accuracy). Therefore, the ohmic value of the load resistor must be greater than 1000 times the maximum ohmic value of the multiplexer switch (when closed), or

$$1000 \times 500 \ \Omega = 500 \ \text{k}\Omega \qquad (13\text{-}5)$$

One way to minimize the voltage drop across the switch is to use an op-amp follower on the output as shown in Fig. 13-7.

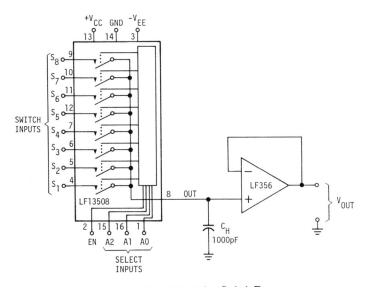

Figure 13-7   Minimizing Switch Drop

The op-amp in Fig. 13-7 is a FET type 741 (LF13741) with a typical open loop input resistance of $10^{12} \ \Omega$. Adding the capacitor C_H converts the circuit into a selectable sample and hold, which samples when the enable input line is high and holds when it is low.

When the ideal switch is in the off position, the resistance across the switch contacts is infinite and no current flows. The multiplexer switch "off" characteristics are rated in the terms of leakage current, which range from about 100 pA to 50 nA and is a function of the voltage developed across the switch (FET). If the voltage developed across the switch becomes too great, then the FET may turn on.

With the multiplexer switch on (closed), we want a high value for the load resistor in order to minimize the voltage drop across the switch. However, a high value of load resistance will cause an increase in the output voltage due to the leakage current. Thus, both cases have to be considered when the load resistor is selected.

Referring to the sample and hold circuit of Fig. 13-7, we must consider the drop in the capacitor voltage caused by the leakage current during the hold time. The time for the multiplexer switch to open and close after the circuit is enabled is about 0.1 to 2 μs. This switching time must be taken into account and allowed for before the output can be sampled.

The maximum analog input voltage for the LF13508 must not exceed the applied positive and negative supply voltages or a maximum peak-to-peak swing of 36 V, which is a device maximum rating. The highest value of current allowed through a LF13508 is 10 mA.

13-9 MULTIPLEXER EXAMPLE

Suppose we wish to convert eight analog input signals into digital for processing by a microcomputer. To do this we can use a multiplexer, sample hold, and an A/D converter connected as shown in Fig. 13-8.

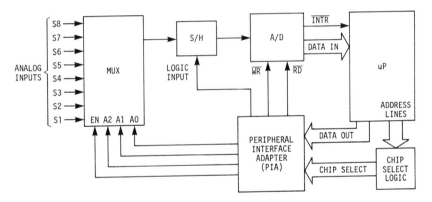

Figure 13-8   Data Conversion System

The PIA has been included to latch (hold) the data, which momentarily appears on the microcomputer data lines. At the start of a conversion sequence, the PIA is addressed via the chip select logic, and an eight-bit word appearing on the microcomputer data lines is latched into the PIA. The data received from the microcomputer will provide the required states on the seven PIA output lines (eighth data line not used). If the data word is such that the multiplexer select inputs (A_0, A_1, and A_2) are low and the enable high, the analog data from the S_1 input will be connected to the input of the sample and hold. With the logic input to the sample and hold high, the S/H capacitor starts charging to the analog input voltage. After allowing the time for the analog signal to be acquired by the S/H, the microcomputer changes the data out of the PIA to put the S/H in the hold mode (logic input low) and puts the A/D $\overline{WR}$ line high. This causes the A/D to convert the analog input signal

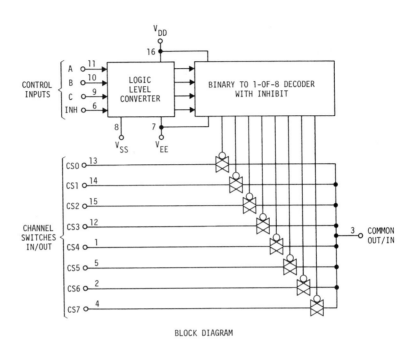

BLOCK DIAGRAM

TRUTH TABLE

CONTROL INPUTS				CHANNEL SWITCH "ON"
INHIBIT	C	B	A	
0	0	0	0	CS0
0	0	0	1	CS1
0	0	1	0	CS2
0	0	1	1	CS3
0	1	0	0	CS4
0	1	0	1	CS5
0	1	1	0	CS6
0	1	1	1	CS7
1	X	X	X	NONE

X = DON'T CARE
1 = LOGIC HIGH
0 = LOGIC LOW

Figure 13-9 CD4051 Analog Multiplexer/Demultiplexer

stored on the S/H capacitor to a digital word. At the completion of conversion, the A/D interrupts the microcomputer. When the microcomputer services the interrupt, it puts the A/D $\overline{RD}$ and $\overline{WR}$ lines low and reads the data from the A/D. It also changes the multiplexer select inputs to cause the second analog input S_2 to be coupled into the S/H, and the cycle repeats.

13-10 MULTIPLEXING AND DEMULTIPLEXING

A device that has the capability for multiplexing and demultiplexing is the CD4051. This is a CMOS device and has the internal circuitry and truth table as shown in Fig. 13-9.

This device is operated as a multiplexer when analog data is applied to any one of eight channel switch (CS1–CS8) inputs. When analog data is applied to the common input, the device operates as a demultiplexer. Notice that this device has an inhibit function that is used in lieu of the enable function for the LF13508. Thus, in order for data to pass through the multiplexer/demultiplexer, the inhibit input line must be low.

13-11 TYPICAL DEMULTIPLEXING OPERATION

A typical demultiplexer operation is shown in Fig. 13-10. As illustrated, the computer provides a digital output to the D/A converter and select (A, B, and C) and inhibit signals to the demultiplexer device via the PIA. The D/A converter transforms the digital input signal to an analog output signal that is applied to the common input of the demultiplexer. The demultiplexer routes the analog input signal to one of the eight output lines as determined by the logic states on the select input lines when the inhibit signal is low. The D/A converter must have an accuracy consistent with the system requirements, and the internal voltage drop of the demultiplexer must be taken into account.

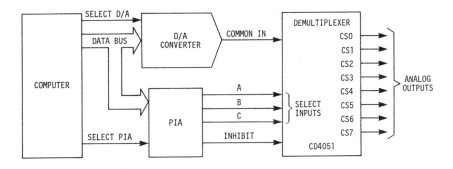

Figure 13-10 Demultiplexer System

13-12 CHAPTER 13—SUMMARY

- A typical data acquisition system consists of a multiplexer, sample and hold, A/D convertor, computer, DAC, and a demultiplexer.

- A sample and hold circuit samples an analog voltage and holds the sampled voltage steady while processing of the voltage is performed.

- The smaller the hold capacitor, the faster the input voltage can be sampled; however, a small hold capacitor means more voltage droop during the hold time.

- The more samples taken of the input waveform, the better the waveform is represented.

- Accuracy of the voltage sampled has to be consistent with the overall data acquisition accuracy requirement; for example, an S/H accuracy of 11 bits is required for an overall system requirement of 8 bits.

- Multiplexers take several input signals and convert them into a signal time shared output signal.

- A demultiplexer does the reverse of providing several outputs from a single time shared input.

- Ideal mux and demux devices should have internal switches that have infinite resistance when open and 0 Ω when closed.

- Because actual mux and demux switches are not ideal, the output circuit loading must be considered.

13-13 CHAPTER 13—REVIEW PROBLEMS

1. Determine the minimum acquisition time for an LF398 if the hold capacitor if 0.1 μF.

2. If the hold capacitor is 0.1 μF, the voltage held is 6 V, and the hold time is 20 ms, find the change in the output voltage. Assume a capacitor leakage resistance of 50 MΩ.

3. Draw the complete sample and hold circuit using an LF398. Assume a 0.01 μF hold capacitor and that sampling occurs with a low logic level. Specify the input logic pulse requirements.

4. Design a voltmeter probe using an LF398. Sample to be taken by activation of a push-button switch. Input voltage range from 0 to +10 V.

5. Determine the maximum permissible gain error if the LF398 is being used with an ADC0804.

6. Show the block diagram for an A/D conversion circuit with a LF398 and a ADC0804. The termination of the hold time is to be controlled from the ADC0804.

7. Show the block diagram for a system which will take one of eight analog inputs and process it in a computer. Output to be one of eight analog signals.

8. Compute the maximum "on" voltage drop across an LF13508 multiplexer with a 1 MΩ load resistor if the input voltage is 1 V.

9. Compute the voltage across a 1 MΩ load resistor with the LF13508 multiplexer turned off if the input voltage is 1 V.

10. If the voltage on a 0.01 μF hold capacitor is 5 V with leakage resistance of 1000 MΩ, compute the drop in voltage after 100 μs of hold time if;
    a) An LF398 S/H is used.
    b) An LF13508 (Fig. 13-7) is used. Assume 50 nA leakage from the mux.

11. What effect does the LF356 bias current have on the result of Part (b) in Problem 10?

12. A data acquisition system consists of a mux, S/H, and an A/D. What must occur before valid data is available at the A/D?

13-14 CHAPTER 13—LABORATORY EXPERIMENTS

Experiment 13A

Purpose: To demonstrate the operation of a sample and hold circuit with a sine wave input and a DC input.

Requirements:

1. Connect up the LF398 device as shown in Fig. 13-11 and couple a low-frequency sine wave into the analog input. Connect a pulse generator to the logic input at a frequency 10 times greater than the sine wave frequency. Determine width of input pulse consistent with acquisition time requirement. Sketch output waveform and label sample time, acquisition time, and hold time. (*Hint:* Adjust pulse frequency to lock in with sine wave frequency.)

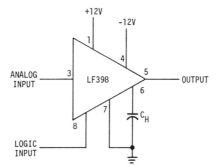

Figure 13-11   Lab Experiment 13A

2. Build a voltmeter probe using an LF398 (reference Review Problem 4) and demonstrate operation.

Measure:

For Requirement 1:

1. Sample time
2. Acquisition time
3. Hold time
4. Gain error

For Requirement 2:

1. Gain error
2. Droop over 10 seconds

Experiment 13B

Purpose: To demonstrate the operation of an analog multiplexer.

Requirements: Connect a voltage divider to generate eight inputs to the CD4051 and use a 1-MΩ resistor for the output load. Use a 555 as a 1-kHz clock and a 7993 counter to provide the select inputs. All circuits to operate off +5 V.

Measure:

1. The "on" voltage drop across the demultiplexer.
2. The voltage across the 1-MΩ resistor with all switches closed.
3. The rise time of the output signal.
4. The switch time delay.

14

Specialized Devices

OBJECTIVES

At the completion of this chapter you should know the following:
1. The theory of operation for a switched capacitor filter (SCF).
2. How to apply an SCF.
3. How a speech synthesizer operates.
4. Applications for a digital gain set device.

14-1 INTRODUCTION

This chapter is reserved for those devices that have a more specialized application than those previously described. Though more limited in capability, they still play an important role as linear integrated circuit components.

There is no functional similarity between the three devices described in this chapter, so they can be studied independently.

The chapter starts with a description of a switched capacitor filter circuit; this is followed by a presentation of a voice synthesis device; finally, a programmable gain set chip is described.

14-2 SWITCHED CAPACITOR FILTER

We looked at various op-amp active filter networks in an earlier chapter. The band pass, band stop, low pass, and high pass were covered; in all these configurations an op-amp was used along with precision resistors and capacitors. Using precision components adds to the cost of equipment as well as making it more difficult to get replacements. This is particularly true with precision capacitors. Manufacturers of linear integrated circuits realized this problem and developed the switched capacitor active filter (SCF) chip, which does away with the need for external capacitors.

Since a key element of the SCF is the integrator, let us first look at how we can filter with this circuit. Figure 14-1 shows an integrator used for a low pass filter. In operation, as the input frequency increases, the reactance of the feedback capacitor decreases, which reduces the gain and the output signal drops.

In part B of Fig. 14-1, we see that a high pass filter can be generated by using the integrator with a summing amplifier (A_1) and feeding the output of the integrator

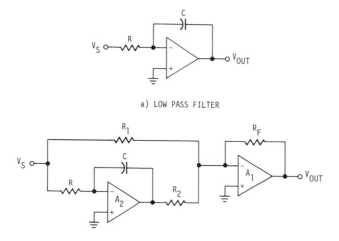

a) LOW PASS FILTER

b) HIGH PASS FILTER   **Figure 14-1**   Integrator Used as a Filter

in to the input of A_1. As the input frequency is increased, less negative feedback occurs from the output of the integrator and the output of A_1 increases. By rearranging the feedback and using a second integrator, a band pass or band stop filter can be generated. Filters that use integrators and summing amplifiers are called *state variable* types.

With a switched capacitor filter, the integrator input resistor is replaced by two switches and a capacitor as shown in Fig. 14-2.

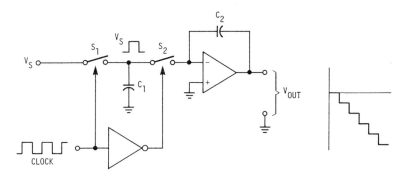

Figure 14-2   Basic Switched Capacitor Circuit

The clock signal controls the switches S_1 and S_2. On a positive clock pulse S_1 closes and C_1 quickly charges to the input voltage V_S (assumes V_S is from a low-resistance source). When the clock signal goes low, S_1 opens and S_2 closes. The charge (q) on C_1, which is equal to V_SC_1, is now all quickly transferred to C_2. The voltage (and charge) on C_1 goes to zero because of the virtual ground at the op-amp input. This transfer of charge to C_2 on each clock cycle causes the voltage on C_2 to increase in steps. With a positive input voltage, the output will step to a more negative voltage.

Comparing the output signal to the conventional integrator, we get a stepped slope rather than a steady slope. But notice that if the clock frequency is increased (more charge transfers), the slope increases. The average current which flows into C_1 and then into C_2 is:

$$I = \frac{q}{t}$$

$$= \frac{V_S C_1}{T}$$

$$= V_S C_1 f_{CLK} \qquad (14\text{-}1)$$

where T is the clock period and f_{CLK} the clock frequency. This current flow can be thought of resulting from a fictional resistor R_1 where:

$$R_1 = \frac{V_S}{I}$$

$$= \frac{1}{C_1 \, f_{CLK}}$$

This effective resistor represents the charging and discharging of C_1, and its value is controlled by the clock frequency. We can change the response (slope) of the integrator by simply changing the clock. The filter characteristics, then, are also under control of the clock.

In a switched capacitor filter chip, the capacitors C_1 and C_2 are formed in the semiconductor material and are therefore closely matched. There is no longer a concern, then, with the drift of external capacitors.

The actual integrator used in the SCF is a non-inverting type. This is achieved by reversing the connection of C_1 when it is connected to C_2 as shown in Fig. 14-3. Both sections of S_1 are closed first, charging C_1, but when S_{2A} and S_{2B} close, the polarity of C_1 is reversed and is connected to the op-amp.

As before, S_1 and S_2 are controlled by opposite states of the input clock. The clock frequency is much higher than the signal frequency, so the steps in the output of the integrator can be filtered out with a simple RC low pass filter.

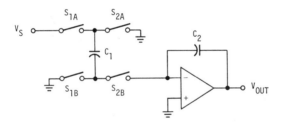

Figure 14-3 Non-Inverting Integrator

14-3 A SWITCHED CAPACITOR FILTER DEVICE

A typical SCF is the MF10. This device can be used as a filter for frequencies up to about 20 kHz with a maximum clock frequency of 1 MHz. The filter center, or 3 dB down, frequency is set by dividing the clock by 50 or 100 (depending on the voltage on a control pin). Filter accuracy is as good as the frequency stability of the clock. If we use a crystal-controlled clock, we will have a very accurate filter. The MF10 contains two separate second-order filters, each as shown in Fig. 14-4.

As can be seen from the figure, there are five different frequency responses available from three outputs, notch (N), all pass (AP), high pass (HP), band pass (BP) and low pass (LP).

Input signals can enter at Pin 4 for a inverted output or Pin 5 (S_1) for a non-inverted output. When using Pin 5, the driving impedance should be less than 1 kΩ.

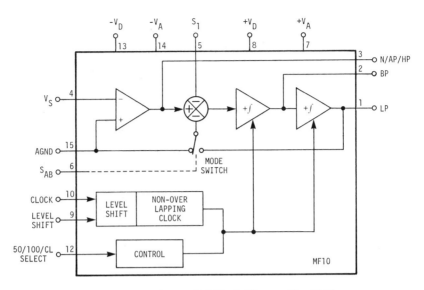

Figure 14-4   MF10 Block Diagram (One Half)

The signal passes through a summing junction where two of the inputs $(-)$ are subtracted from a third $(+)$.

Taking Pin 6 (S_{AB}) low will cause the mode switch to move to a position that will ground the input to the summing junction. With Pin 6 high, the output of the second integrator will be connected to the summing junction. We will see later how this can change the operating mode.

The level shift (Pin 9) allows for various input clock levels. With a TTL input clock, this pin should be grounded, but if the input clock excursion swings to a negative supply value, Pin 9 should be tied to this supply.

14-4 SCF FILTER CONFIGURATIONS

There are several possible circuit configurations using the MF10. We shall look at two which will give us the four filter characteristics. The first configuration shown in Fig. 14-5 is called mode 1 by the manufacturer and provides notch, band pass, and low pass outputs by using three external resistors.

We need to define the filter responses in terms of in-band gain, cutoff frequency, center frequency, Q, etc. Figure 14-6 shows the responses along with the equations for the mode 1 configurations (see p. 271).

Note that the equations for the low pass are somewhat more involved than those of the notch and band pass. Let us look at an example using these equations.

EXAMPLE 14-1

A band pass filter is required with a center frequency of 3 kHz (f_O), a bandwidth of 300 Hz (Q of 10), and a gain of -10. For least clock ripple in the output signal, we

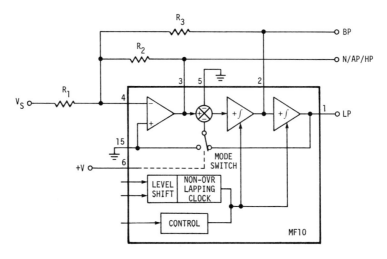

Figure 14-5 Mode 1 Notch, Band Pass, Low Pass Filters

would like the clock frequency as high as possible, so we choose 100:1. From Part b of Fig. 14-6:

$$f_{CLK} = 100 f_O$$

$$= 300 \text{ kHz}$$

The required bandwidth of 300 HZ requires a Q of 10. Let us pick 10 kΩ for R_2, then:

$$R_3 = Q R_2$$

$$= 100 \text{ k}\Omega$$

With the required gain of -10 (H_{OBP}), we can solve for R_1

$$R_1 = -R_3/H_{OBP}$$

$$= 10 \text{ k}\Omega$$

So, with just two 10-kΩ and one 100-kΩ resistors, we have a second order filter. The low pass output is taken from Pin 1 (See Fig. 14-5); band pass response is also available at Pin 2 and notch at Pin 3.

Solution.   To obtain a high pass filter, we can use the mode 3 configuration shown in Fig. 14-7.

The difference between this circuit and the mode 1 type is that Pin 6 is tied to the minus supply rather than the positive supply, causing the lower input to the summing junction to be grounded. Also, an additional resistor (R_4) is used.

We will just consider the equations for the high pass for the mode 3 circuit. Figure 14-8 shows the response and the equations.

Again, we will work an example to reinforce the concept of the high pass filter.

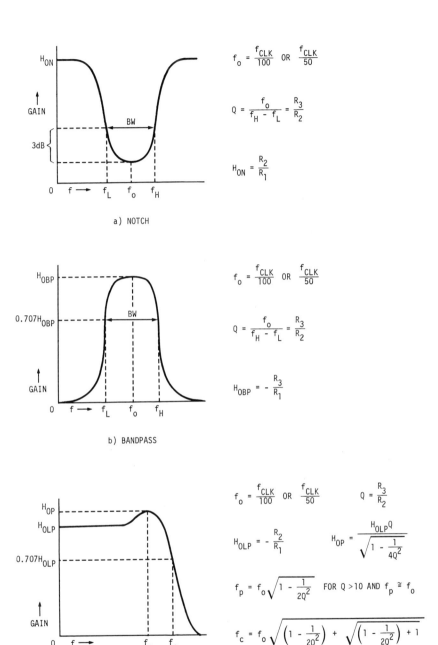

Figure 14-6 Mode 1 Responses and Design Equations

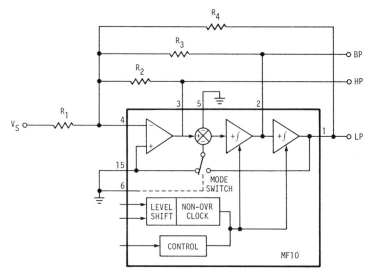

Figure 14-7 Mode 3 High Pass, Band Pass and Low Pass

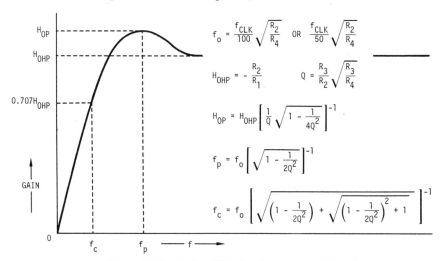

$$f_o = \frac{f_{CLK}}{100} \sqrt{\frac{R_2}{R_4}} \quad OR \quad \frac{f_{CLK}}{50} \sqrt{\frac{R_2}{R_4}}$$

$$H_{OHP} = -\frac{R_2}{R_1} \qquad Q = \frac{R_3}{R_2} \sqrt{\frac{R_3}{R_4}}$$

$$H_{OP} = H_{OHP} \left[\frac{1}{Q} \sqrt{1 - \frac{1}{4Q^2}} \right]^{-1}$$

$$f_p = f_o \left[\sqrt{1 - \frac{1}{2Q^2}} \right]^{-1}$$

$$f_c = f_o \left[\sqrt{\left(1 - \frac{1}{2Q^2}\right) + \sqrt{\left(1 - \frac{1}{2Q^2}\right)^2 + 1}} \right]^{-1}$$

Figure 14-8 Mode 3 High Pass Response and Equations

EXAMPLE 14-2

A high pass filter is required with a cutoff frequency (f_C) of 1 kHz and a gain of -5 (H_{OHP}). If we want little peaking of the response curve (H_{OP} close to H_{OHP}), we can select $Q = 1$.

Choosing a 5-kΩ resistor for R_1, we can solve for R_2:

$$R_2 = H_{OHP} R_1$$

$$= 25 \ k\Omega$$

Solving for f_O:

$$f_O = f_C \sqrt{\left(1 - \frac{1}{2Q^2}\right) + \sqrt{\left(1 - \frac{1}{2Q^2}\right)^2 + 1}}$$

Using $Q = 1$ and $f_c = 1$ kHz

$$f_0 = 1.272 \text{ kHz}$$

If we want the clock frequency 100 times greater than f_O, then:

$$f_{CLK} = 127.2 \text{ kHz}$$

We can determine the peak gain H_{OP}:

$$H_{OP} = \frac{H_{OHP} \, Q}{\sqrt{1 - \dfrac{1}{4Q^2}}}$$

$$= -5.77$$

Finally we can find the frequency (f_P) at which the peak gain occurs:

$$f_P = \frac{f_O}{\sqrt{1 - \dfrac{1}{2Q^2}}}$$

$$= 1.8 \text{ kHz}$$

Being a second-order filter, the slope in the rejection region is 40 dB per decade (12 dB per octave).

Since the MF 10 filter cutoff/center frequency is controlled by the clock frequency, we can move the filter frequency by simply changing the clock. Using a programmable counter (PC) between the clock oscillator and the filter gives us the capability of changing the filter frequency by varying the countdown factor (N) of the PC.

With the system (Fig. 14-9), we can select 16 different clock frequencies for the filter. If we add another programmable counter, we can expand to 256 frequencies. This gives us a tremendous versatility in filtering. If we are using the filters for detecting control tones (3 of 8 tones), and desire to change the frequencies for security reasons, we just program new data out of the microcomputer to the counter. A similar action would have to happen at the tone transmitters in order for the tones and filters to track. Another application could be an audio equalizer consisting of several CSR's which are under computer control. Automatic adjustment of the equalizers could occur to compensate for changes in room acoustics sensed by a monitor microphone.

14-5 SPEECH SYNTHESIS

We are entering an age of voice communication between humans and machines. An automobile status monitoring system informs us, via voice communication, that our

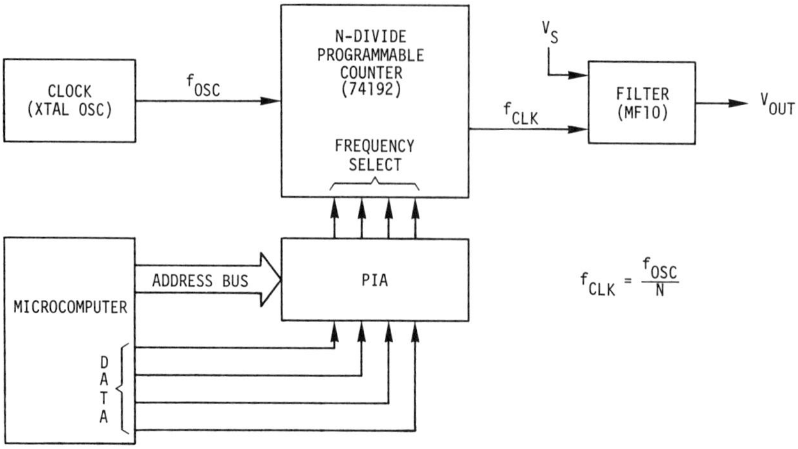

Figure 14-9 Computer Selection of Filter Frequency

seat belts are not fastened or our oil level is too low. Communicating in the other direction, voice recognition circuits determine when a particular human command is given and act on the instruction to start or stop some process. In this section we shall look at a device that can produce a set of distinct vocal sentences through a process called *speech synthesis.*

Speech consists of phonemes, which are the basic sounds uttered by a person. Combining these sounds in various ways forms words. Phonemes fall into two categories—voiced sounds and unvoiced sounds. The voiced sounds are generated by the vocal cords, while the unvoiced sounds are formed by air passing (hissing) through the mouth, which creates the *sh*-type sounds. Unvoiced sounds are less dependent on the speaker and therefore are easier to generate by a speech synthesizer. Actually, a noise source is used to provide the required hissing sound. An impulse source is used for the voiced sound, and both the voiced and unvoiced sources are passed through band pass filters. By controlling the amplitude out of the filter, a required sound can be generated.

Another approach to speech synthesis is to sample and digitize a voice waveform. The data is compressed to eliminate silent intervals and symmetrical redundancy. Use of adaptive delta modulation where differences from the previous sampled value are stored rather than the actual value reduces the amount of data (no change, no data). The stored data for an average word for a male voice is about 1000 bits.

14-6 A SPEECH SYNTHESIZER IC

A speech synthesis device called a DIGITALKER is provided by the National Semiconductor Corporation. This device uses the compression-type synthesis described. It consists of a speech processor chip (SPC), the MM54104, and a ROM containing the stored speech data. The manufacturer has some stored programs available that provide a series of sentences and others that provide words from which sentences can be formed. There is also the option of special ROM's being provided from customer-supplied tapes.

Since the heart of the DIGITALKER system is the MM54104, we will look at the more detailed operation of this device.

The block diagram for the MM54104 is shown in Fig. 14-10.

The function of each input/output line is as follows:

—The eight-bit start address calls up the required word or sentence and speech data from the ROM is received on an eight-line bus.

—The $\overline{WR}$ (write) command initiates the cycle by latching the start address into a register. On the rising edge of this signal, the start address is processed.

—Command Select (CMS) specifies two commands to the chip CMS function.

0   Reset interrupt and start speech sequence.

1   Reset interrupt only.

—Chip Select ($\overline{CS}$) performs the usual function of selecting a given chip when several are used in a system.

—The interrupt signal (INTR) goes high at the completion of the required speech sequence. It is reset by the CMS signal.

—Speech ROM address is the 14-line bus that selects speech data from memory. The MM54104 has the capability of selecting from 128K memory locations.

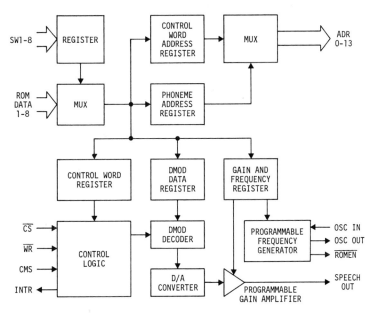

Figure 14-10   MM54104 Block Diagram

—The oscillator IN and OUT terminals are used to connect an external 4.0 MHz crystal for the internal clock, or an external oscillator can be connected to the OSC IN terminal.

ROM enable ($\overline{\text{ROMEN}}$) this line can control power for static speech ROM's.

In operation, when the $\overline{\text{WR}}$ line goes high, the start address is loaded into the control word register. The chip uses this address to fetch the control word from ROM (via the multiplexer) for the first block of speech data. The control word contains waveform information, repeat information, and the ROM address for the raw speech data. This speech address is loaded into the phoneme address register and is passed through the multiplexer to the ROM. The speech data is fetched from the ROM and stored in the speech data register. Under supervision of the control logic, delta demodulation occurs and the digital speech word is converted to an analog signal. Finally, amplitude and frequency (pitch) information is used to modulate the amplitude of the output analog speech signal.

Since digital processing is a sampled data system, the output waveform occurs in small steps which require filtering. Also, as a part of processing, the original data was subjected to pre-emphasis. To smooth and de-emphasize the output waveform, it is recommended that a single section (20 dB per decade) low pass filter be used with a cutoff frequency of about 200 Hz. For low-frequency components of a male voice, a cutoff frequency of 100 Hz is suggested, or 300 Hz for children's voices.

Output from the speech synthesizer ranges from simple statements like "Stop" to more complex sentence forms such as "Fuel remaining is six point five gallons."

Control of the synthesizer can be by hardware or software control logic. A circuit for hardware control is given in Fig. 14-11.

Positioning the eight switches for the starting address set up the required voice word (or sentence), and momentarily depressing the push button enters the address and starts the output voice sequence. If the $\overline{\text{WR}}$ line is momentarily grounded while an output voice message is in process, the output will be interrupted and a new sequence started. Table 14-1 gives the vocabulary set for a DG1056/DT1057 DIGITALKER.

With software logic control of the chip, the microcomputer, data lines connect to the SW 1 through eight lines. Microcomputer address lines enter address decoding logic and cause the CS and WR to be valid to start the sequence. At the completion of the voice output, the INTR line goes high, signaling the microcomputer that the chip is ready for new address data. The microcomputer can acknowledge by switching the CMS line. A possible interconnect diagram is shown in Fig. 14-12.

With the speech synthesizer under computer control, we have an excellent example of machine-to-human communications. The computer can verbally pass along routine status information or it can signal when some human action should be taken.

Applications for devices like the DIGITALKER exist wherever voice is the preferred method of communication. The advantage of audio alarms is that they are

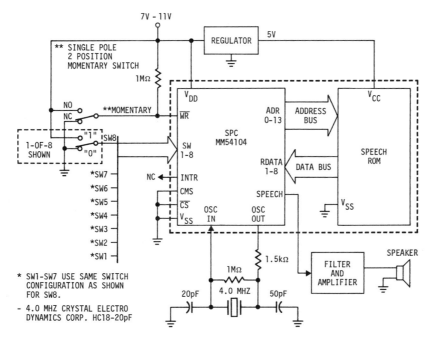

Figure 14-11   Digitalker™ Under Hardware Control

three-dimensional in nature: You don't have to be facing in the right direction to detect the alarm. Speech has the additional advantage of immediately indicating the action that should be taken.

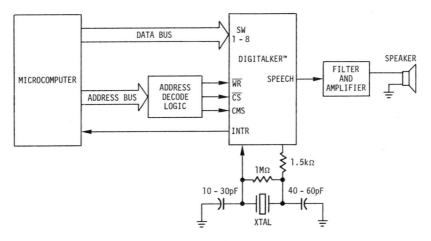

Figure 14-12   Digitalker™ Under Computer Control

Table 14-1 DT1056/DT1057* Master Word List

Word	8-Bit Binary Address SW8 SW1	Word	8-Bit Binary Address SW8 SW1	Word	8-Bit Binary Address SW8 SW1
Abort	00000000	Farad	00101100	Per	01011000
Add	00000001	Fast	00101101	Pico	01011001
Adjust	00000010	Faster	00101110	Place	01011010
Alarm	00000011	Fifth	00101111	Press	01011011
Alert	00000100	Fire	00110000	Pressure	01011100
All	00000101	First	00110001	Quarter	01011101
Ask	00000110	Floor	00110010	Range	01011110
Assistance	00000111	Forward	00110011	Reach	01011111
Attention	00001000	From	00110100	Receive	01100000
Brake	00001001	Gas	00110101	Record	01100001
Button	00001010	Get	00110110	Replace	01100010
Buy	00001011	Going	00110111	Reverse	01100011
Call	00001100	Half	00111000	Room	01100100
Caution	00001101	Hello	00111001	Safe	01100101
Change	00001110	Help	00111010	Secure	01100110
Circuit	00001111	Hertz	00111011	Select	01100111
Clear	00010000	Hold	00111100	Send	01101000
Close	00010001	Incorrect	00111101	Service	01101001
Complete	00010010	Increase	00111110	Side	01101010
Connect	00010011	Intruder	00111111	Slow	01101011
Continue	00010100	Just	01000000	Slower	01101100
Copy	00010101	Key	01000001	Smoke	01101101
Correct	00010110	Level	01000010	South	01101110
Date	00010111	Load	01000011	Station	01101111
Day	00011000	Lock	01000100	Switch	01110000
Decrease	00011001	Meg	01000101	System	01110001
Deposit	00011010	Mega	01000110	Test	01110010
Dial	00011011	Micro	01000111	Th (Note 2)	01110011
Divide	00011100	More	01001000	Thank	01110100
Door	00011101	Move	01001001	Third	01110101
East	00011110	Nano	01001010	This	01110110
Ed (Note 1)	00011111	Need	01001011	Total	01110111
Ed (Note 1)	00100000	Next	01001100	Turn	01111000
Ed (Note 1)	00100001	No	01001101	Use	01111001
Ed (Note 1)	00100010	Normal	01001110	Uth (Note 3)	01111010
Emergency	00100011	North	01001111	Waiting	01111011
End	00100100	Not	01010000	Warning	01111100
Enter	00100101	Notice	01010001	Water	01111101
Entry	00100110	Ohms	01010010	West	01111110
Er	00100111	Onward	01010011	Switch	01111111
Evacuate	00101000	Open	01010100	Window	10000000
Exit	00101001	Operator	01010101	Yes	10000001
Fail	00101010	Or	01010110	Zone	10000010
Failure	00101011	Pass	01010111		

*DT1056 is a complete kit including MM54104 SPC; DT1057 is SSR5 and SSR6 speech ROMs only.

Note 1: "ED" is a suffix that can be used to make any present tense word become a past tense word. The way we say "ED", however, does vary from one word to the next. For that reason, we have offered 4 different "ED" sounds. It is suggested that each "ED" be tested with the desired word for best quality results. Address 31 "ED" or 32 "ED" should be used with words ending in "T" or "D", such as exit or load. Address 34 "ED"

Table 14-1 cont'd.

should be used with words ending with soft sounds such as ask. Address 33 "ED" should be used with all other words.

Note 2: "TH" is a suffix that can be added to words like six, seven, eight to form adjective words like sixth, seventh, eighth.

Note 3: "UTH" is a suffix that can be added to words like twenty, thirty, forty to form adjective words like twentieth, thirtieth, fortieth, etc.

Note 4: Address 130 is the last legal address in this particular word list. Exceeding address 130 will produce pieces of unintelligible invalid speech data.

14-7 DIGITAL GAIN SET

The final specialized device we will study allows remote changing of an amplifier's gain by use of a three-line digital input. This gives computer capability to automatically adjust a circuit's gain to compensate for input signal changes. Imagine an oscilloscope with a comparator in the vertical deflection circuit that senses if the viewed signal is too low. A status signal from the comparator can cause the gain to be increased until the viewed signal is within the correct amplitude range. Two devices with the gain set capability are the LF13006 and the LF13007. With their three-line digital input, eight combination of gain are possible. A block diagram for the LF13006 is shown in Fig. 14-13.

The three control lines are connected to a data bus, and data is transferred into the latches when both the chip select ($\overline{CS}$) and write ($\overline{WR}$) lines are low. The three-

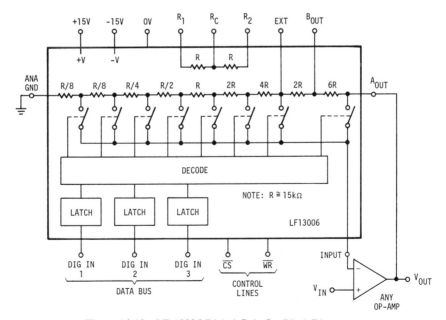

Figure 14-13   LF 13006 Digital Gain Set Block Diagram

TABLE 14-2   GAIN TABLE

	Gain			
	LF13006		LF13007	
Digital Input	A_{OUT}	B_{OUT}	A_{OUT}	B_{OUT}
000	1	1	1	1
001	2	1.25	1.25	1
010	4	2.5	2	1.6
011	8	5	5	4
100	16	10	10	8
101	32	20	20	16
110	64	40	50	40
111	128	80	100	80

line data from the latches is decoded to activate *one* of the eight control switches on the voltage divider. Notice that when a switch closes, it sets up a R_1 resistor to the left of the switch and a R_F resistor to the right of the switch. If, for example, the fourth switch from the left is closed (100 input code), we have an R_1 resistor of value R and an R_F value of $15R$. Since the signal input to the op-amp is on the non-inverting side, we have a gain of $(15R/R + 1)$, or 16. The gain for all possible digital inputs is given in Table 14-2 for both the LF13006 and the LF13007.

Connecting the A_{OUT} line to the output of the op-amp for the LF13006 provides a binary relationship of gains. The same connection for the LF13007 gives a gain relationship that is typical of an oscilloscope's vertical input. Additional relationships are provided by connecting the B_{OUT} to the output of the op-amp rather than the A_{OUT} terminal.

The resistors used in the divider provide a 0.5% gain accuracy, and the total divider resistance is 15 kΩ. Two additional uncommitted resistors are available to add greater versatility to the device.

All logic inputs are at TTL levels; i.e., greater than 2 V signifies a 1, while less than 0.8 V is interpreted as a 0. The maximum time for a gain change to occur is 200 ns after a $\overline{WR}$ transition.

14-8 GAIN SET APPLICATIONS

A further application for the digital gain set is shown in Fig. 14-14.

This circuit provides a capability of inverting (plus to minus or minus to plus) an amplifier by changing the input code. With 000 input code, the signal is fed into both the inverting and non-inverting sides of the op-amp. The gain via the inverting path is $(-R/R)$ or -1, while the gain through the non-inverting side is $(R/R + 1)$ or $+2$. The

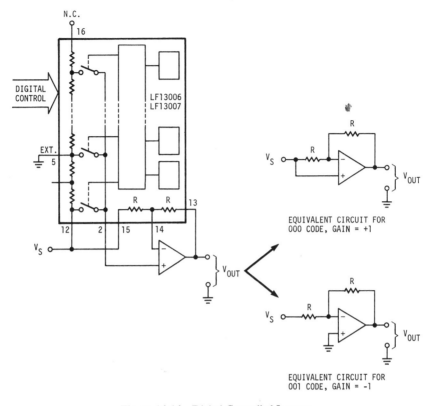

Figure 14-14   Digital Controlled Inverter

net gain is the sum of these two gains, which is $+1$. If the code is now changed to 001, the circuit becomes a simple inverting amplifier with a gain of $-R/R$ or -1.

Another application for the device is a programmable current source, which is illustrated in Fig. 14-15. Current flow out of the circuit is dependent on the amount of voltage drop across the 120-Ω resistor. Since U_2 is a follower, the voltage at the right side of the resistor is fixed at 1.2 V below point A_{OUT}. The voltage at the left side of the resistor is the same as the input to the follower U_1. This voltage varies with the position of the movable tap on the voltage divider within the digital gain set. With an input code of 000, the voltage divider tap is connected to A_{OUT} and the full 1.2 V is across the 120-Ω resistor, providing an output current of 10 mA. If we are using the LF13006 and now the input code is 011, the tap is seven eighths down the divider, which gives a voltage across the 120-Ω resistor of 1.2 V/8 or 0.15 V. The current with this voltage is 1.25 mA. If we take the gain number from Table 14.2 for a given input and put it in the following equation, we can directly determine the current.

$$I_{OUT} = \frac{12 \text{ V}}{120 \ \Omega \times \text{GAIN} \#} \tag{14-1}$$

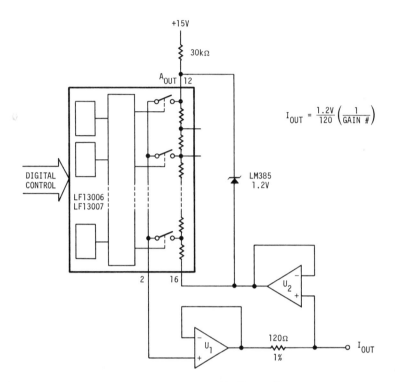

$$I_{OUT} = \frac{1.2V}{120}\left(\frac{1}{GAIN\ \#}\right)$$

Figure 14-15   Programmable Current Source

With the LF13006 we can get a binary selection of current from this source.

These examples have shown some of the versatility of the gain set device. It can be used in any application where digital control of resistance, voltage, or current is desired.

14-9 CHAPTER 14—SUMMARY

- The advantages of a switched capacitor filter are that no external capacitors are required and the filter frequency can be easily changed.
- An SCF can provide low pass, high pass, band pass, or band stop filter characteristics.
- Changing the clock frequency to an SCF moves the filter cutoff or center frequency.
- The state variance filter configuration is used within the SCF.

- A speech synthesizer generates speech from stored data and can provide words or complete sentences.

- Speech can be stored as basic sounds (phonemes) or as samples of a given speech waveform (compression method).

- A speech synthesizer outputs a word or sentence in response to a specific digital data input. Changing the data word changes the voice output message.

- Digitally controlling the gain of an op-amp is a key use for the digital gain set integrated circuit.

- The gain set device can also be used in any application where computer control of resistance is required.

14-10 CHAPTER 14—EXERCISE PROBLEMS

1. Give two advantages of the switched capacitor filter over the conventional op-amp type active filter.

2. State the simplest method to increase the center frequency of an SCF by a factor of 2.

3. Using the mode 1 configuration for the MF10, find the external resistors and the clock frequency (100 to 1) required for a notch filter of center frequency 60 Hz, bandwidth 6 Hz, and in-band gain of 5. Let $R_1 = 1$ kΩ.

4. A low pass filter is required with a cutoff frequency of 3.5 kHz and an inband gain (H_{OLP}) of 10. Use the MF10 in the mode 1 configuration with a $Q = 2$ and find:
   a) Peak gain
   b) Peak frequency
   c) Clock frequency (50 to 1)
   d) All resistor values (Assume $R_1 = 10$ k)

5. Design a high pass filter circuit with a cutoff frequency of 500 Hz, a response with $Q = 1$, and an inband gain of 4. Determine the peak response and all circuit component values. The clock frequency is 100 kHz and $R_1 = 10$ kΩ.

6. The output of a speech synthesizer is in discrete steps corresponding to the digital data. How is the data smoothed?

7. Why is pre-emphasis used on voice data?

8. Determine the specific address codes to provide the following voice message from a DT1056 DIGITALKER: "Clear system alarm record pressure."

9. Describe the connections and determine the input code to provide an op-amp gain of 15 with a LF13006 digital gain set.

10. Show an op-amp Schmitt trigger circuit with hysteresis which can be changed by a digital input.

11. Sketch a circuit that allows digital control of the time internal of a 555 timer.

12. How many gain combinations can be achieved with two digital gain set devices and six data lines?

14-11 CHAPTER 14—LABORATORY EXPERIMENTS

Experiment 14A

Purpose: To demonstrate operation of a switched capacitor filter (MF10).

Requirements: Connect up the MF10 in a mode 1 configuration for a band pass filter with a center frequency of 400 Hz, Q of 2, and a peak gain of 5.

Measure:

1. Center frequency
2. Bandwidth, and determine Q
3. Peak gain

Compare with required values.

Experiment 14B

Purpose: To demonstrate digital gain control using a LF13006 DGS.

Requirements: Connect a 741 op-amp to the LF13006 using the circuit in Fig. 14-13. Use ± 12-V power supplies.

Measure:

1. Gain for each digital input code.
2. Change op-amp output connection from A_{OUT} to B_{OUT} and repeat step 1.

Compare data with predicted data.

15

Device Specification

OBJECTIVES

At the conclusion of this chapter, you should know the following:
1. Why specifications are generated for linear circuits
2. The key parts of a data sheet
3. Significance of "typical" values
4. Why maximum ratings must be considered
5. The various mechanical configurations used for linear circuits
6. Device part number suffixes

15.1 INTRODUCTION

Manufacturers' specifications define the electrical and mechanical features of a particular linear integrated circuit. These specifications serve as a basis for applying the devices in a particular circuit design.

If the manufacturer doesn't specify a particular characteristic of an IC, there is no guarantee that it will be the same from device to device. This means that equipment may no longer operate if a failed IC is replaced. Equipment designers will get around this problem by writing their own specifications, which will include the critical values. In this chapter we will look at a typical data sheet, become aware of the significance of maximum ratings and tolerances, and also look at the different types of packages available.

15.2 TYPICAL DATA SHEET

A data sheet contains the key device characteristic values under given conditions. The conditions can be operation at room temperature (25°C), or at elevated or lower temperatures. Power supply voltages can cause changes in the specifications, so these have to be known along with the specific circuit load conditions. The operating frequency can also have an effect on the device characteristics.

We can look at these effects in greater detail by going through an actual data sheet. The data sheet for the LM2878 dual 5-W power audio amplifier will be used and is shown in Fig. 15-1. The manufacturer suggests this amplifier can be used for a stereo phonograph, an output stage in an AM-FM receiver, a power amplifier or comparator, or as a servo amplifier.

We shall go down through each item on the data sheet—but first, we should note that the specifications are given with the following conditions;

1. V_s (V_{cc}) = 22 V
2. T_{amb} (ambient temperature) = 25°C
3. R_1 (load resistor) = 8 Ω
4. A_v (voltage gain) = 50

If the conditions are different from these values, the specifications could change.

The first item listed in Fig. 15-1 is the total power supply current with no output power. This is also called the standby current and is used to determine how much drain there is on the power supply (or battery) when the amplifier is not in use. If we are concerned with battery life, we must assume that the amplifier is drawing the maximum standby current of 50 mA. Using the typical value of 10 mA could result in the battery running down before we predicted it would.

Absolute Maximum Ratings

Supply Voltage	35V
Input Voltage (Note 1)	± 0.7V
Operating Temperature (Note 2)	0°C to 70°C
Storage Temperature	− 65°C to 150°C
Junction Temperature	150°C
Lead Temperature (Soldering, 10 seconds)	300°C

Electrical Characteristics V_S = 22V, T_A = 25°C, R_L = 8Ω, A_V = 50 (34 dB) unless otherwise specified.

Parameter	Conditions	Min	Typ	Max	Units
Total Supply Current	P_O = 0W		10	50	mA
Operating Supply Voltage		6		32	V
Output Power/Channel	f = 1 kHz, THD = 10%	5	5.5		W
Distortion	f = 1 kHz, R_L = 8Ω				
	P_O = 50 mW		0.20		%
	P_O = 0.5W		0.15		%
	P_O = 2W		0.14		%
Output Swing	R_L = 8Ω		V_S − 6V		Vp-p
Channel Separation	C_{BYPASS} = 50 μF, C_{IN} = 0.1 μF	− 50	− 70		dB
	f = 1 kHz, Output Referred				
	V_O = 4 Vrms				
PSRR Power Supply Rejection Ratio	C_{BYPASS} = 50 μF, C_{IN} = 0.1 μF	− 50	− 60		dB
	f = 120 Hz, Output Referred				
	V_{ripple} = 1 Vrms				
PSRR Negative Supply	Measured at DC, Input Referred		− 60		dB
Common-Mode Range	Split Supplies ± 15V, Pin 1		± 13.5		V
	Tied to Pin 11				
Input Offset Voltage			10		mV
Noise	Equivalent Input Noise				
	R_S = 0, C_{IN} = 0.1 μF				
	BW = 20 − 20 kHz		2.5		μV
	CCIR·ARM		3.0		μV
	Output Noise Wideband		0.8		mV
	R_S = 0, C_{IN} = 0.1 μF, A_V = 200				
Open Loop Gain	R_S = 51Ω, f = 1 kHz, R_L = 8Ω		70		dB
Input Bias Current			100		nA
Input Impedance	Open Loop		4		MΩ
DC Output Voltage	V_S = 22V	10	11	12	V
Slew Rate			2		V/μs
Power Bandwidth	3 dB Bandwidth at 2.5W		65		kHz
Current Limit			1.5		A

Note 1: ± 0.7V applies to audio applications; for extended range, see Application Hints.

Note 2: For operation at ambient temperature greater than 25°C, the LM2878 must be derated based on a maximum 150°C junction temperature using a thermal resistance which depends upon device mounting techniques.

Figure 15-1 LM2878 Data Sheet

Operating supply voltage of 6 to 32 V determines the possible range for the power source—but remember that the rest of the values listed in the data sheet are for a supply voltage of 22 V.

The output power for each channel has a typical value of 5.5 W, but a minimum value of 5 W. Again we must assume the worst case and use the 5-W value. We must also consider the 10% THD (third harmonic distortion). If this is too high, we must operate at a lower output power level.

The distortion versus power data shows that at low power levels (less than 2 W) the distortion actually increases as the output power level is decreased. This is because distortion at the crossover point of a sine wave has a fixed value, so that when the output voltage is reduced, the distortion becomes a greater percentage of the output.

A typical output swing of the supply voltage minus 6 V is listed using a load resistor of 8 Ω. With the supply voltage of 22 V, this provides an output peak-to-peak voltage swing of 16 V. Converting this to RMS by dividing by 2.83 gives 5.66 V. If we square this value and divide by the load resistor, we can find the power out (P_O):

$$P_o = \frac{(5.66 \text{ V})^2}{8 \text{ } \Omega}$$

$$= 4 \text{ W}$$

This is less than the minimum value of 5 W specified, so we run into our first inconsistency in manufacturers' specifications. The other problem we have is that the output swing listed was "typical," but what is the worst case? This could mean a power output of less than 4 W at a power supply of 22 V.

With two amplifiers on the same chip, there is a possibility that the signal being amplified by one amplifier could cross-couple into the other. The specified channel separation is with a 4-VRMS, 1-kHz signal at the output of one amplifier. A minimum separation of 50 dB (316) means the coupled voltage at the output of the second amplifier could be as high as 4 V/316 or 12.6 mV rms.

Power supply rejection was defined earlier when we covered op-amps. A minimum value of 50 dB (316) for the LM2878 can be compared with the 77 dB (7079) for a 741 op-amp. If a negative supply is used, the *typical* rejection is the same as with a positive supply (60 dB). However, no minimum value is listed for this situation, so it is impossible to determine the worst case situation.

The value of 10 mV given for the offset voltage is a typical value, and maximum offset is not provided. No offset pins are provided on the amplifier, so if offset is a problem in a DC amplifier configuration, an opposing DC voltage should be injected into the input as shown in Fig. 15-2.

Connecting the 5-kΩ pot between the positive and negative supplies allows for cancellation of offset of either polarity.

The noise specification is first given with reference to the input because output noise depends on the circuit gain. Using a source resistor (R_s) of zero eliminates noise

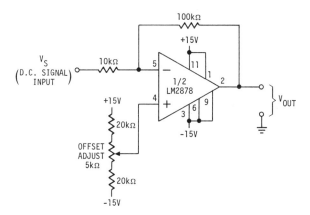

Figure 15-2   Offset Adjust Circuit for LM2878

from this resistor, so we will only see amplifier noise. In the earlier section on op-amp noise, we saw that increasing the bandwidth increased the noise (by a square root factor). The 2.5-μV input noise value listed covers the audio bandwidth of 20 to 20 kHz. Output noise of typically 800 μV is given for an amplifier gain of 200. This is listed as wide-band noise, but the bandwidth is not listed. Using the given 2.5-μV figure with a bandwidth of 20 kHz, a bandwidth of 50 kHz is computed using an input referred noise of 800/200 or 4 μV (remembering again the square root factor on bandwidth). CCIR/ARM is a weighted method of noise measurement developed by Ray Dolby and others, with the response peaking up in the 6-kHz region.

Open loop gain of 70 dB corresponds to a voltage ratio of 3162. This is low compared to the typical op-amp gain of 100 dB (100,000). It will be remembered that low open loop gain means less accurate closed loop gain and a reduction in the gain bandwidth product.

The input bias current of 100 nA is close to the 80 nA figure for a 741 op-amp, while the input resistance of 4 MΩ is twice that of the 741.

An internal equal resistance voltage divider connected to the non-inverting input sets up the DC output voltage. With a 22-V supply, the divider provides a nominal 11 V at the output, which gives the maximum output voltage swing. The possible plus or minus 1-variation means a corresponding reduction in output voltage swing.

The slew rate of 2 V/μs listed is four times better than a 741 op-amp.

Power bandwidth is the frequency range over which the output is within 3 dB (half power) of the maximum value of 5 W. The 65-kHz typical value listed easily exceeds the audio range.

Finally, the current limit of 1.5 A protects both the load and amplifier from excessive dissipation.

What we should learn from going through this typical spec sheet is that not all parameters are well defined, and if a minimum or maximum is not provided then we have no guarantee what the value will be. A way around this problem is to generate a specification sheet which defines the range of variation of all critical parameters. If the IC manufacturer agrees to these specs, they become the basis for incoming acceptance of the devices.

15-3 GRAPHS

In addition to the spec sheet, manufacturers provide curves of the various parameters as a function of voltage, temperature frequency, etc. Figure 15-3 shows graphs for the LM2878.

The first graph shown in the figure relates device dissipation to ambient temperatures, but also includes the effect of various heat sinks. It should be clear that device dissipation is not necessarily the same as output power. The positioning of the heat sink (horizontal or vertical) is not given.

Looking at the curves of power supply rejection, we can see that rejection is quite poor at low frequency (below 1 kHz) and with low values of power supply bypass capacitors. Also, reducing the input coupling capacitor decreases the rejection. Beyond a 10-V supply voltage, the rejection is quite constant.

Channel separation is best with an operating frequency of about 600 Hz. It is interesting that from the $C_{IN} = 0.1$ μF curve the separation reads 78 dB at 1 kHz, whereas the typical listed on the data sheet is 70 dB.

Harmonic distortion below 1% is considered undetectable by the human ear. This level is exceeded, as shown on the graph, with input frequencies greater than 5 kHz if the gain is 50—or greater than 2 kHz with a gain of 200. Power levels greater than 4 W cause a large increase in harmonic distortion.

The open loop gain curve shows that the amplifier has a stable 20 dB per decade slope out to a frequency of 1 MHz.

As expected, power output is a function of a supply voltage. A power output of 6 W is indicated for a power supply of 22 V. The spec sheet indicates 5.5 W for the same voltage (different people generated the graphs).

Finally, the power dissipation versus power out is graphed. Knowing the required power output, this curve can be used in conjunction with the first graph to determine the required heat sink.

15-4 MAXIMUM RATINGS

Actually, the first specifications we should consider when using a device are the absolute maximum ratings, because if we exceed any of these values, the device is no longer able to meet other specifications.

It was brought to the attention of the author that a 723-voltage regulator used in a piece of medical equipment had a very high failure rate. Examination of the circuit revealed that a continuous 50 V was being applied to the input to the regulator, which has an absolute maximum rating of 40 V.

Typical Performance Characteristics

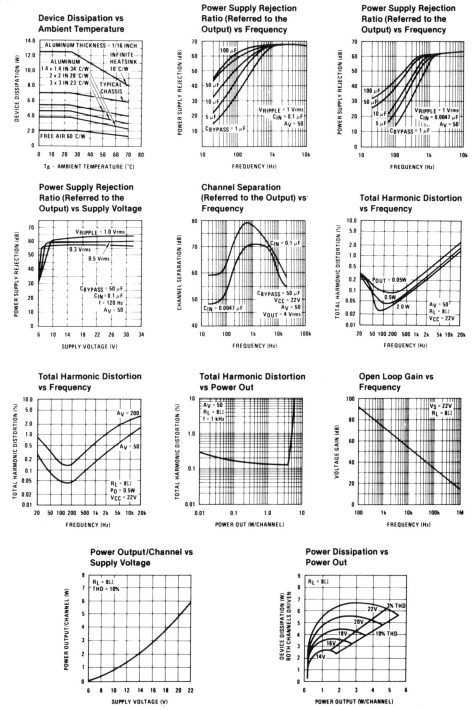

Figure 15-3   LM2878 Graphs

For the LM2878 the maximum input voltage is 35 V. If this is not to be exceeded, we must consider the power supply regulation, voltage spikes on the power supply, and possible power supply failure modes.

A maximum input voltage of ± 0.7 V is allowed without circuit damage. This means care must be taken to avoid high-level inputs. Or, in other words, the gain should be high enough (greater than 15) so that the maximum output power can be achieved without exceeding 0.7-Volts.

The maximum operating temperature range is given as 0°C to 70°C. This is based on 5.5-W internal dissipation and with an infinite heat sink. Practically, we are limited by the heat sink available. For example, using a heat sink that provides a total thermal resistance of 28°C/W ($3 \times 3 \times 1/16$-in. aluminum) and requiring a power *output* of 3 W with 3% THD, we find from the power dissipation versus ambient temperature and power dissipation versus power output curves, that the maximum operating temperature is about 30°C (4.3 W dissipation). From the dissipation versus output curve we also find that we need a required power supply of 18 V.

A storage range of from -65°C to 150°C is specified for a device that is not being operated, while the maximum junction temperature of 150°C controls the amount of allowable dissipation with a given heat sink. Checking numbers from our previous example, we had a junction temperature rise of (150-30)°C. or 120°C. Dividing by the thermal resistance of 28°C/W, we get an allowable power dissipation (P_m) of:

$$P_M = \frac{120°C/W}{28°C}$$

$$= 4.29 \text{ W}$$

which is close to the 4.3 W read from the graph.

The last item under maximum ratings is lead temperature, which is listed as 300°C for a soldering time of 10 sec. This is based on the internal thermal time constant keeping the junction from rising instantaneously to 150°C. But if the 10 sec is exceeded, the junction will rise above 150°C and the chip will be damaged. Good soldering practice of a *hot, clean tinned iron along with clean connections*, for low thermal resistance, will prevent damage during chip installation or removal.

15-5 DEVICE MECHANICAL CONFIGURATIONS

The package used for a given linear integrated circuit depends on such things as the number of circuit connections required, and thermal and sealing requirements.

Circuit connections determine the shape of the device, while thermal considerations could require a metal case, or tab, to insure maximum heat flow out of the device. Sealing is required to prevent moisture from contaminating the semiconductor material and changing its characteristics or causing corrosion of the circuit connections.

The most common packages used are shown in Fig. 15-4, along with the manufacturers' code designators.

The first package is a glass-metal dual in-line package configuration and is designated as a type D. This particular package is higher cost and is seldom used for linear circuits. It is reserved for special low-volume applications because its larger internal cavity can accommodate an oversize semiconductor chip.

The H-type package is a metal can which provides a good seal against moisture and contaminants as well as shielding from electrical interference.

For better temperature stability, the ceramic J package is used. This is because the expansion coefficients of ceramic and the semiconductor material are very similar.

The metal TO-3 three-lead K package gives excellent heat transfer and is therefore used with devices that have high thermal dissipation, i.e., regulators. An aluminum case has lower thermal resistance, but the steel case provides magnetic as well as electrostatic shielding.

Plastic DIP-type N packages are the lowest cost and the most used. Evolution of this configuration has resulted in significant improvement in reliability by better sealing techniques.

The TO-202 is a three-lead transistor-type package. It is mounted to a heat sink through the hole in the metal tab. Mica insulators are available if the tab must be electrically isolated from the chassis. The three leads limit the application of this configuration, but it is used for regulators over a temperature range of 0°C to 125°C. An LM317 regulator in this package has a maximum rating of 7.5 W.

When DIP's have to dissipate high power levels, the SGS-power DIP configuration is used. The attached U-shaped heat conductor is provided with mounting holes so that it can be attached to a larger heat sink.

The TO-220 is a larger version of the TO-202, with increased power dissipation. In this package, the LM317 regulator is rated at 15 W.

Flat-pack configurations are nowhere near as common as the DIP. However, they can be spot-welded or soldered onto the printed circuit board.

Finally, the TO-92 is another low-cost, transistor-type package. When mounted one-eighth inch from a PC board, it has a thermal resistance of 160°C/W. With a maximum temperature of 100°C for this package, the allowance dissipation at 25°C is:

$$P_D = \frac{(100 - 25)°\mathrm{C\,W}}{160°\mathrm{C}}$$

$$= 469 \text{ mW}$$

We can see from this result that this package is restricted to low power levels.

The letter codes listed in Fig. 15-4 are added to the part number to identify the required package. For example, the LM317 National Semiconductor regulator comes in a choice of five packages:

PACKAGE TYPES/ MANUFACTURERS		NSC	SIGNETICS	FAIRCHILD	MOTOROLA	TI	RCA	SILICON GENERAL	AMD	RAYTHEON
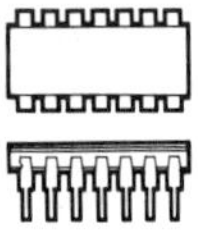 14/16 LEAD GLASS/METAL DIP		D	I	D	L		D	D	D	D, M
 GLASS/METAL FLAT-PACK		F	Q	F	F	F, S	K	F	F	J, F, Q
 TO-5, TO-99, TO-100		H	T, K, L, DB	H	G	L	S*, V1**	T	H	T, H
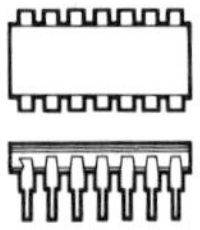 8, 14 AND 16-LEAD LOW-TEMPERATURE CERAMIC DIP		J	F	R, D	U	J				DC, DD
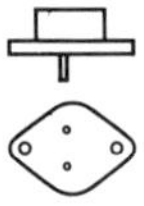 TO-3	(STEEL)	K			KS			K		K
	(ALUMINUM)	KC	DA	K	K	K		K		LK, TK
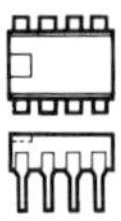 8, 14 AND 16-LEAD PLASTIC DIP		N	V, A, B	T, P	P	P, N	E	M, N	PC	N, DN, DP, MP

* WITH DUAL-IN-LINE FORMED LEADS.
** WITH RADIALLY FORMED LEADS.

Figure 15-4 I.C. Packages

PACKAGE TYPES/ MANUFACTURERS	NSC	SIGNETICS	FAIRCHILD	MOTOROLA	TI	RCA	SILICON GENERAL	AMD	RAYTHEON
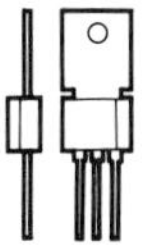 TO-202 (D-40, DURAWATT)	P				KD				
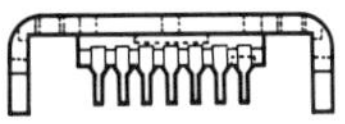 "SGS" TYPE POWER DIP	S		BP						
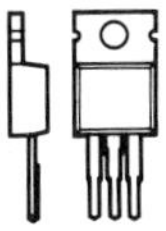 TO-220	T	U	U		KC				
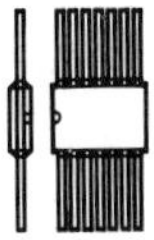 LOW TEMPERATURE GLASS HERMETIC FLAT PACK	W		F	F	W			FM	
 TO-92 (PLASTIC)	Z	S	W	P	LP				

Figure 15-4 (*Continued*)

LM317K Steel   (metal can TO-3)
LM317H   (metal can TO-39)
LM317T   (plastic TO-220)
LM317MP   (plastic TO-202)
LM317MPTB   (plastic TO-202 offset tab)

Other letter codes attached to the part number can indicate the temperature range over which the performance is guaranteed. Let us look at the 741 op-amp, which comes in two operating temperature ranges.

Versions:	*Temperature Ranges:*
LM741 and 741A	$-55°C$ to $+125°C$
LM741C and 741E	$0°C$ to $+70°C$

The 741A and 741E have better electrical characteristics than the 741 and 741C, and the characteristics of the 741 are somewhat better than those of the 741C.

Suppose we want a 741 op-amp in a DIP package which has a minimum voltage offset and can operate from $0°C$ to $70°C$. From the characteristics of the 741 data sheets in Appendix E, we see that we need a 741E, and the DIP requirement makes it a 741EN.

15-6 *INTERCHANGEABILITY*

When a piece of equipment stops operating because of an IC failure, we have to find a replacement part. If the failed part is a 741 op-amp, we can't simply purchase a 741, but we must specify the full part number in order to guarantee electrical and mechanical compatibility. For example, replacing a 741AH with a 741CN will create two problems: First, the 741CN has looser specs; second, the N is a DIP, while the H is a round metal can. But if we had a failed 741CN, we could replace it with a better and mechanically equivalent 741AN or 741EN type if we are willing to pay the extra cost.

Sometimes it is desirable to replace a failed component with a different type. This would be the case if the failure was determined to be caused by exceeding the device maximum rating. Suppose we found that a LM317 voltage regulator failed because the input voltage was greater than the specified 40 V (input-output differential). It would make no sense to replace the failed regulator with the same type. Rather, the LM317HV should be considered because it has a maximum input of 60 V.

If we were trying to fix a radio transmitter on a sinking ship, we couldn't wait until the right part came along. In this extreme example we would find the closest part that would work and put it in the circuit. Ideally, we would select a component equal to or better than the failed part. When the intent is to replace the substituted part when the correct part is available, a tag should be placed on the equipment as a reminder.

We need to know how to read device specifications in order to make an intelligent part selection. It is hoped that this chapter has accomplished this objective.

15-7 CHAPTER 15—SUMMARY

- Specifications set limits on a device so we can predict its performance.

- Without specifications, we would have no guarantee that a replacement part would work.

- A typical data sheet is broken down into the following sections:
  1. Device description
  2. Maximum ratings
  3. Detailed specifications
  4. Performance graphs
  5. Typical applications
- Maximum ratings are usually not conservative, so we should stay below these values.
- Typical values are the average measured values of many devices—but without minimum and maximum values, we have no idea what an individual device will do.
- We must consider device variations with temperature—otherwise, equipment could stop operating with heating or cooling.
- Linear devices come in various mechanical configurations depending on the following requirements:
  1. Power dissipation
  2. Electrical shielding
  3. Space availability
  4. Environmental consideration
  5. Reliability
  6. Cost
- The part number for a device indicates the performance specifications and also the package configuration.
- Specifications on a failed component should always be checked to determine whether the maximum ratings are not being exceeded.

15-8 CHAPTER 15—EXERCISE PROBLEMS

1. List and interpret the maximum ratings for the LM3905 data sheet shown in Appendix E.
2. List and interpret the electrical characteristics for the 741 data sheet shown in Appendix E.
3. What is the problem with typical values?
4. If a part fails in a circuit, what research should be done before a new part is installed?
5. Ordering a LM317K regulator provides which package and power rating?
6. Explain whether a LM329B voltage reference can replace a LM329C.

APPENDIX A

Common Integrals
and Differentials

The following common integral and differential equations are referred to in Chapter 2, "Operational Amplifier Applications."

COMMON INTEGRALS

$$\int dt = t \tag{A-1}$$

$$\int c\,dt = ct \tag{A-2}$$

$$\int t\,dt = \frac{t^2}{2} \tag{A-3}$$

$$\int kt^n\,dt = \frac{kt^{(n+1)}}{n+1} \tag{A-4}$$

$$\int \sin kt\,dt = -\frac{\cos kt}{k} \tag{A-5}$$

$$\int \cos kt\,dt = \frac{\sin kt}{k} \tag{A-6}$$

$$\int e^{kt}\,dt = \frac{e^{kt}}{k} \tag{A-7}$$

COMMON DIFFERENTIALS

$$dt(c) = 0 \tag{A-8}$$

$$dt(t) = 1 \tag{A-9}$$

$$dt(kt^n) = knt^{n-1} \tag{A-10}$$

$$dt(\sin kt) = k(\cos kt) \tag{A-11}$$

$$dt(\cos kt) = -k(\sin kt) \tag{A-12}$$

$$dt(e^{kt}) = ke^{kt} \tag{A-13}$$

APPENDIX B

Switching Regulator Equation Derivation

PART 1: STEP-DOWN SWITCHING REGULATOR EQUATION DERIVATION

The equations for the step-down switching regulator will be derived with reference to Fig. B-1.

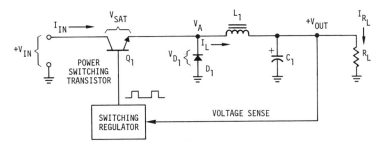

Figure B-1   Step-Down Switching Regulator

When transistor Q_1 is turned on, current flows through inductor L_1 to the load resistor R_L and filter capacitor C_1. Turning off Q_1 causes the field of L_1 to collapse, which will induce a voltage across the inductor that is negative at the diode (CR_1) end and positive at the capacitor end. This causes the diode to conduct, and the load current path is now through R_L, D_1, and L_1. The current flowing through D_1 and L_1 is equal to the nominal DC load current plus ΔI_L due to the charging voltage across the inductor (usually $\Delta I_{L_{p\text{-}p}} = 40\% I_{OUT}$).

In order to derive the equation for the output voltage, we need to list the following relationships.

$$V_{L1} = \frac{L_1 \, di}{dt} \tag{B-1}$$

and

$$\Delta I_{L_1} = \frac{V_{L_1} T}{L_1} \tag{B-2}$$

where V_{L_1} is voltage across inductor ($V_A - V_{OUT}$) and T is the period of the switching frequency. Figure B-2 shows the relationship between the voltage and current waveforms of C_1 during t_{ON} and t_{OFF}.

From Fig. B-2, ΔI_L^+ and ΔI_L^- are determined as follows:

$$\Delta I_{L_1}^+ = \frac{V_A - V_{OUT}}{L_1} t_{ON} \tag{B-3}$$

$$\Delta I_{L_1}^- = \frac{V_{OUT} + V_{CR1}}{L_1} t_{OFF} \tag{B-4}$$

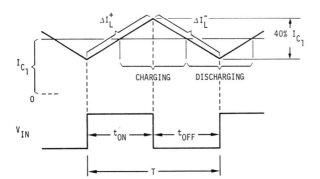

Figure B-2 Current and Voltage Waveforms

Neglecting V_{SAT} and V_{CR1} (then $V_A = V_{IN}$) and setting $I_L^+ = I_L^-$ and solving for V_{OUT}:

$$V_{OUT} = V_{IN}\left(\frac{t_{ON}}{t_{OFF} + t_{ON}}\right) \qquad \text{(B-5)}$$

$$= V_{IN}\left(\frac{t_{ON}}{T}\right)$$

This shows that the output voltage can be changed by varying the Q_1 "on" time. If we assume power out = power in, then $V_{OUT} \times I_{OUT} = V_{IN} \times I_{IN}$ and

$$I_{IN} = I_{OUT}\left(\frac{t_{ON}}{t_{OFF} + t_{ON}}\right) \qquad \text{(B-6)}$$

If we take into account the losses across the transistor switch Q_1 and the diode CR_1, the efficiency η of the circuit is:

$$\eta = \frac{P_{OUT}}{P_{IN}} = \frac{V_{OUT}\, I_{OUT}}{I_{OUT}\, V_{OUT} + V_{SAT}\, I_{IN} + V_{CR1}\, I_{OUT}\left(\dfrac{t_{OFF}}{T}\right)} \qquad \text{(B-7)}$$

Substituting for

$$I_{IN} = I_{OUT}(t_{ON}/T)$$

and letting

$$V_{SAT} = V_{CR_1} = 1 \text{ V}$$

then $\qquad \eta = \dfrac{V_{OUT}\, I_{OUT}}{V_{OUT}\, I_{OUT} + I_{OUT}\left(\dfrac{t_{ON}}{T}\right) + I_{OUT}\left(\dfrac{t_{OFF}}{T}\right)} \qquad \text{(B-8)}$

and recognizing that

$$t_{ON}/T + t_{OFF}/T = 1$$

therefore

$$\eta \cong \frac{V_{OUT}}{V_{OUT} + 1} \tag{B-9}$$

Equation B-9 indicates that the higher the output voltage, the higher the efficiency. Efficiency can be reduced by switching losses in Q_1. Therefore, this transistor should have a fast switching time.

The value of inductor L_1 can be determined by rearranging Eq. B-3 as follows:

$$t_{ON} = \frac{\Delta I_{L_1}^+ \, L_1}{(V_{IN} - V_{OUT})} \tag{B-10}$$

$$t_{OFF} = \frac{\Delta I_{L_1}^- \, L_1}{V_{OUT}} \tag{B-11}$$

Again neglecting V_{SAT} and V_{CR1}:

$$t_{ON} + t_{OFF} = T = \frac{\Delta I_{L_1}^+ \, L_1}{(V_{IN} - V_{OUT})} + \frac{\Delta I_{L_1}^- \, L_1}{V_{OUT}} \tag{B-12}$$

Since

$$\Delta I_L^+ = \Delta I_L^- = 0.4 I_{OUT}$$

then

$$T = \frac{0.4 I_{OUT} \, L_1}{(V_{IN} - V_{OUT})} + \frac{0.4 I_{OUT} L_1}{V_{OUT}} \tag{B-13}$$

Solving for L_1 and substituting $f = 1/T$

$$L_1 = \frac{2.5 V_{OUT}(V_{IN} - V_{OUT})}{I_{OUT} \, V_{IN} \, f} \tag{B-14}$$

In order to determine the value of the filter capacitor C_1, we need to recognize from Fig. B-2 that current is flowing into the capacitor for the last half of t_{ON} and the first half of t_{OFF}, or a time $T/2$. The average current flowing into the capacitor during this time is $\Delta I_{L_1}/4$. The resulting change in voltage across the capacitor is ΔV_{C_1} (or output voltage ripple peak to peak ΔV_{OUT}) is:

$$\Delta V_{OUT} = \frac{\Delta I_{L_1}}{4 \, C_1} \times \frac{T}{2} \tag{B-15}$$

$$= \frac{\Delta I_{L_1} \, T}{8 \, C_1}$$

Since
$$\Delta I_{L_1} = \frac{V_{OUT}\, t_{OFF}}{L_1}$$

again neglecting V_{CR_1}, then $\Delta I_{L_1} = \dfrac{V_{OUT}(T - t_{ON})}{L_1}$ (B-16)

Also from Eq. B-5, $t_{ON} = \dfrac{V_{OUT}\, T}{V_{IN}}$ (B-17)

therefore $V_{OUT_{p-p}} = \dfrac{V_{OUT}\left(T - \dfrac{V_{OUT}\, T}{V_{IN}}\right)T}{8\; C_1\; L_1}$ (B-18)

and $C_1 = \dfrac{(V_{IN} - V_{OUT})V_{OUT}\, T^2}{8\; V_{OUT_{p-p}}\; V_{IN}\; L_1}$ (B-19)

PART 2: STEP-UP SWITCHING REGULATOR EQUATION DERIVATION

Since the circuit configuration is different for the step-up switching regulator, the equations are somewhat different from those derived for the step-down switching regulator. The equations for this regulator will be derived with reference to Fig. B-3.

When transistor Q_1 is on (t_{ON} time), the full input voltage V_{IN} is applied across inductor L_1 and energy is stored in the inductor. Since the collector of Q_1 is near 0 V, the diode CR_1 is reversed biased. The load current through the load resistor R_L is supplied by the capacitor C_1. If we now turn off Q_1, the energy stored in L_1 causes an induced voltage across L_1 which is negative at V_{IN} and positive at CR_1. At CR_1 anode, we have a total positive voltage of V_{IN} plus the induced voltage. This positive

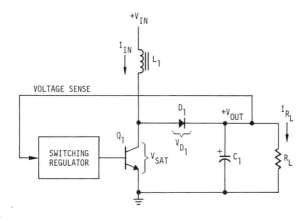

Figure B-3   Step-up Switching Regulator

potential forward biases CR_1 and causes current flow into the load and charges C_1. As with the step-down switching regulator, the current through L_1 consists of a DC component plus some ΔI_{L_1}. The AC component ΔI_{L_1} is again selected to be approximately 40% of I_{L_1}. Figure B-4 shows the L_1 current in relation to Q_1 on and off times.

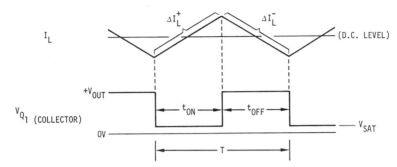

Figure B-4   Current and Voltage Waveforms

In order to develop the equations for the step-up switching regulator, we must first consider the current change ΔI_{L_1} through L_1. Thus,

$$\Delta I_{L_1}^+ = \frac{V_{L_1} \, t_{\text{ON}}}{L_1} \qquad \text{(B-20)}$$

neglecting V_{SAT},

$$\Delta I_{L_1}^+ = \frac{V_{\text{IN}} \, t_{\text{ON}}}{L_1} \qquad \text{(B-21)}$$

and, neglecting V_{CR_1}

$$\Delta I_{L_1}^- \cong \frac{(V_{\text{OUT}} - V_{\text{IN}}) t_{\text{OFF}}}{L_1} \qquad \text{(B-22)}$$

Since

$$\Delta I_{L_1}^+ = \Delta I_{L_1}^-$$

and

$$V_{\text{IN}} t_{\text{ON}} = (V_{\text{OUT}} t_{\text{OFF}}) - (V_{\text{IN}} t_{\text{OFF}})$$

the output voltage is

$$V_{\text{OUT}} = V_{\text{IN}} \left(\frac{t_{\text{ON}} + t_{\text{OFF}}}{t_{\text{OFF}}} \right) = V_{\text{IN}} \left(\frac{T}{t_{\text{OFF}}} \right) \qquad \text{(B-23)}$$

This equation shows that the output voltage is equal to or greater than the input voltage, since:

$$\frac{T}{t_{\text{OFF}}} > 1$$

In calculating input current I_{IN} (DC), which equals L_1 DC current, assuming 100% efficiency:

$$P_{IN} = I_{IN}V_{IN} \tag{B-24}$$

$$P_{OUT} = I_{OUT}V_{OUT} \tag{B-25}$$

$$= I_{OUT}V_{IN}\left(\frac{T}{t_{OFF}}\right)$$

$$P_{OUT} = P_{IN} \tag{B-26}$$

$$I_{OUT}V_{IN}\left(\frac{T}{t_{OFF}}\right) = I_{IN}V_{IN} \tag{B-27}$$

$$I_{IN} = I_{OUT}\left(\frac{T}{t_{OFF}}\right) \tag{B-28}$$

Equation B-28 shows that the input or inductor current is larger than the output current by the factor T/t_{OFF}.

It has been assumed that the power efficiency is 100%, but in actual practice, losses due to the voltage drop across Q_1 and CR_1 will reduce this value. Since the current flows through either Q_1 or CR_1, the losses are not additive. Assuming a 1-V drop across both devices, the power loss will be the same and equal to $I_{IN} \times 1$ V. The actual efficiency will be:

$$\eta = \frac{P_{OUT}}{P_{IN}}$$

$$= \frac{V_{OUT}\ I_{OUT}}{V_{OUT}\ I_{OUT} + I_{IN}(1\ V)}$$

$$= \frac{V_{OUT}\ I_{OUT}}{V_{OUT}\ I_{OUT} + I_{OUT}\left(\dfrac{T}{t_{OFF}}\right)V}$$

$$= \frac{V_{OUT}}{V_{OUT} + \left(\dfrac{T}{t_{OFF}}\right)V} \tag{B-29}$$

This equation assumes only DC losses. However, efficiency is further reduced because of switching losses of Q_1 and CR_1.

The output capacitor C_1 supplies I_{OUT} during t_{ON}. The voltage change on C_1 during this time will be some $\Delta V_{C_1} - \Delta V_{OUT}$, which is the output ripple voltage of the regulator. Thus,

$$\Delta V_{OUT} = \frac{I_{OUT}\ t_{ON}}{C_1} \tag{B-30}$$

or
$$C_1 = \frac{I_{OUT}\, t_{ON}}{\Delta V_{OUT}} \tag{B-31}$$

Since $V_{OUT} = V_{IN}(T/t_{OFF})$ from Eq. B-23

$$t_{OFF} = \frac{V_{IN}\, T}{V_{OUT}} \tag{B-32}$$

where period $T = t_{ON} + t_{OFF} = 1/f.$
    The on time is equal to:

$$t_{ON} = T - t_{OFF} = T - \left(\frac{V_{IN}\, T}{V_{OUT}} \right) \tag{B-33}$$

and C_1 is:
$$C_1 = \frac{I_{OUT} \left(T - \dfrac{V_{IN}\, T}{V_{OUT}} \right)}{\Delta V_{OUT}}$$

$$= \frac{I_{OUT}(V_{OUT} - V_{IN})}{\Delta V_{OUT}\, V_{OUT}\, f} \tag{B-34}$$

where C_1 is in farads, f is the switching frequency, and V_{OUT} is the peak-to-peak ripple voltage.
    Calculation of inductor L_1 by rearranging Eq. B-21 is as follows:

$$L_1 = \frac{V_{IN}\, t_{ON}}{\Delta I_{L_1}^+} \tag{B-35}$$

Since during t_{ON}, V_{IN} is applied across L_1,

$$\Delta I_{L_{1(p-p)}} = 0.4 I_{L_1} = 0.4 I_{IN} = 0.4 I_{OUT} \left(\frac{V_{OUT}}{V_{IN}} \right) \tag{B-36}$$

Therefore
$$L_1 = \frac{V_{IN}\, t_{ON}}{0.4 I_{OUT} \left(\dfrac{V_{OUT}}{V_{IN}} \right)} \tag{B-37}$$

and since t_{ON} is equal to
$$t_{ON} = \frac{T(V_{OUT} - V_{IN})}{V_{OUT}} \tag{B-38}$$

inductor L_1 is equal to
$$L_1 = \frac{2.5\, V_{IN}{}^2(V_{OUT} - V_{IN})}{I_{OUT}\, V_{OUT}{}^2\, f}$$

where L_1 is in henrys, f is the switching frequency.

APPENDIX C

Guidelines for Laboratory Experiments

The purpose of a lab experiment is to reinforce the concepts described in the text. Also, an experiment should teach how devices are applied and how their characteristics can be verified by measurement. "Does the device do what it is supposed to do?" is a question that should be answered by the lab experiment.

The experiments listed at the end of each chapter suggest a particular application or verification of specifications. They are not step-by-step experiments; therefore, a variation in method and results is possible. However, a certain discipline in the approach should be followed in order to prevent damage to components and also to achieve valid results. A suggested generalized procedure is given below:

1. Consult the manufacturer's data sheet to determine the absolute maximum rating for the devices. Try to stay at least 25% (multiply by 0.75) below these values in the experiment.

2. Generate a complete schematic of the required circuit listing all component values, considering item 1 above. If the experiment is conducted in a formal lab situation, the schematic should be presented by the student prior to the start of lab.

3. Wire the circuit from the schematic, checking off on the schematic each component and wire as it is installed. I would suggest using the following color code for hookup wires:
   a) Red—most positive power supply voltage
   b) Orange—less positive power supply voltage (if used)
   c) White—Signal
   d) Black—Power supply return
   e) Brown—Signal return
   f) Blue—most negative supply voltage
   g) Green—less negative supply voltage (if used)

4. Before connecting power to the circuit, verify the power supply voltages. If a signal generator is used, consider whether a coupling capacitor is required to prevent the internal resistance of the generator (typically low) from changing the DC voltage at the circuit input. Also determine whether the generator will be excessively loaded by the input circuit.

5. Think through and determine which wave shape and DC voltage you expect to see at key points in the circuit (this requires an understanding of the circuit operation).

6. Always use an oscilloscope to check circuit operation, because a voltmeter will not indicate unwanted oscillation or distortion. If a signal generator is used, verify that the *same* frequency exists at expected points in the circuit.

7. Take a complete set of circuit measurements and then repeat the measurements. This tends to catch very common measurement errors.

8. Compare measured values with computed or specified values; always compute percent error or deviation. An error as small as 1 mV could be a 100% error.

9. Get in the habit of writing a conclusion that considers and explains deviations of the measured from specified values. Also in the conclusion, list practical applications for the circuit, its advantages and disadvantages.

10. Suggest how the experiment could be improved to meet the objectives of text support and practical application.

APPENDIX D

Resistors and Capacitors

———————————————

Resistors are the most common circuit element used and consist of four basic types:

1. Carbon composition
2. Carbon film
3. Metal wirewound
4. Metal film

Carbon composition resistors are the least expensive type and are available in ¼, ½, 1 or 2 watt ratings. Tolerances of carbon resistors are 5, 10, and 20% (10% most used).

Standard carbon resistor values are listed below for 5% tolerance. The 10% values are limited to those with an asterisk.

OHMS

1.0*	5.1	27*	130	680*	3600	18000*	91000
1.1	5.6*	30	150*	750	3900*	20000	100000*
1.2*	6.2	33*	160	820*	4300	22000*	110000
1.3	6.8*	36	180*	910	4700*	24000	120000*
1.5*	7.5	39*	200	1000*	5100	27000*	130000
1.6	8.2*	43	220*	1100	5600*	30000	150000*
1.8*	9.1	47*	240	1200*	6200	33000*	160000
2.0	10*	51	270*	1300	6800*	36000	180000*
2.2*	11	56*	300	1500*	7500	39000*	200000
2.4	12*	62	330*	1600	8200*	43000	220000*
2.7*	13	68*	360	1800*	9100	47000*	
3.0	15*	75	390*	2000	10000*	51000	
3.3*	16	82*	430	2200*	11000	56000*	
3.6	18*	91	470*	2400	12000*	62000	
3.9*	20	100*	510	2700*	13000	68000*	
4.3	22*	110	569*	3000	15000*	75000	
4.7*	24	120*	620	3300*	16000	82000*	

MEGOHMS

0.24	0.45	0.75	1.3	2.4	4.3	7.5	13.0
0.27*	0.47*	0.82*	1.5*	2.7*	4.7*	8.2*	15.0*
0.30	0.51	0.91	1.6	3.0	5.1	9.1	16.0
0.33*	0.56*	1.0*	1.8*	3.3*	5.6*	10.0*	18.0*
0.36	0.62	1.1	2.0	3.6	6.2	11.0	20.0
0.38*	0.68*	1.2*	2.2*	3.9*	6.8*	12.0*	22.0*

Carbon film resistors come in 5% tolerance values (see chart above) and are smaller and more stable than the composition type.

Wirewound metal resistors are more stable than the carbon types and are available in higher wattage ratings, tighter tolerance, and lower resistance values.

Tolerances are typically 1, 5, or 10%. Standard resistance values for 5% and 10% (asterisk) wirewound resistors are tabulated below.

RESISTANCE VALUE TABLE FOR WIREWOUND RESISTORS

0.10*	0.27*	0.68*	1.8*	4.7*	12*	−3*	82*	220*	560*	1.5K*
0.11	0.30	0.75	2.0	5.1	13	36	91	240	620	1.6K
0.12*	0.33*	0.82*	2.2*	5.6*	15*	39*	100*	270*	680*	1.8K*
0.13	0.36	0.91	2.4	6.2	16	43	110	300	750	2.0K
0.15*	0.39*	1.0*	2.7*	6.8*	18*	47*	120*	330*	820*	
0.16	0.43	1.1	3.0	7.5	20	51	130	360	910	
0.18*	0.47*	1.2*	3.3*	8.2*	22*	56*	150*	390*	1K*	
0.20	0.51	1.3	3.6	9.1*	24	62	160	430	1.1K	
0.22*	0.56*	1.5*	3.9*	10*	27*	68*	180*	470*	1.2K*	
0.24	0.42	1.6	4.3	11	30	75	200	510	1.3K	

Metal film resistors have a 1% tolerance and are very temperature stable compared to the carbon types. They also generate much less electrical noise than carbon and should therefore be used in low noise amplifiers. Standard values are tabulated below:

RESISTANCE VALUE TABLE FOR METAL FILM RESISTORS

1.00	1.21	1.47	1.78	2.15	2.61	3.16	.283	4.64	5.67	6.81	8.25
1.02	1.24	1.50	1.82	2.21	2.67	3.24	3.92	4.75	5.76	6.98	8.45
1.05	1.27	1.54	1.87	2.26	2.74	3.32	4.02	4.87	5.90	7.15	8.66
1.07	1.30	1.58	1.91	2.32	2.80	3.40	4.12	4.99	6.04	7.32	8.87
1.10	1.33	1.62	1.96	2.37	2.87	3.48	4.22	5.11	6.19	7.50	9.09
1.13	1.37	1.65	2.00	2.43	2.94	3.57	4.32	5.23	6.34	7.68	9.31
1.15	1.40	1.69	2.05	2.49	3.01	3.65	4.42	5.36	6.49	7.87	9.53
1.18	1.43	1.74	2.10	2.55	3.09	3.74	4.53	5.49	6.65	8.06	9.76

Capacitors are much more varied than resistors and range from the high capacitance electrolytics to low capacitance ceramic types.

Electrolytic capacitors have wide tolerances such as +75% −10%. They are used when a large capacitor is required (power supply filtering) but the value is not critical. Most electrolytics are polarized which requires consideration of the polarity when they are placed in a circuit.

The smaller value capacitors (mica, ceramic, polystyrene, film etc.) have tighter tolerances which range from 1% to 20% and can provide capacitance values from 1 or 2 picofarads to about 5 microfarads. The actual capacitance value is not marked on some of the smaller capacitors. A numerical code is used similar to the system of precision resistors but with a picofarad baseline. For example, a capacitor marked 103 would have a 10,000 picofarad value (or 0.01 microfarad). Tolerance of the capacitor is indicated by a letter following the number. The letter code is:

$$K = \pm 10\%$$

$$J = \pm 5\%$$

$$G = \pm 2\%$$

Using another example, a capacitor marked 224J would be a 0.22 microfarad $\pm 5\%$ tolerance.

APPENDIX E

Data Sheets

_____

This appendix contains the data sheets for devices discussed in this text. Listing is in numerical order, followed by the manufacturers' symbols.

MF10 Universal Monolithic Dual Switched Capacitor Filter

General Description

The MF10 consists of 2 independent and extremely easy to use, general purpose CMOS active filter building blocks. Each block, together with an external clock and 3 to 4 resistors, can produce various 2nd order functions. Each building block has 3 output pins. One of the outputs can be configured to perform either an allpass, highpass or a notch function; the remaining 2 output pins perform lowpass and bandpass functions. The center frequency of the lowpass and bandpass 2nd order functions can be either directly dependent on the clock frequency, or they can depend on both clock frequency and external resistor ratios. The center frequency of the notch and allpass functions is directly dependent on the clock frequency, while the highpass center frequency depends on both resistor ratio and clock. Up to 4th order functions can be performed by cascading the two 2nd order building blocks of the MF10; higher than 4th order functions can be obtained by cascading MF10 packages. Any of the classical filter configurations (such as Butterworth, Bessel, Cauer and Chebyshev) can be formed.

Features

- Low cost
- 20-pin 0.3" wide package
- Easy to use
- Clock to center frequency ratio accuracy = 0.6%
- Filter cutoff frequency stability directly dependent on external clock quality
- Low sensitivity to external component variation
- Separate highpass (or notch or allpass), bandpass, lowpass outputs
- $f_o \times Q$ range up to 200 kHz
- Operation up to 30 kHz

System Block Diagram

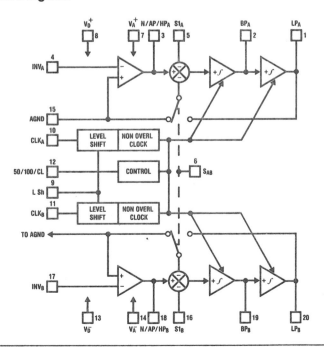

Absolute Maximum Ratings

Supply Voltage	7V
Power Dissipation	500 mW
Operating Temperature	0°C to 70°C
Storage Temperature	150°C
Lead Temperature (Soldering, 10 seconds)	300°C

Electrical Characteristics (Complete Filter) $V_S = \pm 5V$, $T_A = 25°C$

Parameter	Conditions	Min	Typ	Max	Units
Frequency Range	$f_o \times Q < 200$ kHz	20	30		kHz
Clock to Center Frequency Ratio, f_{CLK}/f_o					
MF10BN	Pin 12 High, Q = 10		$49.94 \pm 0.2\%$	$\pm 0.6\%$	
MF10CN	$f_o \times Q < 50$ kHz, Mode 1		$49.94 \pm 0.2\%$	$\pm 1.5\%$	
MF10BN	Pin 12 at Mid Supplies		$99.35 \pm 0.2\%$	$\pm 0.6\%$	
MF10CN	Q = 10, $f_o \times Q < 50$ kHz, Mode 1		$99.35 \pm 0.2\%$	$\pm 1.5\%$	
Q Accuracy (Q Deviation from an Ideal Continuous Filter)					
MF10BN	Pin 12 High, Mode 1		$\pm 2\%$	$\pm 4\%$	
MF10CN	$f_o \times Q < 100$ kHz, $f_o < 5$ kHz		$\pm 2\%$	$\pm 6\%$	
MF10BN	Pin 12 at Mid Supplies		$\pm 2\%$	$\pm 3\%$	
MF10CN	$f_o \times Q < 100$ kHz $f_o < 5$ kHz, Mode 1		$\pm 2\%$	$\pm 6\%$	
f_o Temperature Coefficient	Pin 12 High ($\sim 50{:}1$)		± 10		ppm/°C
	Pin 12 Mid Supplies ($\sim 100{:}1$) $f_o \times Q < 100$ kHz, Mode 1 External Clock Temperature Independent		± 100		ppm/°C
Q Temperature Coefficient	$f_o \times Q < 100$ kHz, Q Setting Resistors Temperature Independent		± 500		ppm/°C
DC Low Pass Gain Accuracy	Mode 1, R1 = R2 = 10k			± 2	%
Crosstalk			50		dB
Clock Feedthrough			10		mV
Maximum Clock Frequency		1	1.5		MHz
Power Supply Current			8	10	mA

Electrical Characteristics (Internal Op Amps) $T_A = 25°C$

Parameter	Conditions	Min	Typ	Max	Units
Supply Voltage		± 4	± 5		V
Voltage Swing (Pins 1, 2, 9, 20)	$V_S = \pm 5V$, $R_L = 5k$				
MF10BN		± 3.8	± 4		V
MF10CN		± 3.2	± 3.7		V
Voltage Swing (Pins 3 and 18)	$V_S = \pm 5V$, $R_L = 3.5k$				
MF10BN		± 3.8	± 4		V
MF10CN		± 3.2	± 3.7		V
Output Short Circuit Current	$V_S = \pm 5V$				
Source			3		mA
Sink			1.5		
Op Amp Gain BW Product			2.5		MHz
Op Amp Slew Rate			7		V/μs

National Semiconductor

LH0070 Series Precision BCD Buffered Reference
LH0071 Series Precision Binary Buffered Reference

General Description

The LH0070 and LH0071 are precision, three terminal, voltage references consisting of a temperature compensated zener diode driven by a current regulator and a buffer amplifier. The devices provide an accurate reference that is virtually independent of input voltage, load current, temperature and time. The LH0070 has a 10.000V nominal output to provide equal step sizes in BCD applications. The LH0071 has a 10.240V nominal output to provide equal step sizes in binary applications.

The output voltage is established by trimming ultra-stable, low temperature drift, thin film resistors under actual operating circuit conditions. The devices are short-circuit proof in both the current sourcing and sinking directions.

The LH0070 and LH0071 series combine excellent long term stability, ease of application, and low cost,

making them ideal choices as reference voltages in precision D to A and A to D systems.

Features

- Accurate output voltage
  - LH0070   10V ±0.01%
  - LH0071   10.24V ±0.01%
- Single supply operation   12.5V to 40V
- Low output impedance   0.1Ω
- Excellent line regulation   0.1 mV/V
- Low zener noise   100 μVp-p
- 3-lead TO-5 (pin compatible with the LM109)
- Short circuit proof
- Low standby current   3 mA

Equivalent Schematic

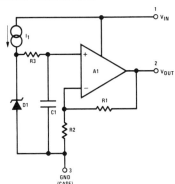

Connection Diagram

TO-5 Metal Can Package

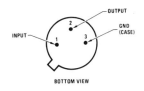

BOTTOM VIEW

Typical Applications

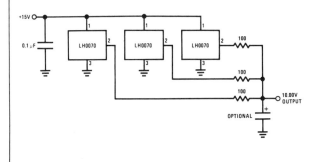

Statistical Voltage Standard

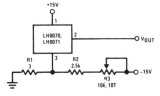

*Note. The output of the LH0070 and LH0071 may be adjusted to a precise voltage by using the above circuit since the supply current of the devices is relatively small and constant with temperature and input voltage. For the circuit shown, supply sensitivities are degraded slightly to 0.01%/V change in V_{OUT} for changes in V_{IN} and V−.

An additional temperature drift of 0.0001%/°C is added due to the variation of supply current with temperature of the LH0070 and LH0071. Sensitivity to the value of R1, R2 and R3 is less than 0.001%/%.

**Output Voltage Fine Adjustment

Absolute Maximum Ratings

Supply Voltage	40V
Power Dissipation (See Curve)	600 mW
Short Circuit Duration	Continuous
Output Current	±20 mA
Operating Temperature Range	$-55°C$ to $+125°C$
Storage Temperature Range	$-65°C$ to $+150°C$
Lead Temperature (Soldering, 10 seconds)	$300°C$

Electrical Characteristics (Note 1)

PARAMETER	CONDITIONS	MIN	TYP	MAX	UNITS
Output Voltage	$T_A = 25°C$				
LH0070			10.000		V
LH0071			10.240		V
Output Accuracy	$T_A = 25°C$				
−0, −1			±0.03	±0.1	%
−2			±0.02	±0.05	%
Output Accuracy					
−0, −1				±0.3	%
−2				±0.2	%
Output Voltage Change With	(Note 2)				
Temperature					
−0				± 0.2	%
−1			±0.02	± 0.1	%
−2			±0.01	±0.04	%
Line Regulation	$13V \leq V_{IN} \leq 33V, T_C = 25°C$				
−0, −1			0.02	0.1	%
−2			0.01	0.03	%
Input Voltage Range		12.5		40	V
Load Regulation	$0 mA \leq I_{OUT} \leq 5 mA$		0.01	0.03	%
Quiescent Current	$13V \leq V_{IN} \leq 33V, I_{OUT} = 0 mA$	2	3	5	mA
Change In Quiescent Current	$\Delta V_{IN} = 20V$ From 13V To 33V		0.75	1.5	mA
Output Noise Voltage	BW = 0.1 Hz To 10 Hz, $T_A = 25°C$		20		μVp-p
Ripple Rejection	f = 120 Hz		0.01		%/Vp-p
Output Resistance			0.2	1	Ω
Long Term Stability	$T_A = 25°C$, (Note 3)				
−0, −1				±0.2	%/yr.
−2				±0.05	%/yr.

Note 1: Unless otherwise specified, these specifications apply for V_{IN} = 15.0V, R_L = 10 kΩ, and over the temperature range of $-55°C \leq T_A \leq +125°C$.

Note 2: This specification is the difference in output voltage measured at T_A = 85°C and T_A = 25°C or T_A = 25°C and T_A = −25°C with readings taken after test chamber and device-under-test stabilization at temperature using a suitable precision voltmeter.

Note 3: This parameter is guaranteed by design and not tested.

![National Semiconductor logo]

LM199/LM299/LM399 Precision Reference

General Description

The LM199/LM299/LM399 are precision, temperature-stabilized monolithic zeners offering temperature coefficients a factor of ten better than high quality reference zeners. Constructed on a single monolithic chip is a temperature stabilizer circuit and an active reference zener. The active circuitry reduces the dynamic impedance of the zener to about 0.5Ω and allows the zener to operate over 0.5 mA to 10 mA current range with essentially no change in voltage or temperature coefficient. Further, a new subsurface zener structure gives low noise and excellent long term stability compared to ordinary monolithic zeners. The package is supplied with a thermal shield to minimize heater power and improve temperature regulation.

The LM199 series references are exceptionally easy to use and free of the problems that are often experienced with ordinary zeners. There is virtually no hysteresis in reference voltage with temperature cycling. Also, the LM199 is free of voltage shifts due to stress on the leads. Finally, since the unit is temperature stabilized, warm up time is fast.

The LM199 can be used in almost any application in place of ordinary zeners with improved performance. Some ideal applications are analog to digital converters,

calibration standards, precision voltage or current sources or precision power supplies. Further in many cases the LM199 can replace references in existing equipment with a minimum of wiring changes.

The LM199 series devices are packaged in a standard hermetic TO-46 package inside a thermal shield. The LM199 is rated for operation from −55°C to +125°C while the LM299 is rated for operation from −25°C to +85°C and the LM399 is rated from 0°C to +70°C.

Features

- Guaranteed 0.0001%/°C temperature coefficient
- Low dynamic impedance − 0.5Ω
- Initial tolerance on breakdown voltage − 2%
- Sharp breakdown at 400μA
- Wide operating current − 500μA to 10 mA
- Wide supply range for temperature stabilizer
- Guaranteed low noise
- Low power for stabilization − 300 mW at 25°C
- Long term stability − 20 ppm

Schematic Diagrams

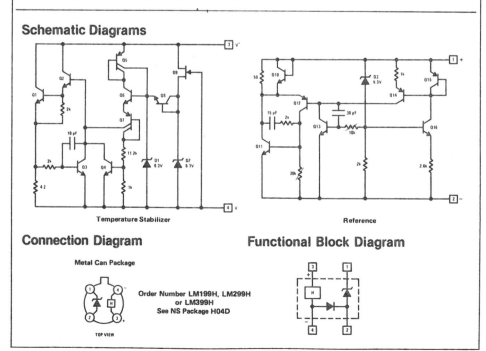

Temperature Stabilizer

Reference

Connection Diagram

Functional Block Diagram

Metal Can Package

Order Number LM199H, LM299H or LM399H
See NS Package H04D

TOP VIEW

Absolute Maximum Ratings

Temperature Stabilizer Voltage	40V
Reverse Breakdown Current	20 mA
Forward Current	1 mA
Reference to Substrate Voltage $V_{(RS)}$ (Note 1)	40V
	$-0.1V$
Operating Temperature Range	
LM199	$-55°C$ to $+125°C$
LM299	$-25°C$ to $+85°C$
LM399	$0°C$ to $+70°C$
Storage Temperature Range	$-55°C$ to $+150°C$
Lead Temperature (Soldering, 10 seconds)	$300°C$

Electrical Characteristics (Note 2)

PARAMETER	CONDITIONS	LM199/LM299			LM399			UNITS
		MIN	TYP	MAX	MIN	TYP	MAX	
Reverse Breakdown Voltage	0.5 mA $\leq I_R \leq 10$ mA	6.8	6.95	7.1	6.6	6.95	7.3	V
Reverse Breakdown Voltage Change With Current	0.5 mA $\leq I \leq 10$ mA		6	9		6	12	mV
Reverse Dynamic Impedance	$I_R = 1$ mA		0.5	1		0.5	1.5	Ω
Reverse Breakdown Temperature Coefficient	$-55°C \leq T_A \leq 85°C$ } LM199		0.00003	0.0001				%/°C
	$85°C \leq T_A \leq 125°C$		0.0005	0.0015				%/°C
	$-25°C \leq T_A \leq 85°C$ LM299		0.00003	0.0001				%/°C
	$0°C \leq T_A \leq 70°C$ LM399					0.00003	0.0002	%/°C
RMS Noise	10 Hz $\leq f \leq 10$ kHz		7	20		7	50	μV
Long Term Stability	Stabilized, $22°C \leq T_A \leq 28°C$, 1000 Hours, $I_R = 1$ mA $\pm0.1\%$		20			20		ppm
Temperature Stabilizer Supply Current	$T_A = 25°C$, Still Air, $V_S = 30V$		8.5	14		8.5	15	mA
	$T_A = -55°C$		22	28				
Temperature Stabilizer Supply Voltage	(Note 3)	9		40	9		40	V
Warm-Up Time to 0.05%	$V_S = 30V$, $T_A = 25°C$		3			3		Seconds
Initial Turn-on Current	$9 \leq V_S \leq 40$, $T_A = 25°C$, (Note 3)		140	200		140	200	mA

Note 1: The substrate is electrically connected to the negative terminal of the temperature stabilizer. The voltage that can be applied to either terminal of the reference is 40V more positive or 0.1V more negative than the substrate.

Note 2: These specifications apply for 30V applied to the temperature stabilizer and $55°C \leq T_A \leq +125°C$ for the LM199; $-25°C \leq T_A \leq +85°C$ for the LM299 and $0°C \leq T_A \leq +70°C$ for the LM399.

Note 3: This initial current can be reduced by adding an appropriate resistor and capacitor to the heater circuit. See the performance characteristic graphs to determine values.

National Semiconductor

Voltage Comparators

LM311 Voltage Comparator

General Description

The LM311 is a voltage comparator that has input currents more than a hundred times lower than devices like the LM306 or LM710C. It is also designed to operate over a wider range of supply voltages: from standard ±15V op amp supplies down to the single 5V supply used for IC logic. Its output is compatible with RTL, DTL and TTL as well as MOS circuits. Further, it can drive lamps or relays, switching voltages up to 40V at currents as high as 50 mA.

- Differential input voltage range: ±30V
- Power consumption: 135 mW at ±15V

Both the input and the output of the LM311 can be isolated from system ground, and the output can drive loads referred to ground, the positive supply or the negative supply. Offset balancing and strobe capability are provided and outputs can be wire OR'ed. Although slower than the LM306 and LM710C (200 ns response time vs 40 ns) the device is also much less prone to spurious oscillations. The LM311 has the same pin configuration as the LM306 and LM710C. See the "application hints" of the LM311 for application help.

Features

- Operates from single 5V supply
- Maximum input current: 250 nA
- Maximum offset current: 50 nA

Auxiliary Circuits **

** Note: Pin connections shown on schematic diagram and typical applications are for TO-5 package.

Typical Applications **

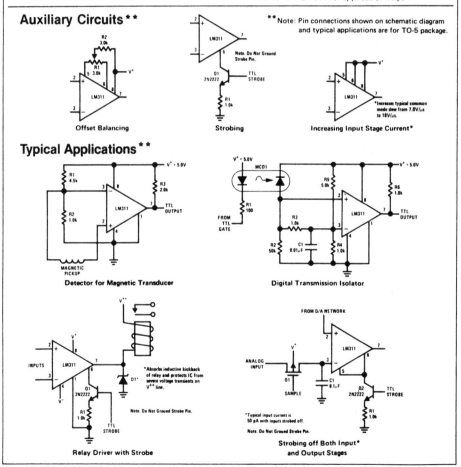

Offset Balancing

Strobing

Increasing Input Stage Current*

Detector for Magnetic Transducer

Digital Transmission Isolator

Relay Driver with Strobe

Strobing off Both Input* and Output Stages

Absolute Maximum Ratings

Total Supply Voltage (V_{84})	36V
Output to Negative Supply Voltage (V_{74})	40V
Ground to Negative Supply Voltage (V_{14})	30V
Differential Input Voltage	±30V
Input Voltage (Note 1)	±15V
Power Dissipation (Note 2)	500 mW
Output Short Circuit Duration	10 sec
Operating Temperature Range	$0°C$ to $70°C$
Storage Temperature Range	$-65°C$ to $150°C$
Lead Temperature (soldering, 10 sec)	$300°C$
Voltage at Strobe Pin	$V^+ -5V$

Electrical Characteristics (Note 3)

PARAMETER	CONDITIONS	MIN	TYP	MAX	UNITS
Input Offset Voltage (Note 4)	$T_A = 25°C$, $R_S \leq 50k$		2.0	7.5	mV
Input Offset Current (Note 4)	$T_A = 25°C$		6.0	50	nA
Input Bias Current	$T_A = 25°C$		100	250	nA
Voltage Gain	$T_A = 25°C$	40	200		V/mV
Response Time (Note 5)	$T_A = 25°C$		200		ns
Saturation Voltage	$V_{IN} \leq -10$ mV, $I_{OUT} = 50$ mA $T_A = 25°C$		0.75	1.5	V
Strobe ON Current	$T_A = 25°C$		3.0		mA
Output Leakage Current	$V_{IN} \geq 10$ mV, $V_{OUT} = 35V$ $T_A = 25°C$, $I_{STROBE} = 3$ mA		0.2	50	nA
Input Offset Voltage (Note 4)	$R_S \leq 50k$			10	mV
Input Offset Current (Note 4)				70	nA
Input Bias Current				300	nA
Input Voltage Range		−14.5	13.8,−14.7	13.0	V
Saturation Voltage	$V^+ \geq 4.5V$, $V^- = 0$ $V_{IN} \leq -10$ mV, $I_{SINK} \leq 8$ mA		0.23	0.4	V
Positive Supply Current	$T_A = 25°C$		5.1	7.5	mA
Negative Supply Current	$T_A = 25°C$		4.1	5.0	mA

Note 1: This rating applies for ±15V supplies. The positive input voltage limit is 30V above the negative supply. The negative input voltage limit is equal to the negative supply voltage or 30V below the positive supply, whichever is less.

Note 2: The maximum junction temperature of the LM311 is 110°C. For operating at elevated temperatures, devices in the TO-5 package must be derated based on a thermal resistance of 150°C/W, junction to ambient, or 45°C/W, junction to case. The thermal resistance of the dual-in-line package is 100°C/W, junction to ambient.

Note 3: These specifications apply for V_S = ±15V and the Ground pin at ground, and $0°C < T_A < +70°C$, unless otherwise specified. The offset voltage, offset current and bias current specifications apply for any supply voltage from a single 5V supply up to ±15V supplies.

Note 4: The offset voltages and offset currents given are the maximum values required to drive the output within a volt of either supply with 1 mA load. Thus, these parameters define an error band and take into account the worst-case effects of voltage gain and input impedance.

Note 5: The response time specified (see definitions) is for a 100 mV input step with 5 mV overdrive.

Note 6: Do not short the strobe pin to ground; it should be current driven at 3 to 5 mA.

Typical Performance Characteristics

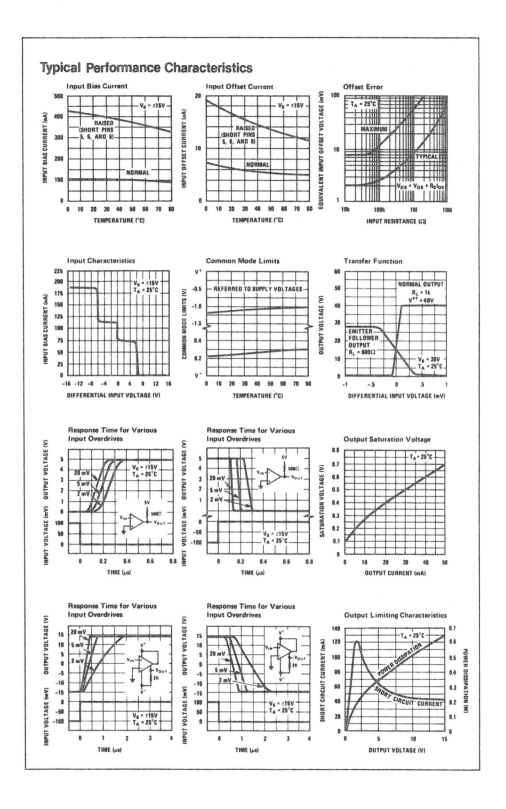

325

Typical Performance Characteristics (Continued)

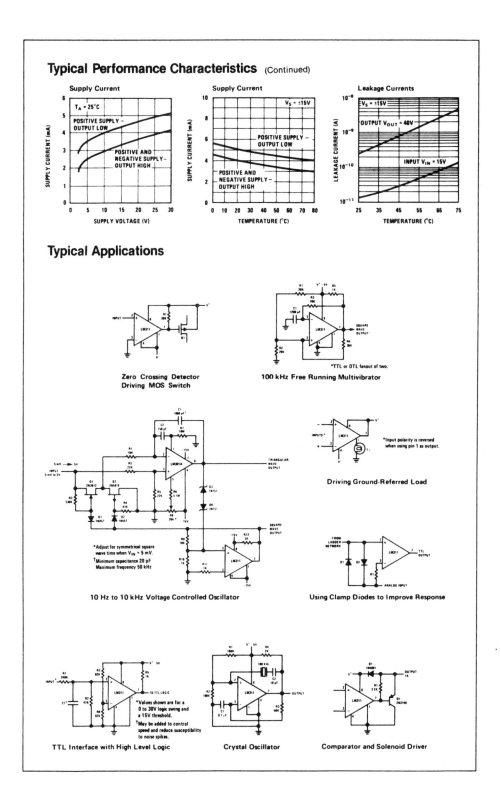

Typical Applications

Zero Crossing Detector
Driving MOS Switch

100 kHz Free Running Multivibrator

*TTL or DTL fanout of two.

10 Hz to 10 kHz Voltage Controlled Oscillator

*Adjust for symmetrical square
wave time when V_{IN} = 5 mV.
†Minimum capacitance 20 pF
Maximum frequency 50 kHz

Driving Ground-Referred Load

*Input polarity is reversed
when using pin 1 as output.

Using Clamp Diodes to Improve Response

TTL Interface with High Level Logic

*Values shown are for a
0 to 30V logic swing and
a 15V threshold.
†May be added to control
speed and reduce susceptibility
to noise spikes.

Crystal Oscillator

Comparator and Solenoid Driver

Typical Applications (Continued)

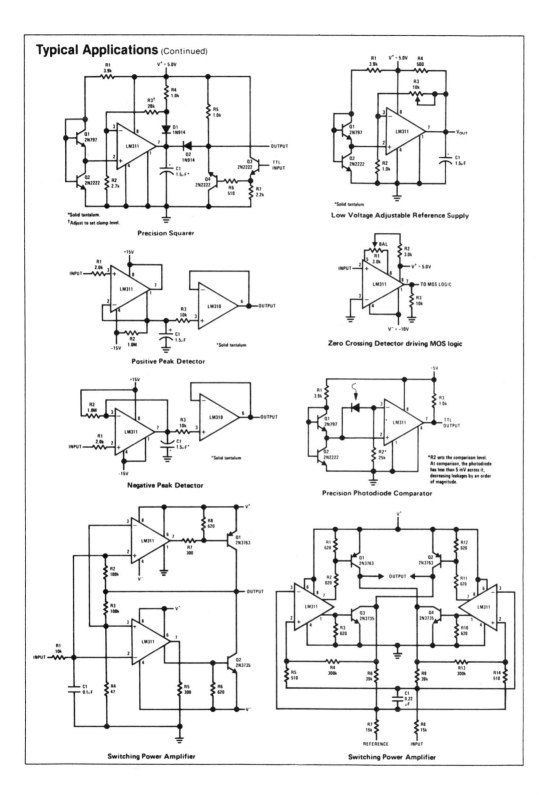

Precision Squarer

*Solid tantalum.
†Adjust to set clamp level.

Low Voltage Adjustable Reference Supply

*Solid tantalum

Positive Peak Detector

*Solid tantalum

Zero Crossing Detector driving MOS logic

Negative Peak Detector

*Solid tantalum

Precision Photodiode Comparator

*R2 sets the comparison level.
At comparison, the photodiode
has less than 5 mV across it,
decreasing leakages by an order
of magnitude.

Switching Power Amplifier

Switching Power Amplifier

Schematic Diagram

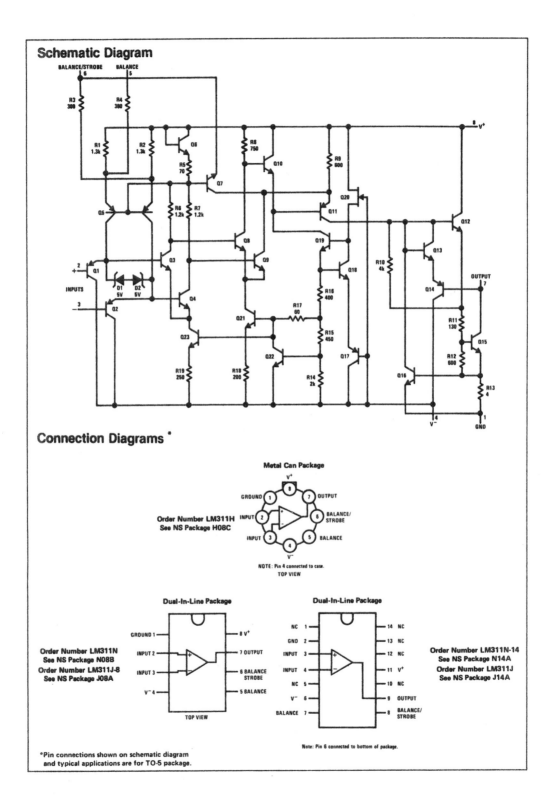

Connection Diagrams *

Metal Can Package

Order Number LM311H
See NS Package H08C

NOTE: Pin 4 connected to case.
TOP VIEW

Dual-In-Line Package

Order Number LM311N
See NS Package N08B
Order Number LM311J-8
See NS Package J08A

TOP VIEW

Dual-In-Line Package

Order Number LM311N-14
See NS Package N14A
Order Number LM311J
See NS Package J14A

Note: Pin 6 connected to bottom of package.

*Pin connections shown on schematic diagram
and typical applications are for TO-5 package.

Application Hints

CIRCUIT TECHNIQUES FOR AVOIDING OSCILLATIONS IN COMPARATOR APPLICATIONS

When a high-speed comparator such as the LM111 is used with fast input signals and low source impedances, the output response will normally be fast and stable, assuming that the power supplies have been bypassed (with 0.1 μF disc capacitors), and that the output signal is routed well away from the inputs (pins 2 and 3) and also away from pins 5 and 6.

However, when the input signal is a voltage ramp or a slow sine wave, or if the signal source impedance is high (1 kΩ to 100 kΩ), the comparator may burst into oscillation near the crossing-point. This is due to the high gain and wide bandwidth of comparators like the LM111. To avoid oscillation or instability in such a usage, several precautions are recommended, as shown in *Figure 1* below.

1. The trim pins (pins 5 and 6) act as unwanted auxiliary inputs. If these pins are not connected to a trim-pot, they should be shorted together. If they are connected to a trim-pot, a 0.01 μA capacitor C1 between pins 5 and 6 will minimize the susceptibility to AC coupling. A smaller capacitor is used if pin 5 is used for positive feedback as in *Figure 1*.

2. Certain sources will produce a cleaner comparator output waveform if a 100 pF to 1000 pF capacitor C2 is connected directly across the input pins.

3. When the signal source is applied through a resistive network, R_s, it is usually advantageous to choose an R_s' of substantially the same value, both for DC and for dynamic (AC) considerations. Carbon, tin-oxide, and metal-film resistors have all been used successfully in comparator input circuitry. Inductive wirewound resistors are not suitable.

4. When comparator circuits use input resistors (eg. summing resistors), their value and placement are particularly important. In all cases the body of the resistor should be close to the device or socket. In other words there should be very little lead length or printed-circuit foil run between comparator and resistor to radiate or pick up signals. The same applies to capacitors, pots, etc. For example, if R_s = 10 kΩ, as little as 5 inches of lead between the resistors and the input pins can result in oscillations that are very hard to damp. Twisting these input leads tightly is the only (second best) alternative to placing resistors close to the comparator.

5. Since feedback to almost any pin of a comparator can result in oscillation, the printed-circuit layout should be engineered thoughtfully. Preferably there should be a groundplane under the LM111 circuitry, for example, one side of a double-layer circuit card. Ground foil (or, positive supply or negative supply foil) should extend between the output and the inputs, to act as a guard. The foil connections for the inputs should be as small and compact as possible, and should be essentially surrounded by ground foil on all sides, to guard against capacitive coupling from any high-level signals (such as the output). If pins 5 and 6 are not used, they should be shorted together. If they are connected to a trim-pot, the trim-pot should be located, at most, a few inches away from the LM111, and the 0.01 μF capacitor should be installed. If this capacitor cannot be used, a shielding printed-circuit foil may be advisable between pins 6 and 7. The power supply bypass capacitors should be located within a couple inches of the LM111. (Some other comparators require the power-supply bypass to be located immediately adjacent to the comparator.)

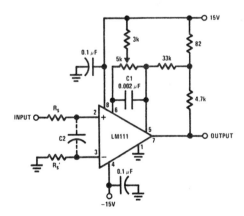

Pin connections shown are for LM111H in 8-lead TO-5 hermetic package

FIGURE 1. Improved Positive Feedback

Application Hints (Continued)

6. It is a standard procedure to use hysteresis (positive feedback) around a comparator, to prevent oscillation, and to avoid excessive noise on the output because the comparator is a good amplifier for its own noise. In the circuit of *Figure 2*, the feedback from the output to the positive input will cause about 3 mV of hysteresis. However, if R_S is larger than 100Ω, such as 50 kΩ, it would not be reasonable to simply increase the value of the positive feedback resistor above 510 kΩ. The circuit of *Figure 3* could be used, but it is rather awkward. See the notes in paragraph 7 below.

7. When both inputs of the LM111 are connected to active signals, or if a high-impedance signal is driving the positive input of the LM111 so that positive feedback would be disruptive, the circuit of *Figure 1* is

ideal. The positive feedback is to pin 5 (one of the offset adjustment pins). It is sufficient to cause 1 to 2 mV hysteresis and sharp transitions with input triangle waves from a few Hz to hundreds of kHz. The positive-feedback signal across the 82Ω resistor swings 240 mV below the positive supply. This signal is centered around the nominal voltage at pin 5, so this feedback does not add to the V_{OS} of the comparator. As much as 8 mV of V_{OS} can be trimmed out, using the 5 kΩ pot and 3 kΩ resistor as shown.

8. These application notes apply specifically to the LM111, LM211, LM311, and LF111 families of comparators, and are applicable to all high-speed comparators in general, (with the exception that not all comparators have trim pins).

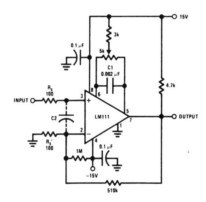

Pin connections shown are for LM111H in 8-lead TO-5 hermetic package

FIGURE 2. Conventional Positive Feedback

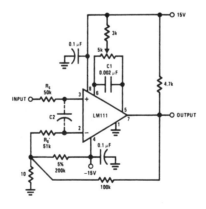

FIGURE 3. Positive Feedback With High Source Resistance

National Semiconductor

LM117/LM217/LM317 3-terminal adjustable regulator

general description

The LM117/LM217/LM317 are adjustable 3-terminal positive voltage regulators capable of supplying in excess of 1.5A over a 1.2V to 37V output range. They are exceptionally easy to use and require only two external resistors to set the output voltage. Further, both line and load regulation are better than standard fixed regulators. Also, the LM117 is packaged in standard transistor packages which are easily mounted and handled.

In addition to higher performance than fixed regulators, the LM117 series offers full overload protection available only in IC's. Included on the chip are current limit, thermal overload protection and safe area protection. All overload protection circuitry remains fully functional even if the adjustment terminal is disconnected.

features

- Adjustable output down to 1.2V
- Guaranteed 1.5A output current
- Line regulation typically 0.01%/V
- Load regulation typically 0.1%
- Current limit constant with temperature
- **100% electrical burn-in**
- Eliminates the need to stock many voltages
- Standard 3-lead transistor package
- 80 dB ripple rejection

Normally, no capacitors are needed unless the device is situated far from the input filter capacitors in which case an input bypass is needed. An optional output capacitor can be added to improve transient response. The adjustment terminal can be bypassed to achieve very high ripple rejections ratios which are difficult to achieve with standard 3-terminal regulators.

Besides replacing fixed regulators, the LM117 is useful in a wide variety of other applications. Since the regulator is "floating" and sees only the input-to-output differential voltage, supplies of several hundred volts can be regulated as long as the maximum input to output differential is not exceeded.

Also, it makes an especially simple adjustable switching regulator, a programmable output regulator, or by connecting a fixed resistor between the adjustment and output, the LM117 can be used as a precision current regulator. Supplies with electronic shutdown can be achieved by clamping the adjustment terminal to ground which programs the output to 1.2V where most loads draw little current.

The LM117K, LM217K and LM317K are packaged in standard TO-3 transistor packages while the LM117H, LM217H and LM317H are packaged in a solid Kovar base TO-5 transistor package. The LM117 is rated for operation from $-55°C$ to $+150°C$, the LM217 from $-25°C$ to $+150°C$ and the LM317 from $0°C$ to $+125°C$. The LM317T and LM317MP, rated for operation over a $0°C$ to $+125°C$ range, are available in a TO-220 plastic package and a TO-202 package, respectively.

For applications requiring greater output current in excess of 3A and 5A, see LM150 series and LM138 series data sheets, respectively. For the negative complement, see LM137 series data sheet.

LM117 Series Packages and Power Capability

DEVICE	PACKAGE	RATED POWER DISSIPATION	DESIGN LOAD CURRENT
LM117	TO-3	20W	1.5A
LM217 LM317	TO-5	2W	0.5A
LM317T	TO-220	15W	1.5A
LM317M	TO-202	7.5W	0.5A

typical applications

1.2V–25V Adjustable Regulator

†Optional—improves transient response

*Needed if device is far from filter capacitors

$$^{\dagger\dagger}V_{OUT} = 1.25V \left(1 + \frac{R2}{R1}\right)$$

Digitally Selected Outputs

*Sets maximum V_{OUT}

5V Logic Regulator with Electronic Shutdown*

*Min output ≈ 1.2V

absolute maximum ratings

Power Dissipation	Internally limited
Input—Output Voltage Differential	40V
Operating Junction Temperature Range	
LM117	-55°C to $+150^\circ$C
LM217	-25°C to $+150^\circ$C
LM317	0°C to $+125^\circ$C
Storage Temperature	-65°C to $+150^\circ$C
Lead Temperature (Soldering, 10 seconds)	300°C

preconditioning

Burn-In in Thermal Limit	100% All Devices

electrical characteristics (Note 1)

PARAMETER	CONDITIONS	LM117/217 MIN	LM117/217 TYP	LM117/217 MAX	LM317 MIN	LM317 TYP	LM317 MAX	UNITS
Line Regulation	$T_A = 25^\circ$C, $3V \leq V_{IN} - V_{OUT} \leq 40V$ (Note 2)		0.01	0.02		0.01	0.04	%/V
Load Regulation	$T_A = 25^\circ$C, $10\,mA \leq I_{OUT} \leq I_{MAX}$							
	$V_{OUT} \leq 5V$, (Note 2)		5	15		5	25	mV
	$V_{OUT} \geq 5V$, (Note 2)		0.1	0.3		0.1	0.5	%
Thermal Regulation	$T_A = 25^\circ$C, 20 ms Pulse		0.03	0.07		0.04	0.07	%/W
Adjustment Pin Current			50	100		50	100	μA
Adjustment Pin Current Change	$10\,mA \leq I_L \leq I_{MAX}$ $2.5V \leq (V_{IN} - V_{OUT}) \leq 40V$		0.2	5		0.2	5	μA
Reference Voltage	$3 \leq (V_{IN} - V_{OUT}) \leq 40V$, (Note 3) $10\,mA \leq I_{OUT} \leq I_{MAX}$, $P \leq P_{MAX}$	1.20	1.25	1.30	1.20	1.25	1.30	V
Line Regulation	$3V \leq V_{IN} - V_{OUT} \leq 40V$, (Note 2)		0.02	0.05		0.02	0.07	%/V
Load Regulation	$10\,mA \leq I_{OUT} \leq I_{MAX}$, (Note 2)							
	$V_{OUT} \leq 5V$		20	50		20	70	mV
	$V_{OUT} \geq 5V$		0.3	1		0.3	1.5	%
Temperature Stability	$T_{MIN} \leq T_j \leq T_{MAX}$		1			1		%
Minimum Load Current	$V_{IN} - V_{OUT} = 40V$		3.5	5		3.5	10	mA
Current Limit	$V_{IN} - V_{OUT} \leq 15V$							
	K and T Package	1.5	2.2		1.5	2.2		A
	H and P Package	0.5	0.8		0.5	0.8		A
	$V_{IN} - V_{OUT} = 40V$							
	K and T Package		0.4			0.4		A
	H and P Package		0.07			0.07		A
RMS Output Noise, % of V_{OUT}	$T_A = 25^\circ$C, $10\,Hz \leq f \leq 10\,kHz$		0.003			0.003		%
Ripple Rejection Ratio	$V_{OUT} = 10V$, $f = 120\,Hz$		65			65		dB
	$C_{ADJ} = 10\mu F$	66	80		66	80		dB
Long-Term Stability	$T_A = 125^\circ$C		0.3	1		0.3	1	%
Thermal Resistance, Junction to Case	H Package		12	15		12	15	$^\circ$C/W
	K Package		2.3	3		2.3	3	$^\circ$C/W
	T Package					4		$^\circ$C/W
	P Package					12		$^\circ$C/W

Note 1: Unless otherwise specified, these specifications apply: -55°C $\leq T_j \leq +150^\circ$C for the LM117, -25°C $\leq T_j \leq +150^\circ$C for the LM217 and 0°C $\leq T_j \leq +125^\circ$C for the LM317; $V_{IN} - V_{OUT} = 5V$ and $I_{OUT} = 0.1A$ for the TO-5 and TO-202 packages and $I_{OUT} = 0.5A$ for the TO-3 package and TO-220 package. Although power dissipation is internally limited, these specifications are applicable for power dissipations of 2W for the TO-5 and TO-202 and 20W for the TO-3 and TO-220. I_{MAX} is 1.5A for the TO-3 and TO-220 package and 0.5A for the TO-5 and TO-202 package.

Note 2: Regulation is measured at constant junction temperature, using pulse testing with a low duty cycle. Changes in output voltage due to heating effects are covered under the specification for thermal regulation.

Note 3: Selected devices with tightend tolerance reference voltage available.

NATIONAL

LM129, LM329 precision reference

general description

The LM129 and LM329 family are precision multi-current temperature compensated 6.9V zener references with dynamic impedances a factor of 10 to 100 less than discrete diodes. Constructed in a single silicon chip, the LM129 uses active circuitry to buffer the internal zener allowing the device to operate over a 0.5 mA to 15 mA range with virtually no change in performance. The LM129 and LM329 are available with selected temperature coefficients of 0.001, 0.002, 0.005 and 0.01%/°C. These new references also have excellent long term stability and low noise.

A new subsurface breakdown zener used in the LM129 gives lower noise and better long term stability than conventional IC zeners. Further the zener and temperature compensating transistor are made by a planar process so they are immune to problems that plague ordinary zeners. For example, there is virtually no voltage shifts in zener voltage due to temperature cycling and the device is insensitive to stress on the leads.

The LM129 can be used in place of conventional zeners with improved performance. The low dynamic impedance simplifies biasing and the wide operating current allows the replacement of many zener types.

The LM129 is packaged in a 2-lead TO-46 package and is rated for operation over a −55°C to +125°C temperature range. The LM329 for operation over 0−70°C is available in both a hermetic TO-46 package and a TO-92 epoxy package.

features

- 0.6 mA to 15 mA operating current
- 0.6Ω dynamic impedance at any current
- Available with temperature coefficients of 0.001%/°C
- 7μV wideband noise
- 5% initial tolerance
- 0.002% long term stability
- Low cost

typical applications

Low Cost 0−25V Regulator

Simple Reference

Adjustable Bipolar Output Reference

absolute maximum ratings

Reverse Breakdown Current	30 mA
Forward Current	2 mA
Operating Temperature Range	
LM129	-55°C to $+125^{\circ}$C
LM329	0°C to $+70^{\circ}$C
Storage Temperature Range	-55°C to $+150^{\circ}$C
Lead Temperature (Soldering, 10 seconds)	300°C

electrical characteristics (Note 1)

PARAMETER	CONDITIONS	LM129A, B, C			LM329B, C, D			UNITS
		MIN	TYP	MAX	MIN	TYP	MAX	
Reverse Breakdown Voltage	$T_A = 25^{\circ}$C, 0.6 mA $\leq I_R \leq 15$ mA	6.7	6.9	7.2	6.55	6.9	7.25	V
Reverse Breakdown Change with Current	$T_A = 25^{\circ}$C, 0.6 mA $\leq I_R \leq 15$ mA		9	14		9	20	mV
Reverse Dynamic Impedance	$T_A = 25^{\circ}$C, $I_R = 1$ mA		0.6	1		0.8	2	Ω
RMS Noise	$T_A = 25^{\circ}$C, 10 Hz $\leq F \leq 10$ kHz		7	20		7	100	μV
Long Term Stability	$T_A = 45^{\circ}$C $\pm0.1^{\circ}$C, $I_R = 1$ mA $\pm0.3\%$		20			20		ppm
Temperature Coefficient	$I_R = 1$ mA							
LM129A			6	10				ppm/$^{\circ}$C
LM129B, LM329B			15	20		15	20	ppm/$^{\circ}$C
LM129C, LM329C			30	50		30	50	ppm/$^{\circ}$C
LM329D						50	100	ppm/$^{\circ}$C
Change In Reverse Breakdown Temperature Coefficient	1 mA $\leq I_R \leq 15$ mA		1			1		ppm/$^{\circ}$C
Reverse Breakdown Change with Current	1 mA $\leq I_R \leq 15$ mA		12			12		mV
Reverse Dynamic Impedance	1 mA $\leq I_R \leq 15$ mA		0.8			1		Ω

Note 1: These specifications apply for -55°C $\leq T_A \leq +125^{\circ}$C for the LM129 and 0°C $\leq T_A \leq +70^{\circ}$C for the LM329 unless otherwise specified.

typical performance characteristics

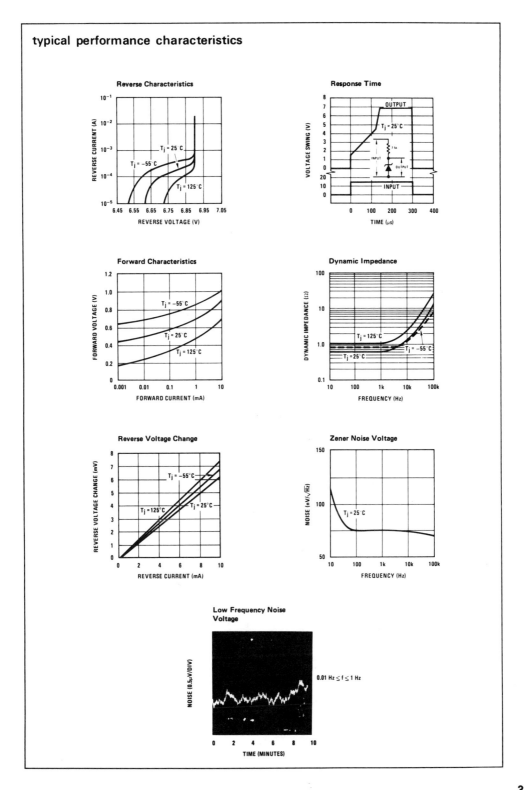

Reverse Characteristics

Response Time

Forward Characteristics

Dynamic Impedance

Reverse Voltage Change

Zener Noise Voltage

Low Frequency Noise Voltage

connection diagrams

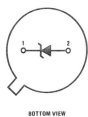

TO-46
Metal Can Package

BOTTOM VIEW

TO-92
Plastic Package

BOTTOM VIEW

physical dimensions

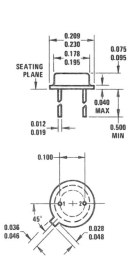

TO-46 Metal Can Package (H)
Order Number LM129AH, LM129BH, LM129CH,
LM329BH, LM329CH or LM329DH

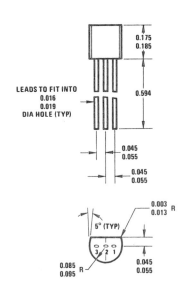

LEADS TO FIT INTO
0.016
0.019
DIA HOLE (TYP)

TO-92 Plastic Package (Z)
Order Number LM329BZ, LM329CZ
or LM329DZ

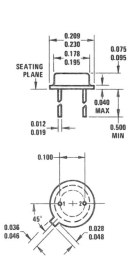

Voltage Comparators

National Semiconductor

LM139/239/339, LM139A/239A/339A, LM2901,LM3302
Low Power Low Offset Voltage Quad Comparators

General Description

The LM139 series consists of four independent precision voltage comparators with an offset voltage specification as low as 2 mV max for all four comparators. These were designed specifically to operate from a single power supply over a wide range of voltages. Operation from split power supplies is also possible and the low power supply current drain is independent of the magnitude of the power supply voltage. These comparators also have a unique characteristic in that the input common-mode voltage range includes ground, even though operated from a single power supply voltage.

Application areas include limit comparators, simple analog to digital converters; pulse, squarewave and time delay generators; wide range VCO; MOS clock timers; multivibrators and high voltage digital logic gates. The LM139 series was designed to directly interface with TTL and CMOS. When operated from both plus and minus power supplies, they will directly interface with MOS logic— where the low power drain of the LM339 is a distinct advantage over standard comparators.

Advantages

- High precision comparators
- Reduced V_{OS} drift over temperature

- Eliminates need for dual supplies
- Allows sensing near gnd
- Compatible with all forms of logic
- Power drain suitable for battery operation

Features

- Wide single supply voltage range or dual supplies

  LM139 series,   2 V_{DC} to 36 V_{DC} or
  LM139A series, LM2901 ± 1 V_{DC} to ± 18 V_{DC}
  LM3302   2 V_{DC} to 28 V_{DC}
                              or ± 1 V_{DC} to ± 14 V_{DC}

- Very low supply current drain (0.8 mA) — independent of supply voltage (2 mW/comparator at +5 V_{DC})
- Low input biasing current   25 nA
- Low input offset current ± 5 nA
  and offset voltage ± 3 mV
- Input common-mode voltage range includes gnd
- Differential input voltage range equal to the power supply voltage
- Low output   250 mV at 4 mA
  saturation voltage
- Output voltage compatible with TTL, DTL, ECL, MOS and CMOS logic systems

Schematic and Connection Diagrams

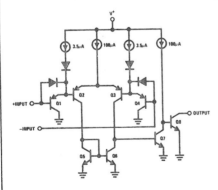

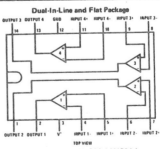

Dual-In-Line and Flat Package

TOP VIEW

Order Number LM139J, LM139AJ,
LM239J, LM239AJ, LM339J,
LM339AJ, LM2901J or LM3302J
See NS Package J14A

Order Number LM339N, LM339AN,
LM2901N or LM3302N
See NS Package N14A

Typical Applications $(V^+ = 5.0 V_{DC})$

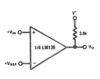

Basic Comparator

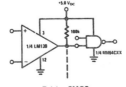

Driving CMOS

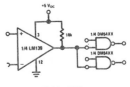

Driving TTL

Absolute Maximum Ratings

	LM139/LM239/LM339 LM139A/LM239A/LM339A LM2901	LM3302
Supply Voltage, V+	36 VDC or ±18 VDC	28 VDC or ±14 VDC
Differential Input Voltage	36 VDC	28 VDC
Input Voltage	−0.3 VDC to +36 VDC	−0.3 VDC to +28 VDC
Power Dissipation (Note 1)		
Molded DIP	570 mW	570 mW
Cavity DIP	900 mW	
Flat Pack	800 mW	
Output Short-Circuit to GND, (Note 2)	Continuous	Continuous
Input Current (V$_{IN}$ < −0.3 V$_{DC}$), (Note 3)	50 mA	50 mA
Operating Temperature Range		−40°C to +85°C
LM339A	0°C to +70°C	
LM239A	−25°C to +85°C	
LM2901	−40°C to +85°C	
LM139A	−55°C to +125°C	
Storage Temperature Range	−65°C to +150°C	−65°C to +150°C
Lead Temperature (Soldering, 10 seconds)	300°C	300°C

Electrical Characteristics (V+ = 5 V$_{DC}$, Note 4)

PARAMETER	CONDITIONS	LM139A			LM239A, LM339A			LM139			LM239, LM339			LM2901			LM3302			UNITS
		MIN	TYP	MAX	MIN	TYP	MAX	MIN	TYP	MAX	MIN	TYP	MAX	MIN	TYP	MAX	MIN	TYP	MAX	
Input Offset Voltage	T$_A$ = 25°C, (Note 9)		±1.0	±2.0		±1.0	±2.0		±2.0	±5.0		±2.0	±5.0		±2.0	±7.0		±3	±20	mV$_{DC}$
Input Bias Current	I$_{IN(+)}$ or I$_{IN(-)}$ with Output in Linear Range, T$_A$ = 25°C, (Note 5)		25	100		25	250		25	100		25	250		25	250		25	500	nA$_{DC}$
Input Offset Current	I$_{IN(+)}$ − I$_{IN(-)}$, T$_A$ = 25°C		±3.0	±25		±5.0	±50		±3.0	±25		±5.0	±50		±5	±50		±3	±100	nA$_{DC}$
Input Common-Mode Voltage Range	T$_A$ = 25°C, (Note 6)	0		V+−1.5	0		V+−1.5	0		V+−1.5	0		V+−1.5	0		V+−1.5	0		V+−1.5	V$_{DC}$
Supply Current	R$_L$ = ∞ on all Comparators, T$_A$ = 25°C; R$_L$ = ∞, V+ = 30V, T$_A$ = 25°C		0.8	2.0		0.8	2.0		0.8	2.0		0.8	2.0		0.8 / 1	2.0 / 2.5		0.8	2	mA$_{DC}$
Voltage Gain	R$_L$ ≥ 15 kΩ, V+ = 15 V$_{DC}$ (To Support Large V$_O$ Swing), T$_A$ = 25°C	50	200		50	200			200			200		25	100		2	30		V/mV
Large Signal Response Time	V$_{IN}$ = TTL Logic Swing, V$_{REF}$ = 1.4 V$_{DC}$, V$_{RL}$ = 5 V$_{DC}$, R$_L$ = 5.1 kΩ, T$_A$ = 25°C		300			300			300			300			300			300		ns
Response Time	V$_{RL}$ = 5 V$_{DC}$, R$_L$ = 5.1 kΩ, T$_A$ = 25°C, (Note 7)		1.3			1.3			1.3			1.3			1.3			1.3		µs
Output Sink Current	V$_{IN(-)}$ ≥ 1 V$_{DC}$, V$_{IN(+)}$ = 0, V$_O$ ≤ 1.5 V$_{DC}$, T$_A$ = 25°C	6.0			6.0			6.0			6.0			6.0			6.0			mA$_{DC}$
Saturation Voltage	V$_{IN(-)}$ ≥ 1 V$_{DC}$, V$_{IN(+)}$ = 0, I$_{SINK}$ ≤ 4 mA, T$_A$ = 25°C		250	400		250	400		250	400		250	400			400		250	500	mV$_{DC}$
Output Leakage Current	V$_{IN(+)}$ ≥ 1 V$_{DC}$, V$_{IN(-)}$ = 0, V$_O$ = 5 V$_{DC}$, T$_A$ = 25°C		0.1			0.1			0.1			0.1			0.1			0.1		nA$_{DC}$

Electrical Characteristics (Continued)

PARAMETER	CONDITIONS	LM139A			LM239A, LM339A			LM139			LM239, LM339			LM2901			LM3302			UNITS
		MIN	TYP	MAX	MIN	TYP	MAX	MIN	TYP	MAX	MIN	TYP	MAX	MIN	TYP	MAX	MIN	TYP	MAX	
Input Offset Voltage	(Note 9)			4.0			4.0			9.0			9.0		9	15			40	mVDC
Input Offset Current	$I_{IN(+)} - I_{IN(-)}$			±100			±150			±100			±150		50	200			300	nADC
Input Bias Current	$I_{IN(+)}$ or $I_{IN(-)}$ with Output in Linear Range			300			400			300			400		200	500			1000	nADC
Input Common-Mode Voltage Range		0		V^+-2.0	0		V^+-2.0	0		V^+-2.0	0		V^+-2.0	0		V^+-2.0	0		V^+-2.0	VDC
Saturation Voltage	$V_{IN(-)} \geq 1\ V_{DC}$, $V_{IN(+)} = 0$, $I_{SINK} \leq 4\ mA$			700			700			700			700		400	700			700	mVDC
Output Leakage Current	$V_{IN(+)} \geq 1\ V_{DC}$, $V_{IN(-)} = 0$, $V_O = 30\ V_{DC}$			1.0			1.0			1.0			1.0			1.0			1.0	µADC
Differential Input Voltage	Keep all V_{IN}'s $\geq 0\ V_{DC}$ (or V^-, if used). (Note 8)			V^+			V^+			36			36	0		V^+			V_{CC}	VDC

Note 1: For operating at high temperatures, the LM339/LM339A, LM2901, LM3302 must be derated based on a 125°C maximum junction temperature and a thermal resistance of 175°C/W which applies for the device soldered in a printed circuit board, operating in a still air ambient. The LM239 and LM139 must be derated based on a 150°C maximum junction temperature. The low bias dissipation and the "ON-OFF" characteristic of the outputs keeps the chip dissipation very small ($P_D \leq 100$ mW), provided the output transistors are allowed to saturate.

Note 2: Short circuits from the output to V^+ can cause excessive heating and eventual destruction. The maximum output current is approximately 20 mA independent of the magnitude of V^+.

Note 3: This input current will only exist when the voltage at any of the input leads is driven negative. It is due to the collector-base junction of the input PNP transistors becoming forward biased and thereby acting as input diode clamps. In addition to this diode action, there is also lateral NPN parasitic transistor action on the IC chip. This transistor action can cause the output voltages of the comparators to go to the V^+ voltage level (or to ground for a large overdrive) for the time duration that an input is driven negative. This is not destructive and normal output states will re-establish when the input voltage, which was negative, again returns to a value greater than −0.3 V_{DC}.

Note 4: These specifications apply for V^+ = 5 V_{DC} and −55°C $\leq T_A \leq$ +125°C, unless otherwise stated. With the LM239/LM239A, all temperature specifications are limited to −25°C $\leq T_A \leq$ +85°C, the LM339/LM339A temperature specifications are limited to 0°C $\leq T_A \leq$ +70°C, and the LM2901, LM3302 temperature range is −40°C $\leq T_A \leq$ +85°C.

Note 5: The direction of the input current is out of the IC due to the PNP input stage. This current is essentially constant, independent of the state of the output so no loading change exists on the reference or input lines.

Note 6: The input common-mode voltage or either input signal voltage should not be allowed to go negative by more than 0.3V. The upper end of the common-mode voltage range is V^+ −1.5V, but either or both inputs can go to +30 V_{DC} without damage.

Note 7: The response time specified is for a 100 mV input step with 5 mV overdrive. For larger overdrive signals 300 ns can be obtained, see typical performance characteristics section.

Note 8: Positive excursions of input voltage may exceed the power supply level. As long as the other voltage remains within the common-mode range, the comparator will provide a proper output state. The low input voltage state must not be less than −0.3 V_{DC} (or 0.3 V_{DC} below the magnitude of the negative power supply, if used).

Note 9: At output switch point, $V_O \cong 1.4\ V_{DC}$, $R_S = 0\Omega$ with V^+ from 5 V_{DC} and over the full input common-mode range (0 V_{DC} to V^+ −1.5 V_{DC}).

Note 10: For input signals that exceed V_{CC}, only the overdriven comparator is affected. With a 5V supply, V_{IN} should be limited to 25V max, and a limiting resistor should be used on all inputs that might exceed the positive supply.

National Semiconductor

LM140A/LM140/LM340A/LM340 Series
3-Terminal Positive Regulators

General Description

The LM140A/LM140/LM340A/LM340 series of positive 3-terminal voltage regulators are designed to provide superior performance as compared to the previously available 78XX series regulator. Computer programs were used to optimize the electrical and thermal performance of the packaged IC which results in outstanding ripple rejection, superior line and load regulation in high power applications (over 15W).

With these advances in design, the LM340 is now guaranteed to have line and load regulation that is a factor of 2 better than previously available devices. Also, all parameters are guaranteed at 1A vs 0.5A output current. The LM140A/LM340A provide tighter output voltage tolerance, ±2% along with 0.01%/V line regulation and 0.3%/A load regulation.

Current limiting is included to limit peak output current to a safe value. Safe area protection for the output transistor is provided to limit internal power dissipation. If internal power dissipation becomes too high for the heat sinking provided, the thermal shutdown circuit takes over limiting die temperature.

Considerable effort was expended to make the LM140-XX series of regulators easy to use and minimize the number of external components. It is not necessary to bypass the output, although this does improve transient response.

Input bypassing is needed only if the regulator is located far from the filter capacitor of the power supply.

Although designed primarily as fixed voltage regulators, these devices can be used with external components to obtain adjustable voltages and currents.

The entire LM140A/LM140/LM340A/LM340 series of regulators is available in the metal TO-3 power package and the LM340A/LM340 series is also available in the TO-220 plastic power package.

Features

- Complete specifications at 1A load
- Output voltage tolerances of ±2% at T_j = 25°C and ±4% over the temperature range (LM140A/LM340A)
- Fixed output voltages available 5, 6, 8, 10, 12, 15, 18 and 24V
- Line regulation of 0.01% of V_{OUT}/V ΔV_{IN} at 1A load (LM140A/LM340A)
- Load regulation of 0.3% of V_{OUT}/A ΔI_{LOAD} (LM140A/LM340A)
- Internal thermal overload protection
- Internal short-circuit current limit
- Output transistor safe area protection

Typical Applications

Fixed Output Regulator

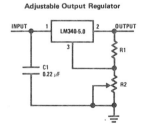

*Required if the regulator is located far from the power supply filter

** Although no output capacitor is needed for stability, it does help transient response. (If needed, use 0.1 μF, ceramic disc)

Adjustable Output Regulator

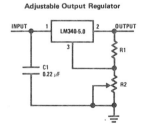

V_{OUT} = 5V + (5V/R1 + I_Q) R2
5V/R1 > 3 I_Q, load regulation (L_r) ≈ [(R1 + R2)/R1] (L_r of LM340-5)

Current Regulator

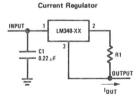

$I_{OUT} = \dfrac{V2\text{-}3}{R1} + I_Q$

ΔI_Q = 1.3 mA over line and load changes

Electrical Characteristics LM140 (Note 2)

$-55°C \leq T_j \leq +150°C$ unless otherwise noted.

OUTPUT VOLTAGE			5V			6V			8V			10V			12V			15V			18V			24V			UNITS
INPUT VOLTAGE (unless otherwise noted)			10V			11V			14V			17V			19V			23V			27V			33V			
PARAMETER		CONDITIONS	MIN	TYP	MAX	MIN	TYP	MAX	MIN	TYP	MAX	MIN	TYP	MAX	MIN	TYP	MAX	MIN	TYP	MAX	MIN	TYP	MAX	MIN	TYP	MAX	
V_O	Output Voltage	$T_j = 25°C$, $5\,mA \leq I_O \leq 1A$	4.8	5	5.2	5.75	6	6.25	7.7	8	8.3	9.6	10	10.4	11.5	12	12.5	14.4	15	15.6	17.3	18	18.7	23.0	24	25.0	V
		$P_D \leq 15W$, $5\,mA \leq I_O \leq 1A$	4.75		5.25	5.7		6.3	7.6		8.4	9.5		10.5	11.4		12.6	14.25		15.75	17.1		18.9	22.8		25.2	V
		$V_{MIN} \leq V_{IN} \leq V_{MAX}$		$(8 \leq V_{IN} \leq 20)$			$(9 \leq V_{IN} \leq 21)$			$(11.5 \leq V_{IN} \leq 23)$			$(13.5 \leq V_{IN} \leq 25)$			$(15.5 \leq V_{IN} \leq 27)$			$(18.5 \leq V_{IN} \leq 30)$			$(22 \leq V_{IN} \leq 33)$			$(28 \leq V_{IN} \leq 38)$		V
ΔV_O	Line Regulation	$I_O \leq 500\,mA$																									
		$T_j = 25°C$, ΔV_{IN}		3	50		3	60		4	80		4	100		4	120		4	150		4	180		6	240	mV
				$(7 \leq V_{IN} \leq 25)$			$(8 \leq V_{IN} \leq 25)$			$(10.5 \leq V_{IN} \leq 25)$			$(12.5 \leq V_{IN} \leq 25)$			$(14.5 \leq V_{IN} \leq 30)$			$(17.5 \leq V_{IN} \leq 30)$			$(21 \leq V_{IN} \leq 33)$			$(27 \leq V_{IN} \leq 38)$		V
		$-55°C \leq T_j \leq +150°C$ ΔV_{IN}			50			60			80			100			120			150			180			240	mV
				$(8 \leq V_{IN} \leq 20)$			$(9 \leq V_{IN} \leq 21)$			$(11 \leq V_{IN} \leq 23)$			$(13 \leq V_{IN} \leq 25)$			$(15 \leq V_{IN} \leq 27)$			$(18.5 \leq V_{IN} \leq 30)$			$(21.5 \leq V_{IN} \leq 33)$			$(28 \leq V_{IN} \leq 38)$		V
ΔV_O		$I_O \leq 1A$																									
		$T_j = 25°C$ ΔV_{IN}			50			60			80			100			120			150			180			240	mV
				$(7.3 \leq V_{IN} \leq 20)$			$(8.35 \leq V_{IN} \leq 21)$			$(10.5 \leq V_{IN} \leq 23)$			$(12.5 \leq V_{IN} \leq 25)$			$(14.6 \leq V_{IN} \leq 27)$			$(17.7 \leq V_{IN} \leq 30)$			$(21 \leq V_{IN} \leq 33)$			$(27.1 \leq V_{IN} \leq 38)$		V
		$-55°C \leq T_j \leq +150°C$ ΔV_{IN}			25			30			40			50			60			75			90			120	mV
				$(8 \leq V_{IN} \leq 12)$			$(9 \leq V_{IN} \leq 13)$			$(11 \leq V_{IN} \leq 17)$			$(14 \leq V_{IN} \leq 20)$			$(16 \leq V_{IN} \leq 22)$			$(20 \leq V_{IN} \leq 26)$			$(24 \leq V_{IN} \leq 30)$			$(30 \leq V_{IN} \leq 36)$		V
ΔV_O	Load Regulation	$T_j = 25°C$, $5\,mA \leq I_O \leq 1.5A$		10	50		12	60		12	80		12	100		12	120		12	150		12	180		12	240	mV
		$250\,mA \leq I_O \leq 750\,mA$			25			30			40			50			60			75			90			120	mV
		$55°C \leq T_j \leq +150°C$, $5\,mA \leq I_O \leq 1A$			50			60			80			100			120			150			180			240	mV
I_Q	Quiescent Current	$I_O \leq 1A$																									
		$T_j = 25°C$			6			6			6			6			6			6			6			6	mA
		$-55°C \leq T_j \leq +150°C$			7			7			7			7			7			7			7			7	mA
ΔI_Q	Quiescent Current Change	$5\,mA \leq I_O \leq 1A$			0.5			0.5			0.5			0.5			0.5			0.5			0.5			0.5	mA
		$T_j = 25°C$, $I_O \leq 1A$ $V_{MIN} \leq V_{IN} \leq V_{MAX}$			0.8			0.8			0.8			0.8			0.8			0.8			0.8			0.8	mA
				$(8 \leq V_{IN} \leq 20)$			$(9 \leq V_{IN} \leq 21)$			$(11.5 \leq V_{IN} \leq 23)$			$(13.5 \leq V_{IN} \leq 25)$			$(15 \leq V_{IN} \leq 27)$			$(18.5 \leq V_{IN} \leq 30)$			$(22 \leq V_{IN} \leq 33)$			$(28 \leq V_{IN} \leq 38)$		V
		$I_O \leq 500\,mA$, $-55°C \leq T_j \leq +150°C$ $V_{MIN} \leq V_{IN} \leq V_{MAX}$			0.8			0.8			0.8			0.8			0.8			0.8			0.8			0.8	mA
				$(8 \leq V_{IN} \leq 25)$			$(9 \leq V_{IN} \leq 25)$			$(11.5 \leq V_{IN} \leq 25)$			$(13.5 \leq V_{IN} \leq 25)$			$(15 \leq V_{IN} \leq 30)$			$(18.5 \leq V_{IN} \leq 30)$			$(22 \leq V_{IN} \leq 33)$			$(28 \leq V_{IN} \leq 38)$		V
V_N	Output Noise Voltage	$T_A = 25°C$, $10\,Hz \leq f \leq 100\,kHz$		40			45			52			70			75			90			110			170		μV
$\Delta V_{IN}/\Delta V_{OUT}$	Ripple Rejection	$f = 120\,Hz$																									
		$I_O \leq 1A$, $T_j = 25°C$ or	68	80		65	78		62	76		61	74		61	72		60	70		59	69		56	66		dB
		$I_O \leq 500\,mA$, $-55°C \leq T_j \leq +150°C$	68			65			62			61			61			60			59			56			dB
		$V_{MIN} \leq V_{IN} \leq V_{MAX}$		$(8 \leq V_{IN} \leq 18)$			$(9 \leq V_{IN} \leq 19)$			$(11.5 \leq V_{IN} \leq 21.5)$			$(13.5 \leq V_{IN} \leq 23.5)$			$(15 \leq V_{IN} \leq 25)$			$(18.5 \leq V_{IN} \leq 28.5)$			$(22 \leq V_{IN} \leq 32)$			$(28 \leq V_{IN} \leq 38)$		V
R_O	Dropout Voltage	$T_j = 25°C$, $I_{OUT} = 1A$		2.0			2.0			2.0			2.0			2.0			2.0			2.0			2.0		V
	Output Resistance	$f = 1\,kHz$		8			9			12			16			18			19			22			28		mΩ
	Short-Circuit Current	$T_j = 25°C$		2.1			2.0			1.9			1.7			1.5			1.2			0.8			0.4		A
	Peak Output Current	$T_j = 25°C$		2.4			2.4			2.4			2.4			2.4			2.4			2.4			2.4		A
	Average TC of V_{OUT}	$0°C \leq T_j \leq +150°C$, $I_O = 5\,mA$		-0.6			-0.7			-1.0			-1.2			-1.5			-1.8			-2.3			-3.0		mV/°C
V_{IN}	Input Voltage Required to Maintain Line Regulation	$T_j = 25°C$, $I_O \leq 1A$	7.3			8.35			10.5			12.5			14.6			17.7			21			27.1			V

Note 2: All characteristics are measured with a capacitor across the input of 0.22 µF and a capacitor across the output of 0.1 µF. All characteristics except noise voltage and ripple rejection ratio are measured using pulse techniques ($t_W \leq 10\,ms$, duty cycle $\leq 5\%$). Output voltage changes due to changes in internal temperature must be taken into account separately.

341

NATIONAL

LF198/LF298/LF398 monolithic sample and hold circuits

general description

The LF198/LF298/LF398 are monolithic sample and hold circuits which utilize BI-FET technology to obtain ultra-high dc accuracy with fast acquisition of signal and low droop rate. Operating as a unity gain follower, dc gain accuracy is 0.002% typical and acquisition time is as low as 6μs to 0.01%. A bipolar input stage is used to achieve low offset voltage and wide bandwidth. Input offset adjust is accomplished with a single pin and does not degrade input offset drift. The wide bandwidth allows the LF198 to be included inside the feedback loop of 1 MHz op amps without having stability problems. Input impedance of $10^{10}\Omega$ allows high source impedances to be used without degrading accuracy.

P-channel junction FET's are combined with bipolar devices in the output amplifier to give droop rates as low as 5 mV/min with a 1μF hold capacitor. The JFET's have much lower noise than MOS devices used in previous designs and do not exhibit high temperature instabilities. The overall design guarantees no feedthrough from input to output in the hold mode even for input signals equal to the supply voltages.

features

- Operates from ±5V to ±18V supplies
- Less than 10μs acquisition time
- TTL, PMOS, CMOS compatible logic input
- 0.5 mV typical hold step at C_h = 0.01μF
- Low input offset
- 0.002% gain accuracy
- Low output noise in hold mode
- Input characteristics do not change during hold mode
- High supply rejection ratio in sample or hold
- Wide bandwidth

Logic inputs on the LF198 are fully differential with low input current, allowing direct connection to TTL, PMOS, and CMOS. Differential threshold is 1.4V. The LF198 will operate from ±5V to ±18V supplies. It is available in an 8-lead TO-5 package.

functional diagram

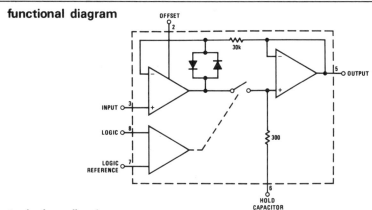

typical applications

Typical Connection

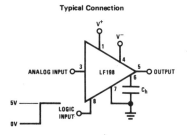

Acquisition Time

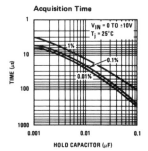

absolute maximum ratings

Supply Voltage	±18V	Input Voltage	Equal to Supply Voltage
Power Dissipation (Package Limitation) (Note 1)	500 mW	Logic To Logic Reference Differential Voltage (Note 2)	+7V, −30V
Operating Ambient Temperature Range		Output Short Circuit Duration	Indefinite
LF198	−55°C to +125°C	Hold Capacitor Short Circuit Duration	10 sec
LF298	−25°C to +85°C	Lead Temperature (Soldering, 10 seconds)	300°C
LF398	0°C to +70°C		
Storage Temperature Range	−65°C to +150°C		

electrical characteristics (Note 3)

PARAMETER	CONDITIONS	LF198/LF298 MIN	LF198/LF298 TYP	LF198/LF298 MAX	LF398 MIN	LF398 TYP	LF398 MAX	UNITS
Input Offset Voltage, (Note 6)	T_j = 25°C		1	3		2	7	mV
	Full Temperature Range			5			10	mV
Input Bias Current, (Note 6)	T_j = 25°C		5	25		10	50	nA
	Full Temperature Range			75			100	nA
Input Impedance	T_j = 25°C		10^{10}			10^{10}		Ω
Gain Error	T_j = 25°C, R_L = 10k		0.002	0.005		0.004	0.01	%
	Full Temperature Range			0.02			0.02	%
Feedthrough Attenuation Ratio at 1 kHz	T_j = 25°C, C_h = 0.01μF	86	96		80	90		dB
Output Impedance	T_j = 25°C, "HOLD" mode		0.5	2		0.5	4	Ω
	Full Temperature Range			4			6	Ω
"HOLD" Step, (Note 4)	T_j = 25°C, C_h = 0.01μF, V_{OUT} = 0		0.5	2.0		1.0	2.5	mV
Supply Current, (Note 6)	$T_j \geq$ 25°C		4.5	5.5		4.5	6.5	mA
Logic and Logic Reference Input Current	T_j = 25°C		2	10		2	10	μA
Leakage Current into Hold Capacitor (Note 6)	T_j = 25°C, (Note 5) Hold Mode		30	100		30	200	pA
Acquisition Time to 0.1%	ΔV_{OUT} = 10V, C_h = 1000 pF		4			4		μs
	C_h = 0.01μF		20			20		μs
Hold Capacitor Charging Current	$V_{IN} - V_{OUT}$ = 2V		5			5		mA
Supply Voltage Rejection Ratio	V_{OUT} = 0	80	110		80	110		dB
Differential Logic Threshold	T_j = 25°C	0.8	1.4	2.4	0.8	1.4	2.4	V

Note 1: The maximum junction temperature of the LF198 is 150°C, for the LF298, 115°C, and for the LF398, 100°C. When operating at elevated ambient temperature, the TO-5 package must be derated based on a thermal resistance (Θ_{jA}) of 150°C/W.

Note 2: Although the differential voltage may not exceed the limits given, the common-mode voltage on the logic pins may be equal to the supply voltages without causing damage to the circuit. For proper logic operation, however, one of the logic pins must always be at least 2V below the positive supply and 3V above the negative supply.

Note 3: Unless otherwise specified, the following conditions apply. Unit is in "sample" mode, V_S = ±15V, T_j = 25°C, −11.5V ≤ V_{IN} ≤ +11.5V, C_h = 0.01μF, and R_L = 10 kΩ. Logic reference voltage = 0V and logic voltage = 2.5V.

Note 4: Hold step is sensitive to stray capacitive coupling between input logic signals and the hold capacitor. 1 pF, for instance, will create an additional 0.5 mV step with a 5V logic swing and a 0.01μF hold capacitor. Magnitude of the hold step is inversely proportional to hold capacitor value.

Note 5: Leakage current is measured at a *junction* temperature of 25°C. The effects of junction temperature rise due to power dissipation or elevated ambient can be calculated by doubling the 25°C value for each 11°C increase in chip temperature. Leakage is guaranteed over full input signal range.

Note 6: These parameters guaranteed over a supply voltage range of ±5 to ±18V.

typical performance characteristics

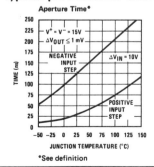

Aperture Time*

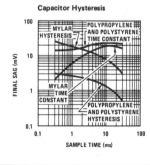

Capacitor Hysteresis

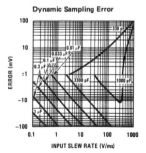

Dynamic Sampling Error

*See definition

typical performance characteristics (con't)

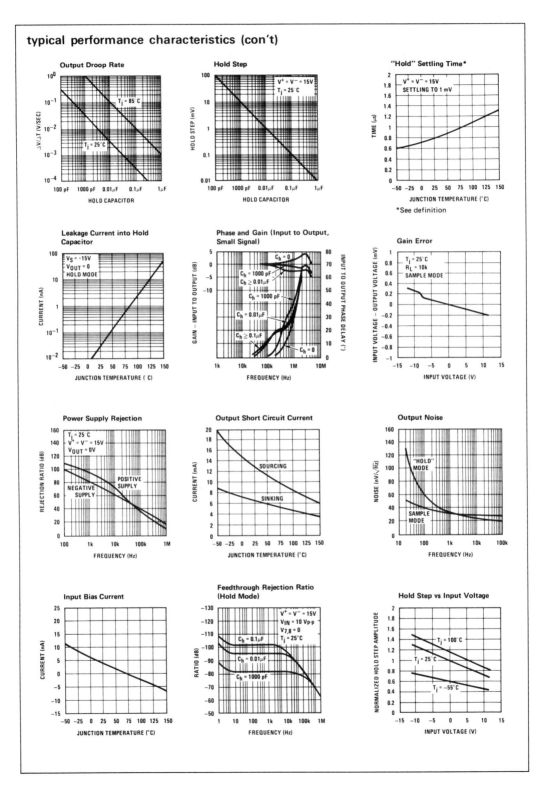

typical applications (con't)

Capacitor Hysteresis Compensation

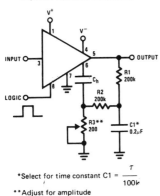

*Select for time constant C1 = $\dfrac{\tau}{100k}$

**Adjust for amplitude

Differential Hold

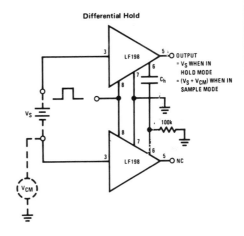

OUTPUT
= V_S WHEN IN HOLD MODE
= (V_S + V_{CM}) WHEN IN SAMPLE MODE

definition of terms

Hold Step: The voltage step at the output of the sample and hold when switching from sample mode to hold mode with a steady (dc) analog input voltage. Logic swing is 5V.

Acquisition Time: The time required to acquire a new analog input voltage with an output step of 10V. Note that acquisition time is not just the time required for the output to settle, but also includes the time required for all internal nodes to settle so that the output assumes the proper value when switched to the hold mode.

Gain Error: The ratio of output voltage swing to input voltage swing in the sample mode expressed as a per cent difference.

Hold Settling Time: The time required for the output to settle within 1 mV of final value after the "hold" logic command.

Dynamic Sampling Error: The error introduced into the held output due to a changing analog input at the time the hold command is given. Error is expressed in mV with a given hold capacitor value and input slew rate. Note that this error term occurs even for long sample times.

Aperture Time: The delay required between "Hold" command and an input analog transition, so that the transition does not affect the held output.

connection diagram

Metal Can Package

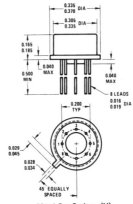

TOP VIEW

physical dimensions

Metal Can Package (H)
Order Number LF198H, LF298H or LF398H

Manufactured under one or more of the following U.S. patents: 3083262, 3189758, 3231797, 3303356, 3317671, 3323071, 3381071, 3408542, 3421025, 3426423, 3440498, 3518750, 3519897, 3557431, 3560765, 3566218, 3571630, 3575609, 3579059, 3593069, 3597640, 3607469, 3617859, 3631312, 3633052, 3638131, 3648071, 3651565, 3693248.

National Semiconductor

LM555/LM555C Timer

General Description

The LM555 is a highly stable device for generating accurate time delays or oscillation. Additional terminals are provided for triggering or resetting if desired. In the time delay mode of operation, the time is precisely controlled by one external resistor and capacitor. For astable operation as an oscillator, the free running frequency and duty cycle are accurately controlled with two external resistors and one capacitor. The circuit may be triggered and reset on falling waveforms, and the output circuit can source or sink up to 200 mA or drive TTL circuits.

Features

- Direct replacement for SE555/NE555
- Timing from microseconds through hours
- Operates in both astable and monostable modes

- Adjustable duty cycle
- Output can source or sink 200 mA
- Output and supply TTL compatible
- Temperature stability better than 0.005% per °C
- Normally on and normally off output

Applications

- Precision timing
- Pulse generation
- Sequential timing
- Time delay generation
- Pulse width modulation
- Pulse position modulation
- Linear ramp generator

Schematic Diagram

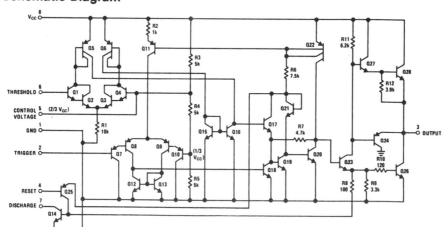

Connection Diagrams

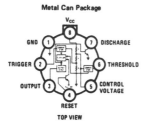

Metal Can Package

TOP VIEW

Order Number LM555H, LM555CH
See NS Package H08C

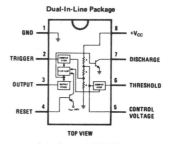

Dual-In-Line Package

TOP VIEW

Order Number LM555CN
See NS Package N08B
Order Number LM555J or LM555CJ
See NS Package J08A

Absolute Maximum Ratings

Supply Voltage	+18V
Power Dissipation (Note 1)	600 mW
Operating Temperature Ranges	
LM555C	0°C to +70°C
LM555	−55°C to +125°C
Storage Temperature Range	−65°C to +150°C
Lead Temperature (Soldering, 10 seconds)	300°C

Electrical Characteristics (T_A = 25°C, V_{CC} = +5V to +15V, unless otherwise specified)

PARAMETER	CONDITIONS	LM555 MIN	LM555 TYP	LM555 MAX	LM555C MIN	LM555C TYP	LM555C MAX	UNITS
Supply Voltage		4.5		18	4.5		16	V
Supply Current	V_{CC} = 5V, R_L = ∞		3	5		3	6	mA
	V_{CC} = 15V, R_L = ∞		10	12		10	15	mA
	(Low State) (Note 2)							
Timing Error, Monostable								
Initial Accuracy			0.5	2		1		%
Drift with Temperature	R_A, R_B = 1k to 100 k,		30			50		ppm/°C
	C = 0.1μF, (Note 3)							
Accuracy over Temperature			1.5	3.0		1.5		%
Drift with Supply			0.05	0.2		0.1		%/V
Timing Error, Astable								
Initial Accuracy			1.5	5		2.25	7	%
Drift with Temperature			90			150		ppm/°C
Accuracy over Temperature			2.5			3.0		%
Drift with Supply			0.15	0.2		0.30	0.5	%/V
Threshold Voltage			0.667			0.667		x V_{CC}
Trigger Voltage	V_{CC} = 15V	4.8	5	5.2		5		V
	V_{CC} = 5V	1.45	1.67	1.9		1.67		V
Trigger Current			0.01	0.5		0.5	0.9	μA
Reset Voltage		0.4	0.5	1	0.4	0.5	1	V
Reset Current			0.1	0.4		0.1	0.4	mA
Threshold Current	(Note 4)		0.1	0.25		0.1	0.25	μA
Control Voltage Level	V_{CC} = 15V	9.6	10	10.4	9	10	11	V
	V_{CC} = 5V	2.9	3.33	3.8	2.6	3.33	4	V
Pin 7 Leakage Output High			1	100		1	100	nA
Pin 7 Sat (Note 5)								
Output Low	V_{CC} = 15V, I_7 = 15 mA		150			180		mV
Output Low	V_{CC} = 4.5V, I_7 = 4.5 mA		70	100		80	200	mV
Output Voltage Drop (Low)	V_{CC} = 15V							
	I_{SINK} = 10 mA		0.1	0.15		0.1	0.25	V
	I_{SINK} = 50 mA		0.4	0.5		0.4	0.75	V
	I_{SINK} = 100 mA		2	2.2		2	2.5	V
	I_{SINK} = 200 mA		2.5			2.5		V
	V_{CC} = 5V							
	I_{SINK} = 8 mA		0.1	0.25				V
	I_{SINK} = 5 mA					0.25	0.35	V
Output Voltage Drop (High)	I_{SOURCE} = 200 mA, V_{CC} = 15V		12.5			12.5		V
	I_{SOURCE} = 100 mA, V_{CC} = 15V	13	13.3		12.75	13.3		V
	V_{CC} = 5V	3	3.3		2.75	3.3		V
Rise Time of Output			100			100		ns
Fall Time of Output			100			100		ns

Note 1: For operating at elevated temperatures the device must be derated based on a +150°C maximum junction temperature and a thermal resistance of +45°C/W junction to case for TO-5 and +150°C/W junction to ambient for both packages.

Note 2: Supply current when output high typically 1 mA less at V_{CC} = 5V.

Note 3: Tested at V_{CC} = 5V and V_{CC} = 15V.

Note 4: This will determine the maximum value of $R_A + R_B$ for 15V operation. The maximum total ($R_A + R_B$) is 20 MΩ.

Note 5: No protection against excessive pin 7 current is necessary providing the package dissipation rating will not be exceeded.

Applications Information

MONOSTABLE OPERATION

In this mode of operation, the timer functions as a one-shot (*Figure 1*). The external capacitor is initially held discharged by a transistor inside the timer. Upon application of a negative trigger pulse of less than 1/3 V_{CC} to pin 2, the flip-flop is set which both releases the short circuit across the capacitor and drives the output high.

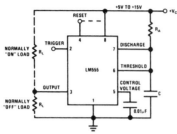

FIGURE 1. Monostable

The voltage across the capacitor then increases exponentially for a period of $t = 1.1\ R_A C$, at the end of which time the voltage equals 2/3 V_{CC}. The comparator then resets the flip-flop which in turn discharges the capacitor and drives the output to its low state. *Figure 2* shows the waveforms generated in this mode of operation. Since the charge and the threshold level of the comparator are both directly proportional to supply voltage, the timing internal is independent of supply.

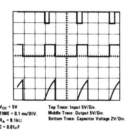

V_{CC} = 5V
TIME = 0.1 ms/DIV.
R_A = 9.1kΩ
C = 0.01μF

Top Trace: Input 5V/Div.
Middle Trace: Output 5V/Div.
Bottom Trace: Capacitor Voltage 2V/Div.

FIGURE 2. Monostable Waveforms

During the timing cycle when the output is high, the further application of a trigger pulse will not effect the circuit. However the circuit can be reset during this time by the application of a negative pulse to the reset terminal (pin 4). The output will then remain in the low state until a trigger pulse is again applied.

When the reset function is not in use, it is recommended that it be connected to V_{CC} to avoid any possibility of false triggering.

Figure 3 is a nomograph for easy determination of R, C values for various time delays.

NOTE: In monostable operation, the trigger should be driven high before the end of timing cycle.

ASTABLE OPERATION

If the circuit is connected as shown in *Figure 4* (pins 2 and 6 connected) it will trigger itself and free run as a

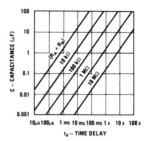

FIGURE 3. Time Delay

multivibrator. The external capacitor charges through $R_A + R_B$ and discharges through R_B. Thus the duty cycle may be precisely set by the ratio of these two resistors.

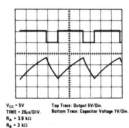

FIGURE 4. Astable

In this mode of operation, the capacitor charges and discharges between 1/3 V_{CC} and 2/3 V_{CC}. As in the triggered mode, the charge and discharge times, and therefore the frequency are independent of the supply voltage.

Figure 5 shows the waveforms generated in this mode of operation.

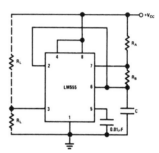

V_{CC} = 5V
TIME = 20μs/DIV.
R_A = 3.9 kΩ
R_B = 3 kΩ
C = 0.01μF

Top Trace: Output 5V/Div.
Bottom Trace: Capacitor Voltage 1V/Div.

FIGURE 5. Astable Waveforms

The charge time (output high) is given by:
$$t_1 = 0.693\ (R_A + R_B)\ C$$

And the discharge time (output low) by:
$$t_2 = 0.693\ (R_B)\ C$$

Thus the total period is:
$$T = t_1 + t_2 = 0.693\ (R_A + 2R_B)\ C$$

Typical Performance Characteristics

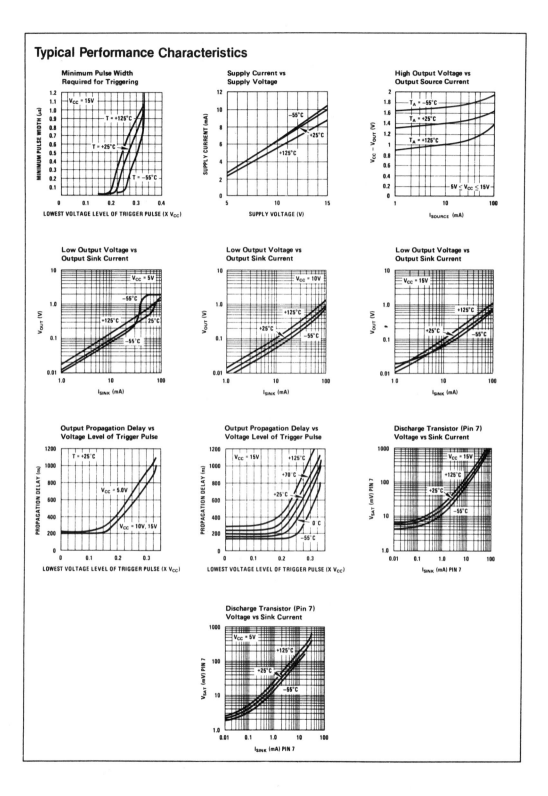

349

DESCRIPTION

The NE564 is a versatile, high guaranteed frequency Phase Locked Loop designed for operation up to 50MHz. As shown in the block diagram, the NE564 consists of a VCO, limiter, phase comparator, and post detection processor.

APPLICATIONS

- High speed modems
- FSK receivers and transmitters
- Frequency synthesizers
- Signal generators
- Various satcom/TV systems

FEATURES

- Operation with single 5V supply
- TTL compatible inputs and outputs
- Guaranteed operation to 50MHz
- External loop gain control
- Reduced carrier feedthrough
- No elaborate filtering needed in FSK applications
- Can be used as a modulator
- Variable loop gain (Externally Controlled)

PIN CONFIGURATION

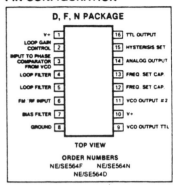

D, F, N PACKAGE

V+	1	16	TTL OUTPUT
LOOP GAIN CONTROL	2	15	HYSTERISIS SET
INPUT TO PHASE COMPARATOR FROM VCO	3	14	ANALOG OUTPUT
LOOP FILTER	4	13	FREQ. SET CAP.
LOOP FILTER	5	12	FREQ. SET CAP.
FM / RF INPUT	6	11	VCO OUTPUT #2
BIAS FILTER	7	10	V+
GROUND	8	9	VCO OUTPUT TTL

TOP VIEW

ORDER NUMBERS
NE/SE564F   NE/SE564N
NE/SE564D

ABSOLUTE MAXIMUM RATINGS

PARAMETER			RATING	UNIT
V+	Supply voltage			V
	Pin 1		14	
	Pin 10		6	
P_D	Power dissipation		400	mW
T_A	Operating temperature	NE	0 to 70	°C
	Operating temperature	SE	−55 to +125	
t_stg	Storage temperature		−65 to 150	°C

FUNCTIONAL DESCRIPTION

The NE564 is a monolithic phase locked loop with a post detection processor. The use of Schottky clamped transistors and optimized device geometries extends the frequency of operation to greater than 50MHz. In addition to the classical PLL applications, the NE564 can be used as a modulator with a controllable frequency deviation.

The output voltage of the PLL can be written as shown in the following equation:

$$V_O = \frac{(f_{in} - f_o)}{K_{vco}} \qquad \text{Equation 1}$$

K_{VCO} = conversion gain of the VCO (see figure 7)
f_{in} = frequency of the input signal
f_o = free running frequency of the VCO

The process of recovering FSK signals involves the conversion of the PLL output into logic compatible signals. For high data rates, a considerable amount of carrier will be present at the output of the PLL due to the wideband nature of the loop filter. To avoid the use of complicated filters, a comparator with hysterisis or Schmitt trigger is required. With the conversion gain of the VCO fixed, the output voltage as given by Equation 1 varies according to the frequency deviation of f_{in} from f_o. Since this differs from system to system, it is necessary that the hysterisis of the Schmitt trigger be capable of being changed, so that it can be optimized for a particular system. This is accomplished in the 564 by varying the voltage at pin 15 which results in a change of the hysterisis of the Schmitt trigger.

For FSK signals, an important factor to be considered is the drift in the free running frequency of the VCO itself. If this changes due to temperature, according to Equation 1 it will lead to a change in the dc levels of the

BLOCK DIAGRAM

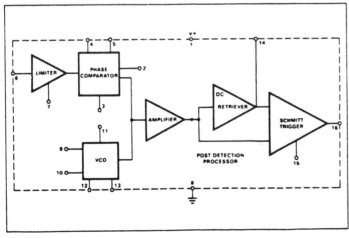

ELECTRICAL CHARACTERISTICS $V_{CC} = 5V$, $T_A = 25°C$, $f_o = 5MHz$, $I_B = 400\mu A$ unless otherwise specified

PARAMETER	TEST CONDITIONS	SE564			NE564			UNIT
		Min	Typ	Max	Min	Typ	Max	
VCO frequency	$C_1 = 6pF$	50	65		45	60		MHz
Lock range	Input $\geq$ 200mVrms $T_A = 25°C$	40	70		40	70		% of f_o
	= 125°C	20	30					
	= −55°C	70	90					
	= 0°C				50	70		
	= 70°C				25	40		
Capture range	Input $\geq$ 200mVrms, $R_2 = R_3 = 27\Omega$	20	30		20	30		% of f_o
VCO frequency drift with temperature	$f_o = 5MHz$, $T_A = −55°C$ to 125°C		400	1000				PPM/°C
	= 0°C to 70°C					400	1250	
	$f_o = 500KHz$, $T_A = −55°C$ to 125°C		250	500				
	= 0°C to 70°C					400	850	
VCO free running frequency	$f_o = \dfrac{1}{25R_CC_1}$, $C_1 = 80pF$ $R_C = 100\Omega$ "Internal"	4	5	6	3.5	5	7	MHz
VCO frequency change with supply voltage	$V_{CC} = 4.5V$ to 5.5V		3	8		3	8	% of f_o
Demodulated output voltage	Modulation frequency: 1KHz $f_o = 5MHz$, input deviation:							
	2% T = 25°C	18	24		18	24		mVrms
	1% T = 25°C	8	14		8	14		mVrms
	= 0°C				9.0	13		mVrms
	= −55°C	6	10					mVrms
	= 70°C				11.0	15		mVrms
	= 125°C	12	16					mVrms
Distortion	Deviation: 1% to 8%		1			1		%
Signal to noise ratio	Std. condition, 1% to 10% dev.		40			40		dB
AM rejection	Std. condition, 30% AM		35			35		dB
Demodulated Output at operating voltage	Modulation frequency: 1KHz $f_o = 5MHz$, input deviation: 1%							
	$V_{CC} = 4.5V$	7	12		7	12		mVrms
	$V_{CC} = 5.5V$	8	14		8	14		mVrms
Supply current	$V_{CC} = 5V$ I_1, I_{10}		45	60		45	60	mA
Output "1" output leakage current "0" output voltage	$V_{OUT} = 5V$, Pin 16, 9		1	20		1	20	μA
	$I_{OUT} = 2mA$, Pin 16, 9		0.3	0.6		0.3	0.6	V
	$I_{OUT} = 6mA$, Pin 16, 9		0.4	0.8		0.4	0.8	V

PLL output, and consequently to errors in the digital output signal. This is especially true for narrow band signals where the deviation in f_{in} itself may be less than the change in f_o due to temperature. This effect can be eliminated if the dc or average value of the signal is retrieved and used as the reference to the comparator. In this manner, variations in the dc levels of the PLL output do not affect the FSK output.

VCO Section

Due to its inherent high frequency performance, an emitter coupled oscillator is used in the VCO. In the circuit, shown in the equivalent schematic, transistors Q_{21} and Q_{23} with current sources Q_{25}–Q_{26} form the basic oscillator. The free running frequency of the oscillator is shown in the following equation:

$$f_o = \frac{1}{16R_CC_1} \qquad \text{Equation 2}$$

$R_C = R_{19} = R_{20} = 100\Omega$ (INTERNAL)
C_1 = external frequency setting capacitor

Variation of V_d (phase detector output voltage) changes the frequency of the oscillator. As indicated by Equation 2, the frequency of the oscillator has a negative temperature coefficient due to the positive temperature coefficient of the monolithic resistor. To compensate for this, a current I_R with negative temperature coefficient is introduced to achieve a low frequency drift with temperature.

Phase Comparator Section

The phase comparator consists of a double balanced modulator with a limiter amplifier to improve AM rejection. Schottky clamped

vertical PNPs are used to obtain TTL level inputs. The loop gain can be varied by changing the current in Q_4 and Q_{15} which effectively changes the gain of the differential amplifiers. This can be accomplished by introducing a current at pin 2.

Post Detection Processor Section

The post detection processor consists of a unity gain transconductance amplifier and comparator. The amplifier can be used as a dc retriever for demodulation of FSK signals, and as a post detection filter for linear FM demodulation. The comparator has adjustable hysterisis so that phase jitter in the output signal can be eliminated.

As shown in the equivalent schematic, the dc retriever is formed by the transductance

![National Semiconductor logo] **National Semiconductor**

LM565/LM565C Phase Locked Loop

General Description

The LM565 and LM565C are general purpose phase locked loops containing a stable, highly linear voltage controlled oscillator for low distortion FM demodulation, and a double balanced phase detector with good carrier suppression. The VCO frequency is set with an external resistor and capacitor, and a tuning range of 10:1 can be obtained with the same capacitor. The characteristics of the closed loop system—bandwidth, response speed, capture and pull in range—may be adjusted over a wide range with an external resistor and capacitor. The loop may be broken between the VCO and the phase detector for insertion of a digital frequency divider to obtain frequency multiplication.

The LM565H is specified for operation over the −55°C to +125°C military temperature range. The LM565CH and LM565CN are specified for operation over the 0°C to +70°C temperature range.

Features

- 200 ppm/°C frequency stability of the VCO

- Power supply range of ±5 to ±12 volts with 100 ppm/% typical
- 0.2% linearity of demodulated output
- Linear triangle wave with in phase zero crossings available
- TTL and DTL compatible phase detector input and square wave output
- Adjustable hold in range from ±1% to > ±60%.

Applications

- Data and tape synchronization
- Modems
- FSK demodulation
- FM demodulation
- Frequency synthesizer
- Tone decoding
- Frequency multiplication and division
- SCA demodulators
- Telemetry receivers
- Signal regeneration
- Coherent demodulators.

Schematic and Connection Diagrams

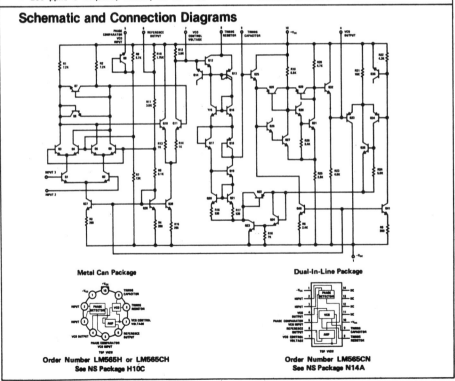

Metal Can Package

Order Number LM565H or LM565CH
See NS Package H10C

Dual-In-Line Package

Order Number LM565CN
See NS Package N14A

Absolute Maximum Ratings

Supply Voltage	±12V
Power Dissipation (Note 1)	300 mW
Differential Input Voltage	±1V
Operating Temperature Range LM565H	$-55°C$ to $+125°C$
LM565CH, LM565CN	$0°C$ to $70°C$
Storage Temperature Range	$-65°C$ to $+150°C$
Lead Temperature (Soldering, 10 sec)	$300°C$

Electrical Characteristics (AC Test Circuit, $T_A = 25°C$, $V_C = ±6V$)

PARAMETER	CONDITIONS	LM565			LM565C			UNITS
		MIN	TYP	MAX	MIN	TYP	MAX	
Power Supply Current			8.0	12.5		8.0	12.5	mA
Input Impedance (Pins 2, 3)	$-4V < V_2, V_3 < 0V$	7	10			5		$k\Omega$
VCO Maximum Operating Frequency	$C_o = 2.7$ pF	300	500		250	500		kHz
Operating Frequency Temperature Coefficient			-100	300		-200	500	ppm/$°C$
Frequency Drift with Supply Voltage			0.01	0.1		0.05	0.2	%/V
Triangle Wave Output Voltage		2	2.4	3	2	2.4	3	V_{p-p}
Triangle Wave Output Linearity			0.2	0.75		0.5	1	%
Square Wave Output Level		4.7	5.4		4.7	5.4		V_{p-p}
Output Impedance (Pin 4)			5			5		$k\Omega$
Square Wave Duty Cycle		45	50	55	40	50	60	%
Square Wave Rise Time			20	100		20		ns
Square Wave Fall Time			50	200		50		ns
Output Current Sink (Pin 4)		0.6	1		0.6	1		mA
VCO Sensitivity	$f_o = 10$ kHz	6400	6600	6800	6000	6600	7200	Hz/V
Demodulated Output Voltage (Pin 7)	±10% Frequency Deviation	250	300	350	200	300	400	mV_{pp}
Total Harmonic Distortion	±10% Frequency Deviation		0.2	0.75		0.2	1.5	%
Output Impedance (Pin 7)			3.5			3.5		$k\Omega$
DC Level (Pin 7)		4.25	4.5	4.75	4.0	4.5	5.0	V
Output Offset Voltage $\lvert V_7 - V_6 \rvert$			30	100		50	200	mV
Temperature Drift of $\lvert V_7 - V_6 \rvert$			500			500		$\mu V/°C$
AM Rejection		30	40			40		dB
Phase Detector Sensitivity K_D		0.6	.68	0.9	0.55	.68	0.95	V/radian

Note 1: The maximum junction temperature of the LM565 is 150$°C$, while that of the LM565C and LM565CN is 100$°C$. For operation at elevated temperatures, devices in the TO-5 package must be derated based on a thermal resistance of 150$°C$/W junction to ambient or 45$°C$/W junction to case. Thermal resistance of the dual-in-line package is 100$°C$/W.

Typical Performance Characteristics

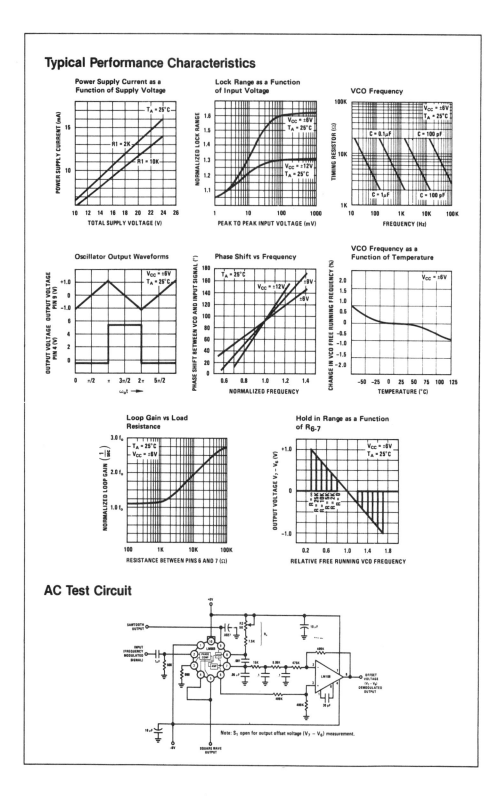

Power Supply Current as a Function of Supply Voltage

Lock Range as a Function of Input Voltage

VCO Frequency

Oscillator Output Waveforms

Phase Shift vs Frequency

VCO Frequency as a Function of Temperature

Loop Gain vs Load Resistance

Hold in Range as a Function of R6-7

AC Test Circuit

Note: S1 open for output offset voltage (V7 − V6) measurement.

LM565/LM565C

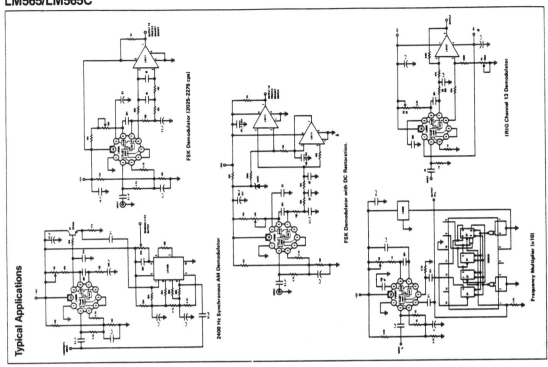

Typical Applications

LM565/LM565C

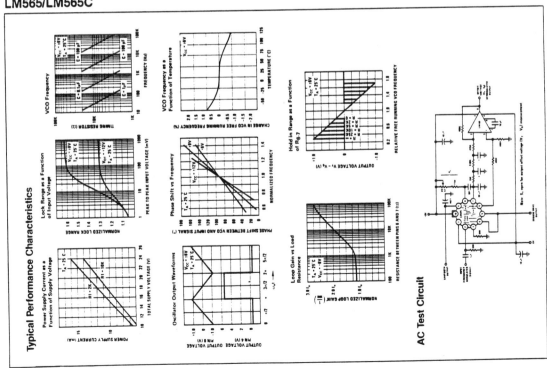

Typical Performance Characteristics

National Semiconductor

Industrial/Automotive/Functional Blocks/Telecommunications

LM566/LM566C Voltage Controlled Oscillator

General Description

The LM566/LM566C are general purpose voltage controlled oscillators which may be used to generate square and triangular waves, the frequency of which is a very linear function of a control voltage. The frequency is also a function of an external resistor and capacitor.

The LM566 is specified for operation over the –55°C to +125°C military temperature range. The LM566C is specified for operation over the 0°C to +70°C temperature range.

Features

- Wide supply voltage range: 10 to 24 volts
- Very linear modulation characteristics

- High temperature stability
- Excellent supply voltage rejection
- 10 to 1 frequency range with fixed capacitor
- Frequency programmable by means of current, voltage, resistor or capacitor.

Applications

- FM modulation
- Signal generation
- Function generation
- Frequency shift keying
- Tone generation

Schematic and Connection Diagrams

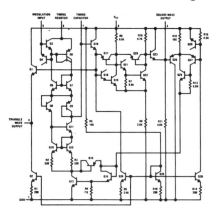

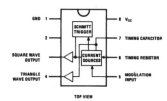

Dual-In-Line Package

Order Number LM566CN
See NS Package N08B

Typical Application

**1 kHz and 10 kHz TTL Compatible
Voltage Controlled Oscillator**

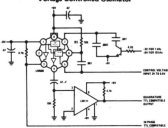

Applications Information

The LM566 may be operated from either a single supply as shown in this test circuit, or from a split (±) power supply. When operating from a split supply, the square wave output (pin 4) is TTL compatible (2 mA current sink) with the addition of a 4.7 kΩ resistor from pin 3 to ground.

A .001 µF capacitor is connected between pins 5 and 6 to prevent parasitic oscillations that may occur during VCO switching.

$$f_O = \frac{2(V^+ - V_5)}{R_1 C_1 V^+}$$

where

$2K < R_1 < 20K$

and V_5 is voltage between pin 5 and pin 1

Absolute Maximum Ratings

Power Supply Voltage	26V
Power Dissipation (Note 1)	300 mW
Operating Temperature Range LM566	$-55°$C to $+125°$C
LM566C	$0°$C to $70°$C
Lead Temperature (Soldering, 10 sec)	$300°$C

Electrical Characteristics V_{CC} = 12V, T_A = 25°C, AC Test Circuit

PARAMETER	CONDITIONS	LM566			LM566C			UNITS
		MIN	TYP	MAX	MIN	TYP	MAX	
Maximum Operating Frequency	R0 = 2k C0 = 2.7 pF		1			1		MHz
Input Voltage Range Pin 5		3/4 V_{CC}		V_{CC}	3/4 V_{CC}		V_{CC}	
Average Temperature Coefficient of Operating Frequency			100			200		ppm/°C
Supply Voltage Rejection	10–20V		0.1	1		0.1	2	%/V
Input Impedance Pin 5		0.5	1		0.5	1		MΩ
VCO Sensitivity	For Pin 5, From 8–10V, f_O = 10 kHz	6.4	6.6	6.8	6.0	6.6	7.2	kHz/V
FM Distortion	±10% Deviation		0.2	0.75		0.2	1.5	%
Maximum Sweep Rate		800	1		500	1		MHz
Sweep Range			10:1			10:1		

Output Impedance
  Pin 3
  Pin 4
			50
			50
						50
						50
								Ω
								Ω

Square Wave Output Level	R_{L1} = 10k	5.0	5.4		5.0	5.4		Vp-p
Triangle Wave Output Level	R_{L2} = 10k	2.0	2.4		2.0	2.4		Vp-p
Square Wave Duty Cycle		45	50	55	40	50	60	%
Square Wave Rise Time			20			20		ns
Square Wave Fall Time			50			50		ns
Triangle Wave Linearity	+1V Segment at 1/2 V_{CC}		0.2	0.75		0.5	1	%

Note 1: The maximum junction temperature of the LM566 is 150°C, while that of the LM566C is 100°C. For operating at elevated junction temperatures, devices in the TO-5 package must be derated based on a thermal resistance of 150°C/W. The thermal resistance of the dual-in-line package is 100°C/W.

Typical Performance Characteristics

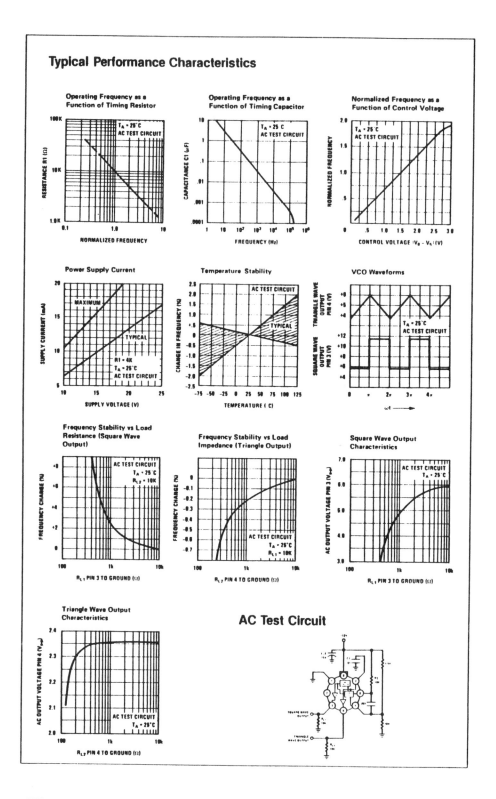

Operating Frequency as a Function of Timing Resistor

Operating Frequency as a Function of Timing Capacitor

Normalized Frequency as a Function of Control Voltage

Power Supply Current

Temperature Stability

VCO Waveforms

Frequency Stability vs Load Resistance (Square Wave Output)

Frequency Stability vs Load Impedance (Triangle Output)

Square Wave Output Characteristics

Triangle Wave Output Characteristics

AC Test Circuit

DESCRIPTION

The SE/NE567 tone and frequency decoder is a highly stable phase-locked loop with synchronous AM lock detection and power output circuitry. Its primary function is to drive a load whenever a sustained frequency within its detection band is present at the self-biased input. The bandwidth center frequency, and output delay are independently determined by means of four external components.

FEATURES

- Wide frequency range (.01Hz to 500kHz)
- High stability of center frequency
- Independently controllable bandwidth (up to 14 percent)
- High out-band signal and noise rejection
- Logic-compatible output with 100mA current sinking capability
- Inherent Immunity to false signals
- Frequency adjustment over a 20 to 1 range with an external resistor
- Military processing available

APPLICATIONS

- Touch Tone® decoding
- Carrier current remote controls
- Ultrasonic controls (remote TV, etc.)
- Communications paging
- Frequency monitoring and control
- Wireless Intercom
- Precision oscillator

PIN CONFIGURATIONS

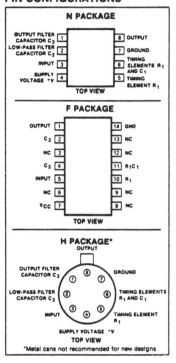

ABSOLUTE MAXIMUM RATINGS

PARAMETER	RATING	UNIT
Operating temperature		
NE567	0 to +70	°C
SE567	–55 to +125	°C
Operating voltage	10	V
Positive voltage at input	0.5 + Vs	V
Negative voltage at input	–10	Vdc
Output voltage (collector of output transistor)	15	Vdc
Storage temperature	–65 to +150	°C
Power dissipation	300	mW

BLOCK DIAGRAM

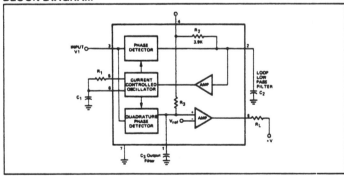

Signetics

DC ELECTRICAL CHARACTERISTICS (V+ = 5.0V; T_A = 25°C unless otherwise specified.)

PARAMETER	TEST CONDITIONS	SE567 Min	SE567 Typ	SE567 Max	NE567 Min	NE567 Typ	NE567 Max	UNIT
CENTER FREQUENCY[1]								
Highest center frequency (f_o)		100	500		100	500		kHz
Center frequency stability[2]	–55 to +125°C		35±140			35±140		ppm/°C
	0 to +70°C		35±60			35±60		ppm/°C
Center frequency shift with supply voltage	f_o = 100kHz		0.5	1		0.7	2	%/V
DETECTION BANDWIDTH								
Largest detection bandwidth	f_o = 100kHz	12	14	16	10	14	18	% of f_o
Largest detection bandwidth skew			2	4		3	6	% of f_o
Largest detection bandwidth— variation with temperature	V_i = 300mVrms		±0.1			±0.1		%/°C
Largest detection bandwidth— variation with supply voltage	V_i = 300mVrms		±2			±2		%/V
INPUT								
Input resistance			20			20		kΩ
Smallest detectable input voltage (V_i)	I_L = 100mA, f_i = f_o		20	25		20	25	mVrms
Largest no-output input voltage	I_L = 100mA, f_i = f_o	10	15		10	15		mVrms
Greatest simultaneous outband signal to inband signal ratio			+6			+6		dB
Minimum input signal to wideband noise ratio	B_n = 140kHz		–6			–6		dB
OUTPUT								
Fastest on-off cycling rate			f_o/20			f_o/20		
"1" output leakage current			0.01	25		0.01	25	μA
"0" output voltage	I_L = 30mA		0.2	0.4		0.2	0.4	V
	I_L = 100mA		0.6	1.0		0.6	1.0	V
Output fall time[3]	R_L = 50Ω		30			30		ns
Output rise time[3]	R_L = 50Ω		150			150		ns
GENERAL								
Operating voltage range		4.75		9.0	4.75		9.0	V
Supply current quiescent			6	8		7	10	mA
Supply current—activated	R_L = 20kΩ		11	13		12	15	mA
Quiescent power dissipation			30			35		mW

NOTES
1. Frequency determining resistor R_1 should be between 1 and 20kΩ.
2. Applicable over 4.75 to 5.75 volts. See graphs for more detailed information.
3. Pin 8 to Pin 1 feedback R_L network selected to eliminate pulsing during turn-on and turn-off.

TYPICAL PERFORMANCE CHARACTERISTICS

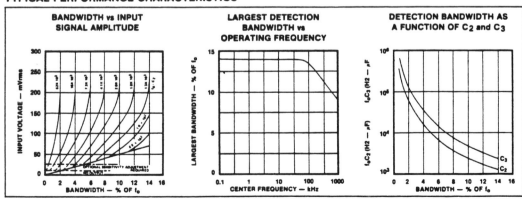

LM723/LM723C Voltage Regulator

General Description

The LM723/LM723C is a voltage regulator designed primarily for series regulator applications. By
itself, it will supply output currents up to 150 mA;
but external transistors can be added to provide
any desired load current. The circuit features extremely low standby current drain, and provision
is made for either linear or foldback current limiting. Important characteristics are:

- 150 mA output current without external pass
  transistor
- Output currents in excess of 10A possible by
  adding external transistors

- Input voltage 40V max
- Output voltage adjustable from 2V to 37V
- Can be used as either a linear or a switching
  regulator.

The LM723/LM723C is also useful in a wide range
of other applications such as a shunt regulator, a
current regulator or a temperature controller.

The LM723C is identical to the LM723 except
that the LM723C has its performance guaranteed
over a 0°C to 70°C temperature range, instead of
−55°C to +125°C.

Schematic and Connection Diagrams *

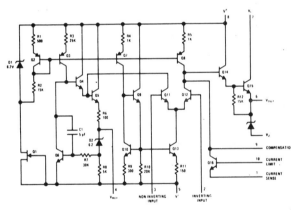

Dual-In-Line Package

Order Number LM723CN
See NS Package N14A
Order Number LM723J or LM723CJ
See NS Package J14A

Metal Can Package

Note: Pin 5 connected to case.
TOP VIEW
Order Number LM723H or LM723CH
See NS Package H10C

Equivalent Circuit*

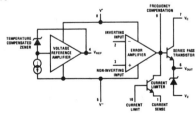

*Pin numbers refer to metal can package.

Absolute Maximum Ratings

Pulse Voltage from V⁺ to V⁻ (50 ms)	50V
Continuous Voltage from V⁺ to V⁻	40V
Input-Output Voltage Differential	40V
Maximum Amplifier Input Voltage (Either Input)	7.5V
Maximum Amplifier Input Voltage (Differential)	5V
Current from V_Z	25 mA
Current from V_{REF}	15 mA
Internal Power Dissipation Metal Can (Note 1)	800 mW
Cavity DIP (Note 1)	900 mW
Molded DIP (Note 1)	660 mW
Operating Temperature Range LM723	-55°C to $+125^\circ$C
LM723C	0°C to $+70^\circ$C
Storage Temperature Range Metal Can	-65°C to $+150^\circ$C
DIP	-55°C to $+125^\circ$C
Lead Temperature (Soldering, 10 sec)	300°C

Electrical Characteristics (Note 2)

PARAMETER	CONDITIONS	LM723			LM723C			UNITS
		MIN	TYP	MAX	MIN	TYP	MAX	
Line Regulation	V_{IN} = 12V to V_{IN} = 15V		.01	0.1		.01	0.1	% V_{OUT}
	-55°C $\leq T_A \leq +125^\circ$C			0.3				% V_{OUT}
	0°C $\leq T_A \leq +70^\circ$C						0.3	% V_{OUT}
	V_{IN} = 12V to V_{IN} = 40V		.02	0.2		0.1	0.5	% V_{OUT}
Load Regulation	I_L = 1 mA to I_L = 50 mA		.03	0.15		.03	0.2	% V_{OUT}
	-55°C $\leq T_A \leq +125^\circ$C			0.6				%V_{OUT}
	0°C $\leq T_A \leq +70^\circ$C						0.6	%V_{OUT}
Ripple Rejection	f = 50 Hz to 10 kHz, C_{REF} = 0		74			74		dB
	f = 50 Hz to 10 kHz, C_{REF} = 5 μF		86			86		dB
Average Temperature Coefficient of Output Voltage	-55°C $\leq T_A \leq +125^\circ$C		.002	.015				%/$^\circ$C
	0°C $\leq T_A \leq +70^\circ$C					.003	.015	%/$^\circ$C
Short Circuit Current Limit	R_{SC} = 10Ω, V_{OUT} = 0		65			65		mA
Reference Voltage		6.95	7.15	7.35	6.80	7.15	7.50	V
Output Noise Voltage	BW = 100 Hz to 10 kHz, C_{REF} = 0		20			20		μVrms
	BW = 100 Hz to 10 kHz, C_{REF} = 5 μF		2.5			2.5		μVrms
Long Term Stability			0.1			0.1		%/1000 hrs
Standby Current Drain	I_L = 0, V_{IN} = 30V		1.3	3.5		1.3	4.0	mA
Input Voltage Range		9.5		40	9.5		40	V
Output Voltage Range		2.0		37	2.0		37	V
Input-Output Voltage Differential		3.0		38	3.0		38	V

Note 1: See derating curves for maximum power rating above 25°C.

Note 2: Unless otherwise specified, T_A = 25°C, V_{IN} = V⁺ = V_C = 12V, V⁻ = 0, V_{OUT} = 5V, I_L = 1 mA, R_{SC} = 0, C_1 = 100 pF, C_{REF} = 0 and divider impedance as seen by error amplifier $\leq$ 10 kΩ connected as shown in Figure 1. Line and load regulation specifications are given for the condition of constant chip temperature. Temperature drifts must be taken into account separately for high dissipation conditions.

Note 3: L_1 is 40 turns of No. 20 enameled copper wire wound on Ferroxcube P36/22-3B7 pot core or equivalent with 0.009 in. air gap.

Note 4: Figures in parentheses may be used if R1/R2 divider is placed on opposite input of error amp.

Note 5: Replace R1/R2 in figures with divider shown in Figure 13.

Note 6: V⁺ must be connected to a +3V or greater supply.

Note 7: For metal can applications where V_Z is required, an external 6.2 volt zener diode should be connected in series with V_{OUT}.

Maximum Power Ratings

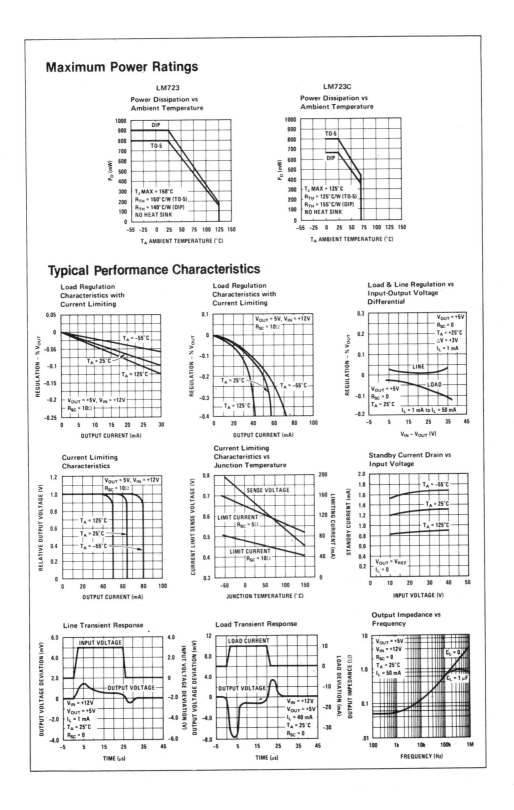

LM723

Power Dissipation vs Ambient Temperature

LM723C

Power Dissipation vs Ambient Temperature

Typical Performance Characteristics

Load Regulation Characteristics with Current Limiting

Load Regulation Characteristics with Current Limiting

Load & Line Regulation vs Input-Output Voltage Differential

Current Limiting Characteristics

Current Limiting Characteristics vs Junction Temperature

Standby Current Drain vs Input Voltage

Line Transient Response

Load Transient Response

Output Impedance vs Frequency

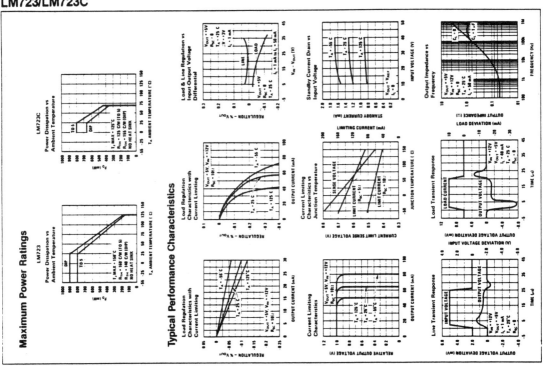

Typical Applications

FIGURE 1. Basic Low Voltage Regulator (VOUT = 2 to 7 Volts)

FIGURE 2. Basic High Voltage Regulator (VOUT = 7 to 37 Volts)

FIGURE 3. Negative Voltage Regulator

FIGURE 4. Positive Voltage Regulator (External NPN Pass Transistor)

Maximum Power Ratings

Typical Performance Characteristics

Operational Amplifiers/Buffers

LM741/LM741A/LM741C/LM741E Operational Amplifier

General Description

The LM741 series are general purpose operational amplifiers which feature improved performance over industry standards like the LM709. They are direct, plug-in replacements for the 709C, LM201, MC1439 and 748 in most applications.

The amplifiers offer many features which make their application nearly foolproof: overload pro-

tection on the input and output, no latch-up when the common mode range is exceeded, as well as freedom from oscillations.

The LM741C/LM741E are identical to the LM741/LM741A except that the LM741C/LM741E have their performance guaranteed over a 0°C to +70°C temperature range, instead of −55°C to +125°C.

Schematic and Connection Diagrams (Top Views)

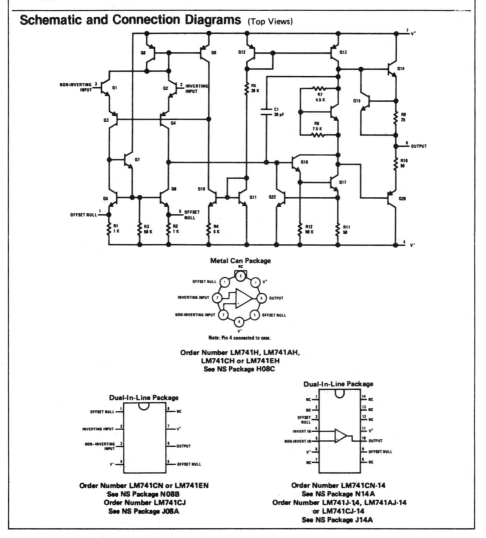

Metal Can Package

Note: Pin 4 connected to case.

Order Number LM741H, LM741AH, LM741CH or LM741EH
See NS Package H08C

Dual-In-Line Package

Order Number LM741CN or LM741EN
See NS Package N08B
Order Number LM741CJ
See NS Package J08A

Dual-In-Line Package

Order Number LM741CN-14
See NS Package N14A
Order Number LM741J-14, LM741AJ-14
or LM741CJ-14
See NS Package J14A

Absolute Maximum Ratings

	LM741A	LM741E	LM741	LM741C
Supply Voltage	±22V	±22V	±22V	±18V
Power Dissipation (Note 1)	500 mW	500 mW	500 mW	500 mW
Differential Input Voltage	±30V	±30V	±30V	±30V
Input Voltage (Note 2)	±15V	±15V	±15V	±15V
Output Short Circuit Duration	Indefinite	Indefinite	Indefinite	Indefinite
Operating Temperature Range	−55°C to +125°C	0°C to +70°C	−55°C to +125°C	0°C to +70°C
Storage Temperature Range	−65°C to +150°C	−65°C to +150°C	−65°C to +150°C	−65°C to +150°C
Lead Temperature (Soldering, 10 seconds)	300°C	300°C	300°C	300°C

Electrical Characteristics (Note 3)

PARAMETER	CONDITIONS	LM741A/LM741E			LM741			LM741C			UNITS
		MIN	TYP	MAX	MIN	TYP	MAX	MIN	TYP	MAX	
Input Offset Voltage	$T_A = 25°C$										
	$R_S \leq 10\ k\Omega$					1.0	5.0		2.0	6.0	mV
	$R_S \leq 50\Omega$		0.8	3.0							mV
	$T_{AMIN} \leq T_A \leq T_{AMAX}$										
	$R_S \leq 50\Omega$			4.0							mV
	$R_S \leq 10\ k\Omega$						6.0			7.5	mV
Average Input Offset Voltage Drift				15							$\mu V/°C$
Input Offset Voltage Adjustment Range	$T_A = 25°C$, $V_S = \pm20V$	±10				±15			±15		mV
Input Offset Current	$T_A = 25°C$		3.0	30		20	200		20	200	nA
	$T_{AMIN} \leq T_A \leq T_{AMAX}$			70		85	500			300	nA
Average Input Offset Current Drift				0.5							$nA/°C$
Input Bias Current	$T_A = 25°C$		30	80		80	500		80	500	nA
	$T_{AMIN} \leq T_A \leq T_{AMAX}$			0.210			1.5			0.8	μA
Input Resistance	$T_A = 25°C$, $V_S = \pm20V$	1.0	6.0		0.3	2.0		0.3	2.0		$M\Omega$
	$T_{AMIN} \leq T_A \leq T_{AMAX}$, $V_S = \pm20V$	0.5									$M\Omega$
Input Voltage Range	$T_A = 25°C$							±12	±13		V
	$T_{AMIN} \leq T_A \leq T_{AMAX}$				±12	±13					V
Large Signal Voltage Gain	$T_A = 25°C$, $R_L \geq 2\ k\Omega$										
	$V_S = \pm20V$, $V_O = \pm15V$	50									V/mV
	$V_S = \pm15V$, $V_O = \pm10V$				50	200		20	200		V/mV
	$T_{AMIN} \leq T_A \leq T_{AMAX}$, $R_L \geq 2\ k\Omega$,										
	$V_S = \pm20V$, $V_O = \pm15V$	32									V/mV
	$V_S = \pm15V$, $V_O = \pm10V$				25			15			V/mV
	$V_S = \pm5V$, $V_O = \pm2V$	10									V/mV
Output Voltage Swing	$V_S = \pm20V$										
	$R_L \geq 10\ k\Omega$	±16									V
	$R_L \geq 2\ k\Omega$	±15									V
	$V_S = \pm15V$										
	$R_L \geq 10\ k\Omega$				±12	±14		±12	±14		V
	$R_L \geq 2\ k\Omega$				±10	±13		±10	±13		V
Output Short Circuit Current	$T_A = 25°C$	10	25	35		25			25		mA
	$T_{AMIN} \leq T_A \leq T_{AMAX}$	10		40							mA
Common-Mode Rejection Ratio	$T_{AMIN} \leq T_A \leq T_{AMAX}$										
	$R_S \leq 10\ k\Omega$, $V_{CM} = \pm12V$				70	90		70	90		dB
	$R_S \leq 50\ k\Omega$, $V_{CM} = \pm12V$	80	95								dB

Electrical Characteristics (Continued)

PARAMETER	CONDITIONS	LM741A/LM741E			LM741			LM741C			UNITS
		MIN	TYP	MAX	MIN	TYP	MAX	MIN	TYP	MAX	
Supply Voltage Rejection Ratio	$T_{AMIN} \leq T_A \leq T_{AMAX}$, $V_S = \pm20V$ to $V_S = \pm5V$										
	$R_S \leq 50\Omega$	86	96								dB
	$R_S \leq 10\,k\Omega$				77	96		77	96		dB
Transient Response	$T_A = 25°C$, Unity Gain										
Rise Time			0.25	0.8		0.3			0.3		μs
Overshoot			6.0	20		5			5		%
Bandwidth (Note 4)	$T_A = 25°C$	0.437	1.5								MHz
Slew Rate	$T_A = 25°C$, Unity Gain	0.3	0.7			0.5			0.5		V/μs
Supply Current	$T_A = 25°C$					1.7	2.8		1.7	2.8	mA
Power Consumption	$T_A = 25°C$										
	$V_S = \pm20V$		80	150							mW
	$V_S = \pm15V$					50	85		50	85	mW
LM741A	$V_S = \pm20V$										
	$T_A = T_{AMIN}$			165							mW
	$T_A = T_{AMAX}$			135							mW
LM741E	$V_S = \pm20V$			150							mW
	$T_A = T_{AMIN}$			150							mW
	$T_A = T_{AMAX}$			150							mW
LM741	$V_S = \pm15V$										
	$T_A = T_{AMIN}$					60	100				mW
	$T_A = T_{AMAX}$					45	75				mW

Note 1: The maximum junction temperature of the LM741/LM741A is 150°C, while that of the LM741C/LM741E is 100°C. For operation at elevated temperatures, devices in the TO-5 package must be derated based on a thermal resistance of 150°C/W junction to ambient, or 45°C/W junction to case. The thermal resistance of the dual-in-line package is 100°C/W junction to ambient.

Note 2: For supply voltages less than ±15V, the absolute maximum input voltage is equal to the supply voltage.

Note 3: Unless otherwise specified, these specifications apply for $V_S = \pm15V$, $-55°C \leq T_A \leq +125°C$ (LM741/LM741A). For the LM741C/LM741E, these specifications are limited to $0°C \leq T_A \leq +70°C$.

Note 4: Calculated value from: BW (MHz) = 0.35/Rise Time(μs).

DAC0800(LMDAC08) 8-bit digital-to-analog converter

general description

The DAC08 is a monolithic 8-bit high-speed current-output digital-to-analog converter (DAC) featuring typical settling times of 100 ns. When used as a multiplying DAC, monotonic performance over a 40 to 1 reference current range is possible. The DAC08 also features high compliance complementary current outputs to allow differential output voltages of 20 Vp-p with simple resistor loads as shown in *Figure 1*. The reference-to-full-scale current matching of better than ±1 LSB eliminates the need for full scale trims in most applications while the nonlinearities of better than ±0.1% over temperature minimizes system error accumulations.

The noise immune inputs of the DAC08 will accept TTL levels with the logic threshold pin, V_{LC}, pin 1 grounded. Simple adjustments of the V_{LC} potential allow direct interface to all logic families. The performance and characteristics of the device are essentially unchanged over the full ±4.5V to ±18V power supply range; power dissipation is only 33 mW with ±5V supplies and is independent of the logic input states.

The DAC0800L, DAC0800LC, DAC0801LC are a direct replacement for the DAC08, DAC08E and DAC08C, respectively.

features

- Fast settling output current 100 ns
- Full scale error ±1 LSB
- Nonlinearity over temperature ±0.1%
- Full scale current drift ±10 ppm/°C
- High output compliance −10V to +18V
- Complementary current outputs
- Interface directly with TTL, CMOS, PMOS and others
- 2 quadrant wide range multiplying capability
- Wide power supply range ±4.5V to ±18V
- Low power consumption 33 mW at ±5V
- Low cost

typical applications

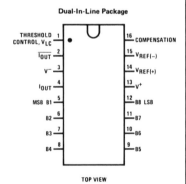

FIGURE 1. ±20 Vp-p Output Digital-to-Analog Converter

connection diagram

Dual-In-Line Package

TOP VIEW

ordering information

NONLINEARITY	TEMPERATURE RANGE	ORDER NUMBERS*					
		D PACKAGE		J PACKAGE		N PACKAGE	
±0.19% FS	−55°C ≤ T_A ≤ +125°C	DAC0800LD	LMDAC08D	DAC0800LJ	LMDAC08J		
±0.19% FS	0°C ≤ T_A ≤ +70°C	DAC0800LCD	LMDAC08ED	DAC0800LCJ	LMDAC08EJ	DAC0800LCN	LMDAC08EN
±0.39% FS	0°C ≤ T_A ≤ +70°C	DAC0801LCD	LMDAC08CD	DAC0801LCJ	LMDAC08CJ	DAC0801LCN	LMDAC08CN

*Note. Devices may be ordered by using either order number.

absolute maximum ratings

Supply Voltage	±18V or 36V
Power Dissipation (Note 1)	500 mW
Reference Input Differential Voltage (V14 to V15)	V^- to V^+
Reference Input Common-Mode Range (V14, V15)	V^- to V^+
Reference Input Current	5 mA
Logic Inputs	V^- to V^- plus 36V
Analog Current Outputs	*Figure 24*
Operating Temperature	0°C to +70°C
Storage Temperature	−65°C to +150°C
Lead Temperature (Soldering, 10 seconds)	300°C

operating conditions

DEVICE	T_{MIN}	T_{MAX}
DAC0800L, LMDAC08	−55°C	+125°C
DAC0800LC, LMDAC08E	0°C	+70°C
DAC0801LC, LMDAC08C	0°C	+70°C

electrical characteristics

(V_S = ±15V, I_{REF} = 2 mA, $T_{MIN} \leq T_A \leq T_{MAX}$ unless otherwise specified. Output characteristics refer to both I_{OUT} and $\overline{I_{OUT}}$.)

	PARAMETER	CONDITIONS	DAC0800L/DAC0800LC MIN	TYP	MAX	DAC0801LC MIN	TYP	MAX	UNITS
	Resolution		8	8	8	8	8	8	Bits
	Monotonicity		8	8	8	8	8	8	Bits
	Nonlinearity				±0.19			±0.39	%FS
t_s	Settling Time	To ±1/2 LSB, All Bits Switched "ON" or "OFF", T_A = 25°C							
		DAC0800L		100	135				ns
		DAC0800LC, DAC0801LC		100	150		100	150	ns
t_{PLH},t_{PH}	Propagation Delay	T_A = 25°C							
	Each Bit			35	60		35	60	ns
	All Bits Switched			35	60		35	60	ns
TCI_{FS}	Full Scale Tempco			±10	±50		±10	±80	ppm/°C
V_{OC}	Output Voltage Compliance	Full Scale Current Change < 1/2 LSB, R_{OUT} > 20 MΩ	−10		18	−10		18	V
I_{FS4}	Full Scale Current	V_{REF} = 10.000V, R14 = 5.000 kΩ R15 = 5.000 kΩ, T_A = 25°C	1.94	1.99	2.04	1.94	1.99	2.04	mA
I_{FSS}	Full Scale Symmetry	$I_{FS4} - I_{FS2}$		±1	±8.0		±2	±16	µA
I_{ZS}	Zero Scale Current			0.2	2.0		0.2	4.0	µA
I_{FSR}	Output Current Range	V^- = −5V	0	2.0	2.1	0	2.0	2.1	mA
		V^- = −7V to −18V	0	2.0	4.2	0	2.0	4.2	mA
	Logic Input Levels								
V_{IL}	Logic "0"	V_{LC} = 0V			0.8			0.8	V
V_{IH}	Logic "1"		2.0			2.0			V
	Logic Input Current	V_{LC} = 0V							
I_{IL}	Logic "0"	$-10V \leq V_{IN} \leq +0.8V$		−2.0	−10		−2.0	−10	µA
I_{IH}	Logic "1"	$2V \leq V_{IN} \leq +18V$		0.002	10		0.002	10	µA
V_{IS}	Logic Input Swing	V^- = −15V	−10		18	−10		18	V
V_{THR}	Logic Threshold Range	V_S = ±15V	−10		13.5	−10		13.5	V
I_{15}	Reference Bias Current			−1.0	−3.0		−1.0	−3.0	µA
dI/dt	Reference Input Slew Rate	*(Figure 24)*	8.0			8.0			mA/µs
$PSSI_{FS+}$	Power Supply Sensitivity	$4.5V \leq V^+ \leq 18V$		0.0001	0.01		0.0001	0.01	%/%
$PSSI_{FS-}$		$-4.5V \leq V^- \leq 18V$ I_{REF} = 1 mA		0.0001	0.01		0.0001	0.01	%/%
	Power Supply Current	V_S = ±5V, I_{REF} = 1 mA							
$I+$				2.3	3.8		2.3	3.8	mA
$I-$				−4.3	−5.8		−4.3	−5.8	mA
		V_S = 5V, −15V, I_{REF} = 2 mA							
$I+$				2.4	3.8		2.4	3.8	mA
$I-$				−6.4	−7.8		−6.4	−7.8	mA
		V_S = ±15V, I_{REF} = 2 mA							
$I+$				2.5	3.8		2.5	3.8	mA
$I-$				−6.5	−7.8		−6.5	−7.8	mA
P_D	Power Dissipation	±5V, I_{REF} = 1 mA		33	48		33	48	mW
		5V, −15V, I_{REF} = 2 mA		108	136		108	136	mW
		±15V, I_{REF} = 2 mA		135	174		135	174	mW

Note 1: The maximum junction temperature of the DAC0800 and DAC0801 is 100°C. For operating at elevated temperatures, devices in the dual-in-line J or D package must be derated based on a thermal resistance of 100°C/W, junction to ambient, 175°C/W for the molded dual-in-line N, package.

block diagram

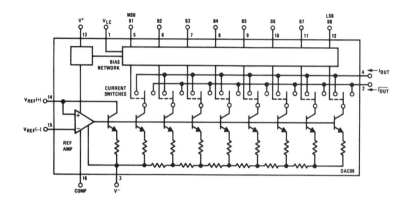

equivalent circuit

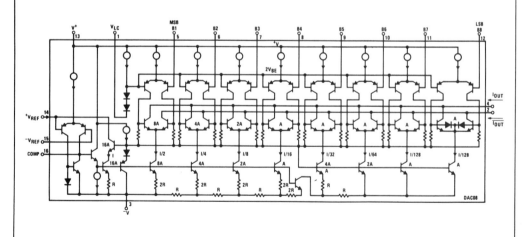

FIGURE 2

typical performance characteristics

Full Scale Current vs Reference Current

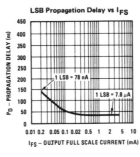

FIGURE 3

LSB Propagation Delay vs I_{FS}

FIGURE 4

Reference Input Frequency Response

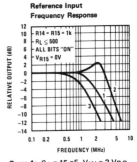

Curve 1: $C_C = 15$ pF, $V_{IN} = 2$ Vp-p centered at 1V.

Curve 2: $C_C = 15$ pF, $V_{IN} = 50$ mVp-p centered at 200 mV.

Curve 3: $C_C = 0$ pF, $V_{IN} = 100$ mVp-p at 0V and applied through 50 Ω connected to pin 14. 2V applied to R14.

FIGURE 5

Reference Amp Common-Mode Range

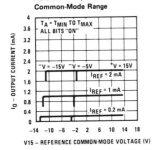

Note. Positive common-mode range is always (V+) − 1.5V.

FIGURE 6

Logic Input Current vs Input Voltage

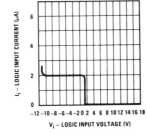

FIGURE 7

$V_{TH} − V_{LC}$ vs Temperature

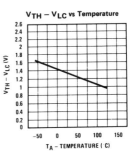

FIGURE 8

Output Current vs Output Voltage (Output Voltage Compliance)

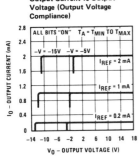

FIGURE 9

Output Voltage Compliance vs Temperature

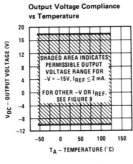

FIGURE 10

Bit Transfer Characteristics

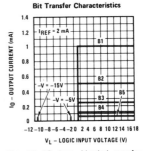

Note. B1—B8 have identical transfer characteristics. Bits are fully switched with less than 1/2 LSB error, at less than ±100 mV from actual threshold. These switching points are guaranteed to lie between 0.8 and 2V over the operating temperature range ($V_{LC} = 0$V).

FIGURE 11

371

**A to D, D to A**

ADC0801, ADC0802, ADC0803, ADC0804 8-Bit μP Compatible A/D Converters

General Description

The ADC0801, ADC0802, ADC0803, ADC0804 are CMOS 8-bit, successive approximation A/D converters which use a modified potentiometric ladder—similar to the 256R products. They are designed to meet the NSC MICROBUS™ standard to allow operation with the 8080A control bus, and TRI-STATE® output latches directly drive the data bus. These A/Ds appear like memory locations or I/O ports to the microprocessor and no interfacing logic is needed.

A new differential analog voltage input allows increasing the common-mode rejection and offsetting the analog zero input voltage value. In addition, the voltage reference input can be adjusted to allow encoding any smaller analog voltage span to the full 8 bits of resolution.

Features

■ MICROBUS (8080A) compatible—no interfacing logic needed

■ Easy interface to all microprocessors, or operates "stand alone"

■ Differential analog voltage inputs

■ Logic inputs and outputs meet T^2L voltage level specifications

■ Works with 2.5V (LM336) voltage reference

■ On-chip clock generator

■ 0V to 5V analog input voltage range with single 5V supply

■ No zero adjust required

■ 0.3" standard width 20-pin DIP package

Key Specifications

■ Resolution	8 bits
■ Total error	±1/4 LSB, ±1/2 LSB and ±1 LSB
■ Conversion time	100 μs
■ Access time	135 ns
■ Single supply	5 V_{DC}
■ Operates ratiometrically or with 5 V_{DC}, 2.5 V_{DC}, or analog span adjusted voltage reference	

Typical Application

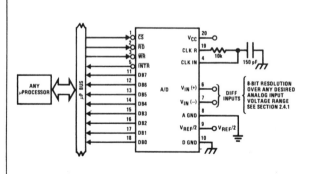

Connection Diagram

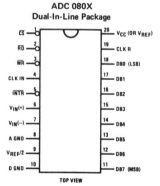

ADC 080X
Dual-In-Line Package

TOP VIEW

Absolute Maximum Ratings (Notes 1 and 2)

Supply Voltage (V_{CC}) (Note 3)	6.5V
Voltage at Any Input	−0.3V to (V_{CC} +0.3V)
Storage Temperature Range	−65°C to +150°C
Package Dissipation at T_A = 25°C	875 mW
Lead Temperature (Soldering, 10 seconds)	300°C

Operating Ratings (Notes 1 and 2)

Temperature Range (Note 1)	$T_{MIN} \leq T_A \leq T_{MAX}$
ADC0801/02/03 LD	−55°C $\leq T_A \leq$ +125°C
ADC0801/02/03/04 LCD	−40°C $\leq T_A \leq$ +85°C
ADC0801/02/03/04 LCN	0°C $\leq T_A \leq$ 70°C
Range of V_{CC} (Note 1)	4.5 V_{DC} to 6.3 V_{DC}

Electrical Characteristics

Converter Specifications:
V_{CC} = 5 V_{DC}, $V_{REF}/2$ = 2.500 V_{DC}, $T_{MIN} \leq T_A \leq T_{MAX}$ and f_{CLK} = 640 kHz unless otherwise stated.

PARAMETER	CONDITIONS	MIN	TYP	MAX	UNITS
ADC0801:					
Total Adjusted Error (Note 8)	With Full-Scale Adj.			±1/4	LSB
ADC0802:					
Total Unadjusted Error	Completely Unadjusted			±1/2	LSB
(Note 8)					
ADC0803:					
Total Adjusted Error (Note 8)	With Full-Scale Adj.			±1/2	LSB
ADC0804:					
Total Unadjusted Error	Completely Unadjusted			±1	LSB
(Note 8)					
$V_{REF}/2$ Input Resistance	Input Resistance at Pin 9	1.0	1.3		kΩ
Analog Input Voltage Range	(Note 4) V(+) or V(−)	Gnd−0.05		V_{CC}+0.05	V_{DC}
DC Common-Mode Rejection	Over Analog Input Voltage Range		±1/16	±1/8	LSB
Power Supply Sensitivity	V_{CC} = 5 V_{DC} ±10% Over Allowed V_{IN}(+) and V_{IN}(−) Voltage Range (Note 4)		±1/16	±1/8	LSB

Electrical Characteristics

Timing Specifications: V_{CC} = 5 V_{DC} and T_A = 25°C unless otherwise noted.

	PARAMETER	CONDITIONS	MIN	TYP	MAX	UNITS
f_{CLK}	Clock Frequency	V_{CC} = 6V, (Note 5)	100	640	1280	kHz
		V_{CC} = 5V	100	640	800	kHz
T_C	Conversion Time	(Note 6)	66		73	$1/f_{CLK}$
CR	Conversion Rate In Free-Running Mode	$\overline{INTR}$ tied to $\overline{WR}$ with $\overline{CS}$ = 0 V_{DC}, f_{CLK} = 640 kHz			8770	conv/s
$t_{W(\overline{WR})L}$	Width of $\overline{WR}$ Input (Start Pulse Width)	$\overline{CS}$ = 0 V_{DC} (Note 7)	100			ns
t_{ACC}	Access Time (Delay from Falling Edge of $\overline{RD}$ to Output Data Valid)	C_L = 100 pF (Use Bus Driver IC for Larger C_L)		135	200	ns
t_{1H}, t_{0H}	TRI-STATE Control (Delay from Rising Edge of $\overline{RD}$ to Hi-Z State)	C_L = 10 pF, R_L = 10k (See TRI-STATE Test Circuits)		125	250	ns
t_{WI}	Delay from Falling Edge of $\overline{WR}$ to Reset of $\overline{INTR}$			300	450	ns
C_{IN}	Input Capacitance of Logic Control Inputs			5	7.5	pF
C_{OUT}	TRI-STATE Output Capacitance (Data Buffers)			5	7.5	pF

Electrical Characteristics

Digital Levels and DC Specifications:
V_{CC} = 5 V_{DC} and $T_{MIN} \leq T_A \leq T_{MAX}$, unless otherwise noted.

PARAMETER		CONDITIONS	MIN	TYP	MAX	UNITS
CONTROL INPUTS [Note: CLK IN (pin 4) is the input of a Schmitt trigger circuit and is therefore specified separately]						
V_{IN} (1)	Logical "1" Input Voltage (Except Pin 4 CLK IN)	V_{CC} = 5.25 V_{DC}	2.0		15	V_{DC}
V_{IN} (0)	Logical "0" Input Voltage (Except Pin 4 CLK IN)	V_{CC} = 4.75 V_{DC}			0.8	V_{DC}
V_{T+}	CLK IN (Pin 4) Positive Going Threshold Voltage		2.7	3.1	3.5	V_{DC}
V_{T-}	CLK IN (Pin 4) Negative Going Threshold Voltage		1.5	1.8	2.1	V_{DC}
V_H	CLK IN (Pin 4) Hysteresis $(V_{T+}) - (V_{T-})$		0.6	1.3	2.0	V_{DC}
I_{IN} (1)	Logical "1" Input Current (All Inputs)	V_{IN} = 5 V_{DC}		0.005	1	μA_{DC}
I_{IN} (0)	Logical "0" Input Current (All Inputs)	V_{IN} = 0 V_{DC}	−1	−0.005		μA_{DC}
I_{CC}	Supply Current (Includes Ladder Current)	f_{CLK} = 640 kHz, T_A = 25°C and $\overline{CS}$ = "1"		1.3	2.5	mA
DATA OUTPUTS AND $\overline{INTR}$						
V_{OUT} (0)	Logical "0" Output Voltage	I_O = 1.6 mA V_{CC} = 4.75 V_{DC}			0.4	V_{DC}
V_{OUT} (1)	Logical "1" Output Voltage	I_O = −360 μA V_{CC} = 4.75 V_{DC}	2.4			V_{DC}
I_{OUT}	TRI-STATE Disabled Output Leakage (All Data Buffers)	V_{OUT} = 0 V_{DC}	−3			μA_{DC}
		V_{OUT} = 5 V_{DC}			3	μA_{DC}
	Output Short Circuit Current	T_A = 25°C				
I_{SOURCE}		V_{OUT} Short to Gnd	4.5	6		mA_{DC}
I_{SINK}		V_{OUT} Short to V_{CC}	9.0	16		mA_{DC}

Note 1: Absolute maximum ratings are those values beyond which the life of the device may be impaired.

Note 2: All voltages are measured with respect to Gnd, unless otherwise specified. The separate A Gnd point should always be wired to the D Gnd.

Note 3: A zener diode exists, internally, from V_{CC} to Gnd and has a typical breakdown voltage of 7 V_{DC}.

Note 4: For $V_{IN}(-) \geq V_{IN}(+)$ the digital output code will be 0000 0000. Two on-chip diodes are tied to each analog input (see block diagram) which will forward conduct for analog input voltages one diode drop below ground or one diode drop greater than the V_{CC} supply. Be careful, during testing at low V_{CC} levels (4.5V), as high level analog inputs (5V) can cause this input diode to conduct—especially at elevated temperatures, and cause errors for analog inputs near full-scale. The spec allows 50 mV forward bias of either diode. This means that as long as the analog V_{IN} does not exceed the supply voltage by more than 50 mV, the output code will be correct. To achieve an absolute 0 V_{DC} to 5 V_{DC} input voltage range will therefore require a minimum supply voltage of 4.950 V_{DC} over temperature variations, initial tolerance and loading.

Note 5: With V_{CC} = 6V, the digital logic interfaces are no longer TTL compatible.

Note 6: With an asynchronous start pulse, up to 8 clock periods may be required before the internal clock phases are proper to start the conversion process.

Note 7: The $\overline{CS}$ input is assumed to bracket the $\overline{WR}$ strobe input and therefore timing is dependent on the $\overline{WR}$ pulse width. An arbitrarily wide pulse width will hold the converter in a reset mode and the start of conversion is initiated by the low to high transition of the $\overline{WR}$ pulse (see timing diagrams).

Note 8: None of these A/Ds requires a zero adjust. However, if an all zero code is desired for an analog input other than 0.0V, or if a narrow full-scale span exists (for example: 0.5V to 4.0V full-scale) the $V_{IN}(-)$ input can be adjusted to achieve this. See section 2.5 and *Figure 19.*

Typical Performance Characteristics

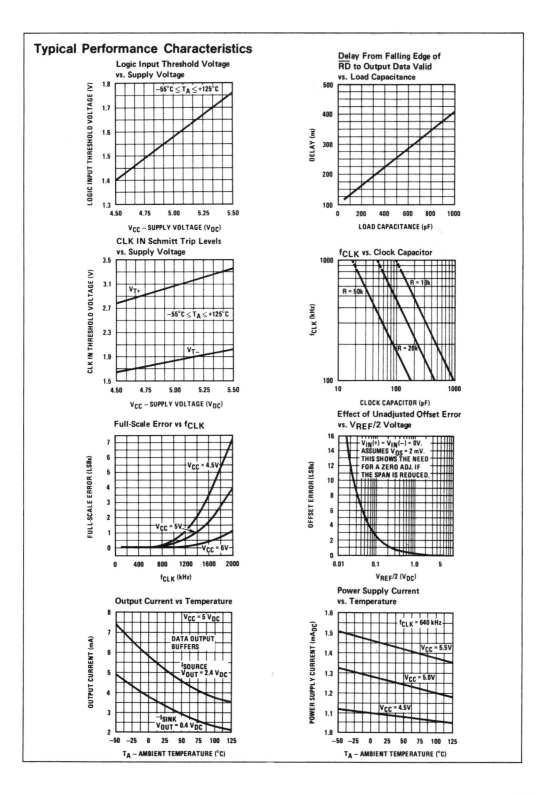

Logic Input Threshold Voltage vs. Supply Voltage

Delay From Falling Edge of $\overline{RD}$ to Output Data Valid vs. Load Capacitance

CLK IN Schmitt Trip Levels vs. Supply Voltage

f_{CLK} vs. Clock Capacitor

Full-Scale Error vs f_{CLK}

Effect of Unadjusted Offset Error vs. $V_{REF}/2$ Voltage

Output Current vs Temperature

Power Supply Current vs. Temperature

TRI-STATE® Test Circuits and Waveforms

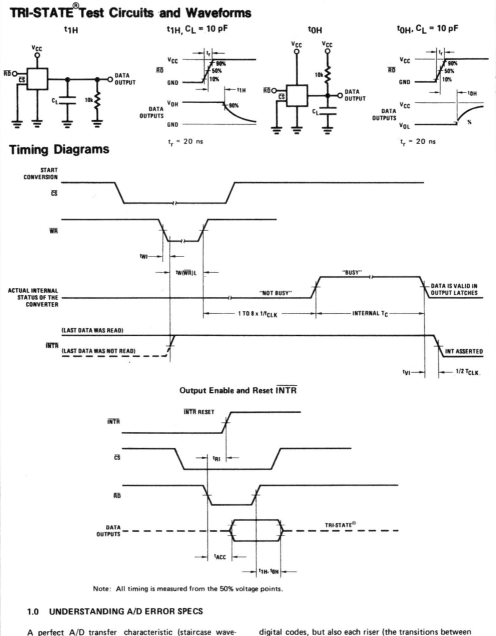

Timing Diagrams

Output Enable and Reset INTR

Note: All timing is measured from the 50% voltage points.

1.0 UNDERSTANDING A/D ERROR SPECS

A perfect A/D transfer characteristic (staircase waveform) is shown in *Figure 1a*. The horizontal scale is analog input voltage and the particular points labeled are in steps of 1 LSB (19.53 mV with 2.5V tied to the $V_{REF}/2$ pin). The digital output codes which correspond to these inputs are shown as D−1, D, and D+1. For the perfect A/D, not only will center-value (A−1, A, A+1, . . .) analog inputs produce the correct output

digital codes, but also each riser (the transitions between adjacent output codes) will be located ±1/2 LSB away from each center-value. As shown, the risers are ideal and have no width. Correct digital output codes will be provided for a range of analog input voltages which extend ±1/2 LSB from the ideal center-values. Each tread (the range of analog input voltage which provides the same digital output code) is therefore 1 LSB wide.

Figure 1b shows worst case error plot for ADC0801. All center-valued inputs are guaranteed to produce the correct output codes and the adjacent risers are guaranteed to be no closer to the center-value points than ±1/4 LSB. In other words, if we apply an analog input equal to the center-value ±1/4 LSB, *we guarantee* that the A/D will produce the correct digital code. The maximum range of the position of the code transition is indicated by the horizontal arrow and it is guaranteed to be no more than 1/2 LSB.

The error curve of *Figure 1c* shows worst case error plot for ADC0802. Here we guarantee that if we apply an analog input equal to the LSB analog voltage center-value the A/D will produce the correct digital code.

Next to each transfer function is shown the corresponding error plot. Many people may be more familiar with error plots than transfer functions. The analog input voltage to the A/D is provided by either a linear ramp or by the discrete output steps of a high resolution DAC. Notice that the error is continuously displayed and includes the quantization uncertainty of the A/D. For example the error at point 1 of *Figure 1a* is +1/2 LSB because the digital code appeared 1/2 LSB in advance of the center-value of the tread. The error plots always have a constant negative slope and the abrupt upside steps are always 1 LSB in magnitude.

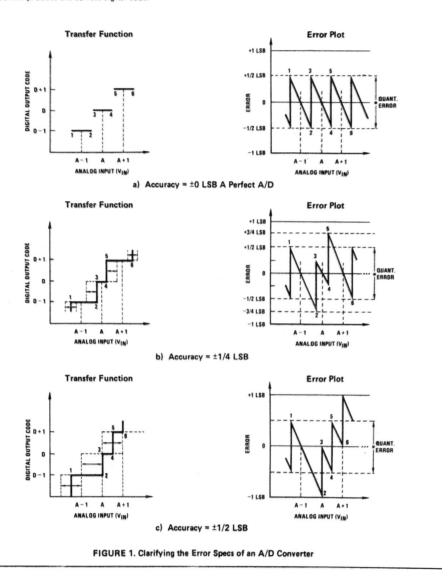

a) Accuracy = ±0 LSB A Perfect A/D

b) Accuracy = ±1/4 LSB

c) Accuracy = ±1/2 LSB

FIGURE 1. Clarifying the Error Specs of an A/D Converter

MICRO-DAC™ DAC0830/0831/0832
8-Bit μP Compatible, Double-Buffered D to A Converters

General Description

The DAC0830 is an advanced CMOS/Si-Cr 8-bit multiplying DAC designed to interface directly with the 8080, 8048, 8085, Z-80, and other popular microprocessors. A deposited silicon-chromium R-2R resistor ladder network divides the reference current and provides the circuit with excellent temperature tracking characteristics (0.05% of Full Scale Range maximum linearity error over temperature). The circuit uses CMOS current switches and control logic to achieve low power consumption and low output leakage current errors. Special circuitry provides TTL logic input voltage level compatibility.

Double buffering allows these DACs to output a voltage corresponding to one digital word while holding the next digital word. This permits the simultaneous updating of any number of DACs.

The DAC0830 series are the 8-bit members of a family of microprocessor-compatible DACs (MICRO-DACs). For applications demanding higher resolution, the DAC1000 series (10-bits) and the DAC1208 and DAC1230 (12-bits) are available alternatives.

BI-FET™ and MICRO DAC™ are trademarks of National Semiconductor Corp.

Features

- Double-buffered, single-buffered or flow-through digital data inputs
- Easy interchange and pin-compatible with 12-bit DAC1230 series
- Direct interface to all popular microprocessors
- Linearity specified with zero and full scale adjust only—NOT BEST STRAIGHT LINE FIT.
- Works with ±10V reference-full 4-quadrant multiplication
- Can be used in the voltage switching mode
- Logic inputs which meet TTL voltage level specs (1.4V logic threshold)
- Operates "STAND ALONE" (without μP) if desired

Key Specifications

- Current settling time   1 μs
- Resolution   8-bits
- Linearity   8, 9, or 10 bits
  (guaranteed over temp.)
- Gain Tempco   0.0002% FS/°C
- Low power dissipation   20 mW
- Single power supply   5 to 15 V_{DC}

Typical Application

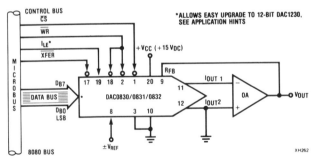

*ALLOWS EASY UPGRADE TO 12-BIT DAC1230, SEE APPLICATION HINTS

Connection Diagram Top View

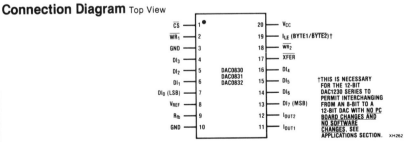

$\overline{CS}$	1	20 — V_{CC}
$\overline{WR}_1$	2	19 — I_{LE} (BYTE1/BYTE2)†
GND	3	18 — $\overline{WR}_2$
DI_3	4	17 — $\overline{XFER}$
DI_2	5	16 — DI_4
DI_1	6	15 — DI_5
DI_0 (LSB)	7	14 — DI_6
V_{REF}	8	13 — DI_7 (MSB)
R_{fb}	9	12 — I_{OUT2}
GND	10	11 — I_{OUT1}

DAC0830 / DAC0831 / DAC0832

†THIS IS NECESSARY FOR THE 12-BIT DAC1230 SERIES TO PERMIT INTERCHANGING FROM AN 8-BIT TO A 12-BIT DAC WITH NO PC BOARD CHANGES AND NO SOFTWARE CHANGES, SEE APPLICATIONS SECTION.   XH262

Absolute Maximum Ratings (Notes 1 and 2)

Supply Voltage (V_{CC})	17 V_{DC}
Voltage at any digital input	V_{CC} to GND
Voltage at V_{REF} input	±25 V
Storage temperature range	−65°C to +150°C
Package dissipation at T_A = 25°C (Note 3)	500 mW
DC voltage applied to I_{OUT1} or I_{OUT2} (Note 4)	−100 mV to V_{CC}
Lead temperature (soldering, 10 seconds)	300°C

Operating Ratings

Temperature Range	$T_{MIN} \leqslant T_A \leqslant T_{MAX}$
Part numbers with 'LCN' suffix	0°C to 70°C
Part numbers with 'LCD' suffix	−40°C to +85°C
Part numbers with 'LD' Suffix	−55°C to +125°C
Voltage at any digital input	V_{CC} TO GND

Electrical Characteristics

V_{REF} = 10.000 V_{DC} unless otherwise noted. Boldface limits apply over temperature, $T_{MIN} \leqslant T_A \leqslant T_{MAX}$. For all other limits T_A = 25°C.

Parameter	Conditions	See Note	V_{CC}=12 V_{DC}±5% to 15 V_{DC}±5%			V_{CC} = 5 V_{DC}±5%			Limit Units
			Typ.	Tested Limit (Note 5)	Design Limit (Note 6)	Typ.	Tested Limit (Note 5)	Design Limit (Note 6)	
Converter Characteristics									
Resolution			8	8		8	8		bits
Linearity Error Max.	Zero and full scale adjusted −10V ≤ V_{REF} ≤ +10V	4, 7 8							
	DAC0830 LD & LCD			**0.05**			**0.05**		% FSR
	DAC0832 LD & LCD			**0.2**			**0.2**		% FSR
	DAC0830 LCN		0.05		**0.05**	0.05		**0.05**	% FSR
	DAC0831 LCN		0.1		**0.1**	0.1		**0.1**	% FSR
	DAC0832 LCN		0.2		**0.2**	0.2		**0.2**	% FSR
Differential Nonlinearity Max.	Zero and full scale adjusted −10V ≤ V_{REF} ≤ +10V	4, 7 8							
	DAC0830 LD & LCD			**0.1**			**0.1**		% FSR
	DAC0832 LD & LCD			**0.4**			**0.4**		% FSR
	DAC0830 LCN		0.1		**0.1**	0.1		**0.1**	% FSR
	DAC0831 LCN		0.2		**0.2**	0.2		**0.2**	% FSR
	DAC0832 LCN		0.4		**0.4**	0.4		**0.4**	% FSR
Monotonicity	−10V ≤ V_{REF} ≤ +10V LD & LCD	4, 7		**8**			**8**		bits
	LCN			8	**8**		8	**8**	bits
Gain Error Max.	Using internal R_{fb} −10V ≤ V_{REF} ≤ +10V	7	±0.2	±1		±0.2	±1		% FS
Gain Error Tempco Max.	Using internal R_{fb}		**0.0002**		**0.0006**	**0.0002**		**0.0006**	% FS/°C
Power Supply Rejection	All digital inputs latched high V_{CC} = 14.5V to 15.5V		0.0002						% FSR/V
	11.5V to 12.5V		0.0006						
	4.5V to 5.5V					0.0130			
Reference Input Max.			15	20		15	20		kΩ
Min.			15	10		15	10		kΩ
Output Feedthrough Error	V_{REF} = 20 V_{P-P}, f = 100kHz All data inputs latched low	9	3			3			mV_{P-P}
Output Leakage Current Max. I_{OUT1}	All data inputs LD & LCD latched low LCN	10		**100** 50	**100**		**100** 50	**100**	nA
I_{OUT2}	All data inputs LD & LCD latched high LCN			**100** 50	**100**		**100** 50	**100**	nA
Output Capacitance I_{OUT1} I_{OUT2}	All data inputs latched low		45 115			45 115			pF
I_{OUT1} I_{OUT2}	All data inputs latched high		130 30			130 30			pF

Electrical Characteristics (Continued) $V_{REF} = 10.000\,V_{DC}$ unless otherwise noted. Boldface limits apply over temperature, $T_{MIN} \leqslant T_A \leqslant T_{MAX}$. For all other limits $T_A = 25°C$.

Parameter		Conditions		See Note	$V_{CC}=12\,V_{DC}\pm5\%$ to $15\,V_{DC}\pm5\%$			$V_{CC}=5\,V_{DC}\pm5\%$			Limit Units
					Typ.	Tested Limit (Note 5)	Design Limit (Note 6)	Typ.	Tested Limit (Note 5)	Design Limit (Note 6)	
Digital and DC Characteristics											
Digital Input Voltages	Max.	Logic Low	LD			**0.8**			**0.6**		V_{DC}
			LCD			**0.8**			**0.8**		
			LCN			1.0	**0.8**		1.0	**0.8**	
	Min.	Logic High	LD & LCD			**2.0**			**2.0**		V_{DC}
			LCN			1.9	**2.0**		1.9	**2.0**	
Digital Input Currents Max.		Digital inputs<0.8V	LD & LCD		−50	**−200**		−50	**−200**		μA_{DC}
			LCN			160	**−200**		−160	**−200**	
		Digital inputs>2.0V	LD & LCD		0.1	**+10**		0.1	**+10**		μA_{DC}
			LCN			+8	**+10**		+8	**+10**	
Supply Current Drain Max.			LD & LCD		1.2	**2.0**		1.2	**2.0**		mA
			LCN			1.7	**2.0**		1.7	**2.0**	
AC Characteristics											
t_S	Current Setting Time	$V_{IL}=0V, V_{IH}=5V$			1.0			1.0			μs
t_W	Write and XFER Pulse Width Min.	$V_{IL}=0V, V_{IH}=5V$		11	100 **180**		320 **320**	375 **500**		600 **900**	
t_{DS}	Data Setup Time Min.	$V_{IL}=0V, V_{IH}=5V$			100 **180**		320 **320**	375 **500**		600 **900**	
t_{DH}	Data Hold Time Min.	$V_{IL}=0V, V_{IH}=5V$			**10**		**50**	**10**		**50**	ns
t_{CS}	Control Setup Time Min.	$V_{IL}=0V, V_{IH}=5V$			110 **200**		320 **320**	400 **500**		650 **900**	
t_{CH}	Control Hold Time Min.	$V_{IL}=0V, V_{IH}=5V$					**10**			**10**	

Note 1: "Absolute Maximum Ratings" are those values beyond which the safety of the device cannot be guaranteed. These specifications are not meant to imply that the devices should be operated at these "Absolute Maximum" limits.

Note 2: All voltages are measured with respect to GND, unless otherwise specified.

Note 3: Max. T_J for the D suffix package is 150°C with $\theta_{JA} = 80°C/W$. Max. T_J for the N suffix package is 125°C with $\theta_{JA} = 120°C/W$.

Note 4: For current switching applications, both I_{OUT1} and I_{OUT2} must go to ground or the "Virtual Ground" of an operational amplifier. The linearity error is degraded by approximately $V_{OS} \div V_{REF}$. For example, if $V_{REF} = 10\,V$ then a 1mV offset, V_{OS}, on I_{OUT1} or I_{OUT2} will introduce an additional 0.01% linearity error.

Note 5: Guaranteed and 100% production tested.

Note 6: Guaranteed, but not 100% production tested. These limits are not used to calculate outgoing quality levels.

Note 7: Guaranteed at $V_{REF} = \pm 10\,V_{DC}$ and $V_{REF} = \pm 1\,V_{DC}$.

Note 8: The unit "FSR" stands for "Full Scale Range." "Linearity Error" and "Power Supply Rejection" specs are based on this unit to eliminate dependence on a particular V_{REF} value and to indicate the true performance of the part. The "Linearity Error" specification of the DAC0830 is "0.05% of FSR (MAX)." This guarantees that after performing a zero and full scale adjustment (see Sections 2.5 and 2.6), the plot of the 256 analog voltage outputs will each be within $0.05\% \times V_{REF}$ of a straight line which passes through zero and full scale.

Note 9: To achieve this low feedthrough in the D package, the user must ground the metal lid. If the lid is left floating, the feedthrough is typically 6mV.

Note 10: A 100nA leakage current with $R_{fb} = 20k$ and $V_{REF} = 10\,V$ corresponds to a zero error of $(100 \times 10^{-9} \times 20 \times 10^3) \times 100/10$ which is 0.02% of FS.

Note 11: The entire write pulse must occur within the valid data interval for the specified t_W, t_{DS}, t_{DH}, and t_S to apply.

DT1056/DT1057 DIGITALKER™ Standard Vocabulary Kit

General Description

The DIGITALKER™ is a speech synthesis system consisting of several N-channel MOS integrated circuits. It contains a speech processor chip (SPC) and speech ROM and when used with external filter, amplifier, and speaker, produces a system which generates high quality speech including the natural inflection and emphasis of the original speech. Male, female, and children's voices can be synthesized.

The SPC communicates with the speech ROM, which contains the compressed speech data as well as the frequency and amplitude data required for speech output. Up to 128k bits of speech data can be directly accessed.

With the addition of an external resistor, on-chip debounce is provided for use with a switch interface.

An interrupt is generated at the end of each speech sequence so that several sequences or words can be cascaded to form different speech expressions.

The DT1056/DT1057 is a standard DIGITALKER kit encoded with 131 separate and useful words (see the Master Word List Table I) and when used with the DT1050 Standard Vocabulary Kit, provides a library of 274 useful words. The words have been assigned discrete addresses, making it possible to output single words or words concatenated into phrases or even sentences.

The "voice" output of the DT1056/DT1057 is a highly intelligible male voice. The vocabulary is chosen so that it is applicable to many products and markets.

Features

- Easily adaptable to DT1050 Standard Vocabulary Kit
- 131 useful words
- COPS™ and MICROBUS™ compatible
- Designed to be easily interfaced to other popular microprocessors
- Natural inflection and emphasis of original speech
- Addresses 128k bits of ROM directly
- TTL compatible
- On-chip switch debounce for interfacing to manual switches independent of a microprocessor
- Interrupt capability for cascading words or phrases
- Crystal controlled or externally driven oscillator
- Available in complete kit (DT1056) or speech ROMs only (DT1057)

Applications

- Telecommunications
- Appliance
- Automotive
- Teaching aids
- Consumer products
- Clocks
- Language translation
- Annunciators

Typical Applications

Minimum Configuration Using Switch Interface

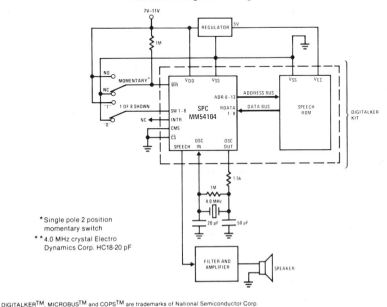

* Single pole 2 position momentary switch
** 4.0 MHz crystal Electro Dynamics Corp. HC18-20 pF

DIGITALKER™, MICROBUS™ and COPS™ are trademarks of National Semiconductor Corp.

Absolute Maximum Ratings*

Storage Temperature Range	$-65°C$ to $+150°C$	Voltage at Any Pin	12V
Operating Temperature Range	0°C to 70°C	Operating Voltage Range, V_{DD}-V_{SS}	7V to 11V
V_{DD}-V_{SS}	12V	Lead Temperature (Soldering, 10 seconds)	300°C

DC Electrical Characteristics* $T_A = 0°C$ to 70°C, $V_{DD} = 7V$–11V, $V_{SS} = 0V$, unless otherwise specified.

Symbol	Parameter	Conditions	Min	Typ	Max	Units
V_{IL}	Input Low Voltage		-0.3		0.8	V
V_{IH}	Input High Voltage		2.0		V_{DD}	V
V_{OL}	Output Low Voltage	$I_{OL} = 1.6$ mA			0.4	V
V_{OH}	Output High Voltage	$I_{OH} = -100$ μA	2.4		5.0	V
V_{ILX}	Clock Input Low Voltage		-0.3		1.2	V
V_{IHX}	Clock Input High Voltage		5.5		V_{DD}	V
I_{DD}	Power Supply Current				45	mA
I_{IL}	Input Leakage				± 10	μA
I_{ILX}	Clock Input Leakage				± 10	μA
V_S	Silence Voltage			$0.45 V_{DD}$		V
V_{OUT}	Peak to Peak Speech Output	$V_{DD} = 11V$		2.0		V
R_{EXT}	External Load on Speech Output	R_{EXT} Connected Between Speech Output and V_{SS}	50			$k\Omega$

AC Electrical Characteristics* $T_A = 0°C$ to 70°C, $V_{DD} = 7V$–11V, $V_{SS} = 0V$, unless otherwise specified.

Symbol	Parameter	Min	Max	Units
t_{aw}	CMS Valid to Write Strobe	350		ns
t_{csw}	Chip Select ON to Write Strobe	310		ns
t_{dw}	Data Bus Valid to Write Strobe	50		ns
t_{wa}	CMS Hold Time after Write Strobe	50		ns
t_{wd}	Data Bus Hold Time after Write Strobe	100		ns
t_{ww}	Write Strobe Width (50% Point)	430		ns
t_{red}	$\overline{ROMEN}$ ON to Valid ROM Data		2	μs
t_{wss}	Write Strobe to Speech Output Delay		410	μs
f_t	External Clock Frequency	3.92	4.08	MHz

Note: Rise and fall times (10% to 90%) of MICROBUS signals should be 50 ns maximum.

*SPC characteristics only. ROM characteristics covered by separate data sheet for MM52164.

Timing Waveforms

Command Sequence

Functional Description (Continued)

TABLE I. DT1056/DT1057* MASTER WORD LIST

Word	8-Bit Binary Address SW 8 — SW 1	Word	8-Bit Binary Address SW 8 — SW 1	Word	8-Bit Binary Address SW 8 — SW 1
ABORT	0 0 0 0 0 0 0 0	FARAD	0 0 1 0 1 1 0 0	PER	0 1 0 1 1 0 0 0
ADD	0 0 0 0 0 0 0 1	FAST	0 0 1 0 1 1 0 1	PICO	0 1 0 1 1 0 0 1
ADJUST	0 0 0 0 0 0 1 0	FASTER	0 0 1 0 1 1 1 0	PLACE	0 1 0 1 1 0 1 0
ALARM	0 0 0 0 0 0 1 1	FIFTH	0 0 1 0 1 1 1 1	PRESS	0 1 0 1 1 0 1 1
ALERT	0 0 0 0 0 1 0 0	FIRE	0 0 1 1 0 0 0 0	PRESSURE	0 1 0 1 1 1 0 0
ALL	0 0 0 0 0 1 0 1	FIRST	0 0 1 1 0 0 0 1	QUARTER	0 1 0 1 1 1 0 1
ASK	0 0 0 0 0 1 1 0	FLOOR	0 0 1 1 0 0 1 0	RANGE	0 1 0 1 1 1 1 0
ASSISTANCE	0 0 0 0 0 1 1 1	FORWARD	0 0 1 1 0 0 1 1	REACH	0 1 0 1 1 1 1 1
ATTENTION	0 0 0 0 1 0 0 0	FROM	0 0 1 1 0 1 0 0	RECEIVE	0 1 1 0 0 0 0 0
BRAKE	0 0 0 0 1 0 0 1	GAS	0 0 1 1 0 1 0 1	RECORD	0 1 1 0 0 0 0 1
BUTTON	0 0 0 0 1 0 1 0	GET	0 0 1 1 0 1 1 0	REPLACE	0 1 1 0 0 0 1 0
BUY	0 0 0 0 1 0 1 1	GOING	0 0 1 1 0 1 1 1	REVERSE	0 1 1 0 0 0 1 1
CALL	0 0 0 0 1 1 0 0	HALF	0 0 1 1 1 0 0 0	ROOM	0 1 1 0 0 1 0 0
CAUTION	0 0 0 0 1 1 0 1	HELLO	0 0 1 1 1 0 0 1	SAFE	0 1 1 0 0 1 0 1
CHANGE	0 0 0 0 1 1 1 0	HELP	0 0 1 1 1 0 1 0	SECURE	0 1 1 0 0 1 1 0
CIRCUIT	0 0 0 0 1 1 1 1	HERTZ	0 0 1 1 1 0 1 1	SELECT	0 1 1 0 0 1 1 1
CLEAR	0 0 0 1 0 0 0 0	HOLD	0 0 1 1 1 1 0 0	SEND	0 1 1 0 1 0 0 0
CLOSE	0 0 0 1 0 0 0 1	INCORRECT	0 0 1 1 1 1 0 1	SERVICE	0 1 1 0 1 0 0 1
COMPLETE	0 0 0 1 0 0 1 0	INCREASE	0 0 1 1 1 1 1 0	SIDE	0 1 1 0 1 0 1 0
CONNECT	0 0 0 1 0 0 1 1	INTRUDER	0 0 1 1 1 1 1 1	SLOW	0 1 1 0 1 0 1 1
CONTINUE	0 0 0 1 0 1 0 0	JUST	0 1 0 0 0 0 0 0	SLOWER	0 1 1 0 1 1 0 0
COPY	0 0 0 1 0 1 0 1	KEY	0 1 0 0 0 0 0 1	SMOKE	0 1 1 0 1 1 0 1
CORRECT	0 0 0 1 0 1 1 0	LEVEL	0 1 0 0 0 0 1 0	SOUTH	0 1 1 0 1 1 1 0
DATE	0 0 0 1 0 1 1 1	LOAD	0 1 0 0 0 0 1 1	STATION	0 1 1 0 1 1 1 1
DAY	0 0 0 1 1 0 0 0	LOCK	0 1 0 0 0 1 0 0	SWITCH	0 1 1 1 0 0 0 0
DECREASE	0 0 0 1 1 0 0 1	MEG	0 1 0 0 0 1 0 1	SYSTEM	0 1 1 1 0 0 0 1
DEPOSIT	0 0 0 1 1 0 1 0	MEGA	0 1 0 0 0 1 1 0	TEST	0 1 1 1 0 0 1 0
DIAL	0 0 0 1 1 0 1 1	MICRO	0 1 0 0 0 1 1 1	TH (NOTE 2)	0 1 1 1 0 0 1 1
DIVIDE	0 0 0 1 1 1 0 0	MORE	0 1 0 0 1 0 0 0	THANK	0 1 1 1 0 1 0 0
DOOR	0 0 0 1 1 1 0 1	MOVE	0 1 0 0 1 0 0 1	THIRD	0 1 1 1 0 1 0 1
EAST	0 0 0 1 1 1 1 0	NANO	0 1 0 0 1 0 1 0	THIS	0 1 1 1 0 1 1 0
ED (NOTE 1)	0 0 0 1 1 1 1 1	NEED	0 1 0 0 1 0 1 1	TOTAL	0 1 1 1 0 1 1 1
ED (NOTE 1)	0 0 1 0 0 0 0 0	NEXT	0 1 0 0 1 1 0 0	TURN	0 1 1 1 1 0 0 0
ED (NOTE 1)	0 0 1 0 0 0 0 1	NO	0 1 0 0 1 1 0 1	USE	0 1 1 1 1 0 0 1
ED (NOTE 1)	0 0 1 0 0 0 1 0	NORMAL	0 1 0 0 1 1 1 0	UTH (NOTE 3)	0 1 1 1 1 0 1 0
EMERGENCY	0 0 1 0 0 0 1 1	NORTH	0 1 0 0 1 1 1 1	WAITING	0 1 1 1 1 0 1 1
END	0 0 1 0 0 1 0 0	NOT	0 1 0 1 0 0 0 0	WARNING	0 1 1 1 1 1 0 0
ENTER	0 0 1 0 0 1 0 1	NOTICE	0 1 0 1 0 0 0 1	WATER	0 1 1 1 1 1 0 1
ENTRY	0 0 1 0 0 1 1 0	OHMS	0 1 0 1 0 0 1 0	WEST	0 1 1 1 1 1 1 0
ER	0 0 1 0 0 1 1 1	ONWARD	0 1 0 1 0 0 1 1	SWITCH	0 1 1 1 1 1 1 1
EVACUATE	0 0 1 0 1 0 0 0	OPEN	0 1 0 1 0 1 0 0	WINDOW	1 0 0 0 0 0 0 0
EXIT	0 0 1 0 1 0 0 1	OPERATOR	0 1 0 1 0 1 0 1	YES	1 0 0 0 0 0 0 1
FAIL	0 0 1 0 1 0 1 0	OR	0 1 0 1 0 1 1 0	ZONE	1 0 0 0 0 0 1 0
FAILURE	0 0 1 0 1 0 1 1	PASS	0 1 0 1 0 1 1 1		

*DT1056 is a complete kit including MM54104 SPC; DT1057 is SSR5 and SSR6 speech ROMs only.

Note 1: "ED" is a suffix that can be used to make any present tense word become a past tense word. The way we say "ED," however, does vary from one word to the next. For that reason, we have offered 4 different "ED" sounds. It is suggested that each "ED" be tested with the desired word for best quality results. Address 31 "ED" or 32 "ED" should be used with words ending in "T" or "D," such as exit or load. Address 34 "ED" should be used with words ending with soft sounds such as ask. Address 33 "ED" should be used with all other words.

Note 2: "TH" is a suffix that can be added to words like six, seven, eight to form adjective words like sixth, seventh, eighth.

Note 3: "UTH" is a suffix that can be added to words like twenty, thirty, forty to form adjective words like thirtieth, fortieth, etc.

Note 4: Address 130 is the last legal address in this particular word list. Exceeding address 130 will produce pieces of unintelligible invalid speech data.

LH1605 5 Amp, High Efficiency Switching Regulator

General Description

The LH1605 is a hybrid switching regulator with high output current capability. It incorporates a temperature-compensated voltage reference, a duty cycle modulator with the oscillator frequency programmable, error amplifier, high current-high voltage output switch, and a power diode. The LH1605 can supply up to 5 A of output current over a wide range of regulated output voltages.

Features

- Step down switching regulator
- Output adjustable from 3.0 to 30V
- 5 A output current
- High efficiency
- Frequency adjustable to 100 kHz
- Standard 8-pin TO-3 package

Block Diagram and Connection Diagram

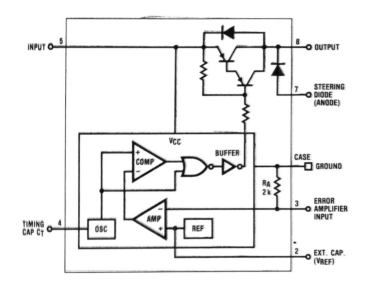

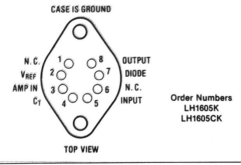

CASE IS GROUND

N.C.
V_REF
AMP IN
C_T
OUTPUT
DIODE
N.C.
INPUT

TOP VIEW

Order Numbers
LH1605K
LH1605CK

Absolute Maximum Ratings

V_{IN}	Input Voltage	35V Max.
I_{OUT}	Output Current	6A
T_J	Operating Temperature	150°C
P_D	Internal Power Dissipation	20W
T_A	Operating Temperature Range	
	LH1605C	−25°C to +85°C
	LH1605	−55°C to +125°C
T_{STG}	Storage Temperature Range	−65°C to +150°C
V_R (V_{8-7})	Steering Diode Reverse Voltage	60V
I_D (I_{7-8})	Steering Diode Forward Current	6A

Electrical Characteristics $T_C = 25°C$, $V_{IN} = 15V$ unless otherwise specified.

Symbol	Characteristics	Conditions	LH1605 Min.	LH1605 Typ.	LH1605 Max.	LH1605C Min.	LH1605C Typ.	LH1605C Max.	Units
V_{OUT}	Output Voltage Range	$V_{IN} \geqslant V_{OUT} + 5V$ $I_{OUT} = 2A$ (Note 2)	3.0		30	3.0		30	V
V_S	Switch Saturation Voltage	$I_C = 5.0A$ $I_C = 2.0A$		1.5 1.0	2.0 1.2		1.5 1.0	2.0 1.2	V
V_F	Steering Diode On Voltage	$I_D = 5.0A$ $I_D = 2.0A$		2.0 1.6	2.8 2.0		2.0 1.6	2.8 2.0	
V_{IN}	Supply Voltage Range		10		35	10		35	
I_R	Steering Diode Reverse Current	$V_R = 25V$		0.1	10.0		0.1	10.0	μA
I_Q	Quiescent Current	$I_{OUT} = 0.2A$ (Note 3) 50% Duty Cycle		30			30		mA
		0% Duty Cycle ($V_3 = 3.0V$)		6			6		
		100% Duty Cycle ($V_3 = 0V$)		46			46		
V_2	Reference Voltage on Pin 2		2.42	2.50	2.58		2.50		V
		$T_{MIN} \leqslant T_A \leqslant T_{MAX}$	2.40	2.50	2.60		2.50		
$\Delta V_2/\Delta T$	V_2 Temperature Coefficient			100			100		ppm/°C
ΔV_2	Line Regulation of Reference Voltage on Pin 2	$10V \leqslant V_{IN} \leqslant 35V$ $T_{MIN} \leqslant T_A \leqslant T_{MAX}$		20	30		20		mV
V_3	Voltage on Pin 3	(Note 4)	2.45	2.50	2.55		2.50		V
		$T_{MIN} \leqslant T_C \leqslant T_{MAX}$	2.42	2.50	2.58		2.50		
V_4	Voltage Swing — Pin 4			3.0			3.0		V
I_4	Charging Current — Pin 4			70			70		μA
$\Delta R_A/\Delta T$	Resistance Temp. Coeff.			75			75		ppm/°C
t_r	Voltage Rise Time	$V_{OUT} = 10V$ $I_{OUT} = 2.0A$ $I_{OUT} = 5.0A$		350 500			350 500		ns
t_f	Voltage Fall Time	$V_{OUT} = 10V$ $I_{OUT} = 2.0A$ $I_{OUT} = 5.0A$		300 400			300 400		
t_s	Storage Time	$V_{OUT} = 10V$		1.5			1.5		μs
t_d	Delay Time	$I_{OUT} = 5.0A$		100			100		ns
P_D	Power Dissipation	$V_{OUT} = 10V$		16			16		W
η	Efficiency	$I_{OUT} = 5.0A$		75			75		%
θ_{JC}	Thermal Resistance			5.0			5.0		°C/W

Note 1: θ_{JA} is typically 30°C/W for natural convection cooling.

Note 2: V_{OUT} and I_{OUT} refer to the output DC voltage and output current of a switching supply after the output LC filter as shown in the Typical Application circuit.

Note 3: Quiescent current depends on the duty cycle of the switching transistor. The average quiescent current may be calculated from known operating parameters.

Note 4: Voltage on pin 3 is tested by applying a +5.0V_{DC} voltage through a precision 2.0kΩ resistor to pin 3. This method combines the error due to the input bias current of the error amplifier, and the tolerance of the 2kΩ resistor from pin 3 to ground.

Note 5: The input offset voltage of the error amplifier is wafer tested to a maximum of 10mV.

LM2878 Dual 5 Watt Power Audio Amplifier

General Description

The LM2878 is a high voltage stereo power amplifier designed to deliver 5W/channel continuous into 8Ω loads. The amplifier is ideal for use with low regulation power supplies due to the absolute maximum rating of 35V and its superior power supply rejection. The LM2878 is designed to operate with a low number of external components, and still provide flexibility for use in stereo phonographs, tape recorders, and AM-FM stereo receivers. The flexibility of the LM2878 allows it to be used as a power operational amplifier, power comparator or servo amplifier. The LM2878 is internally compensated for all gains greater than 10, and comes in an 11-lead single-in-line package (SIP).

Features

- Wide operating range 6V–32V
- 5 W/channel output
- 60 dB ripple rejection, output referred
- 70 dB channel separation, output referred
- Low crossover distortion
- Internal current limiting, short circuit protection
- Internal thermal shutdown

Applications

- Stereo phonographs
- AM-FM radio receivers
- Power op amp, power comparator
- Servo amplifiers

Typical Applications

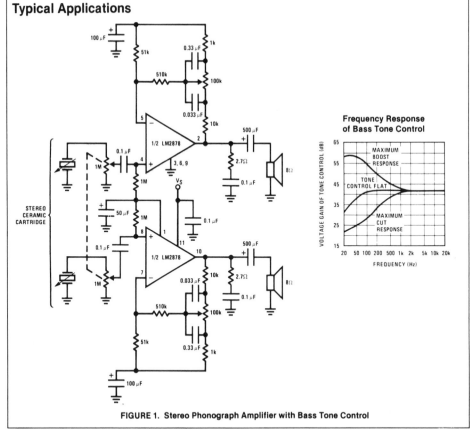

FIGURE 1. Stereo Phonograph Amplifier with Bass Tone Control

Absolute Maximum Ratings

Supply Voltage	35V
Input Voltage (Note 1)	± 0.7V
Operating Temperature (Note 2)	0°C to 70°C
Storage Temperature	− 65°C to 150°C
Junction Temperature	150°C
Lead Temperature (Soldering, 10 seconds)	300°C

Electrical Characteristics $V_S = 22V$, $T_A = 25°C$, $R_L = 8\Omega$, $A_V = 50$ (34 dB) unless otherwise specified.

Parameter	Conditions	Min	Typ	Max	Units
Total Supply Current	$P_O = 0W$		10	50	mA
Operating Supply Voltage		6		32	V
Output Power/Channel	f = 1 kHz, THD = 10%	5	5.5		W
Distortion	f = 1 kHz, $R_L = 8\Omega$				
	$P_O = 50$ mW		0.20		%
	$P_O = 0.5W$		0.15		%
	$P_O = 2W$		0.14		%
Output Swing	$R_L = 8\Omega$		$V_S - 6V$		Vp-p
Channel Separation	$C_{BYPASS} = 50\ \mu F$, $C_{IN} = 0.1\ \mu F$	− 50	− 70		dB
	f = 1 kHz, Output Referred				
	$V_O = 4$ Vrms				
PSRR Power Supply Rejection Ratio	$C_{BYPASS} = 50\ \mu F$, $C_{IN} = 0.1\ \mu F$	− 50	− 60		dB
	f = 120 Hz, Output Referred				
	$V_{ripple} = 1$ Vrms				
PSRR Negative Supply	Measured at DC, Input Referred		− 60		dB
Common-Mode Range	Split Supplies ± 15V, Pin 1 Tied to Pin 11		± 13.5		V
Input Offset Voltage			10		mV
Noise	Equivalent Input Noise $R_S = 0$, $C_{IN} = 0.1\ \mu F$				
	BW = 20 − 20 kHz		2.5		μV
	CCIR·ARM		3.0		μV
	Output Noise Wideband $R_S = 0$, $C_{IN} = 0.1\ \mu F$, $A_V = 200$		0.8		mV
Open Loop Gain	$R_S = 51\Omega$, f = 1 kHz, $R_L = 8\Omega$		70		dB
Input Bias Current			100		nA
Input Impedance	Open Loop		4		MΩ
DC Output Voltage	$V_S = 22V$	10	11	12	V
Slew Rate			2		V/μs
Power Bandwidth	3 dB Bandwidth at 2.5W		65		kHz
Current Limit			1.5		A

Note 1: ± 0.7V applies to audio applications; for extended range, see Application Hints.

Note 2: For operation at ambient temperature greater than 25°C, the LM2878 must be derated based on a maximum 150°C junction temperature using a thermal resistance which depends upon device mounting techniques.

LM2907, LM2917 tachometer-speed switch

general description

The LM2907, LM2917 series are monolithic frequency to voltage converters with a high gain op amp/comparator designed to operate a relay, lamp, or other load when the input frequency reaches or exceeds a selected rate. The tachometer uses a charge pump technique and offers frequency doubling for low ripple, full input protection in two versions (LM2907-8, LM2917-8) and its output swings to ground for a zero frequency input. (continued on page 5)

- Frequency doubling for low ripple
- Tachometer has built-in hysteresis with either differential input or ground referenced input
- Built-in zener on LM2917
- ±0.3% linearity typical
- Ground referenced tachometer is fully protected from damage due to swings above V_{CC} and below ground

advantages

- Output swings to ground for zero frequency input
- Easy to use; $V_{OUT} = f_{IN} \times V_{CC} \times R1 \times C1$
- Only one RC network provides frequency doubling
- Zener regulator on chip allows accurate and stable frequency to current conversion. (LM2917)

applications

- Over/under speed sensing
- Frequency to voltage conversion (tachometer)
- Speedometers
- Breaker point dwell meters
- Hand-held tachometer
- Speed governors
- Cruise control
- Automotive door lock control
- Clutch control
- Horn control
- Touch or sound switches

features

- Ground referenced tachometer input interfaces directly with magnetic variable reluctance pickups
- Op amp/comparator has floating transistor output
- 50 mA sink or source to operate relays, solenoids, meters, or LEDs

block and connection diagrams Dual-In-Line Package, Top Views

LM2907-8

LM2917-8

LM2907

LM2917

absolute maximum ratings (Note 1)

Supply Voltage	28V	Input Voltage Range	
Supply Current (Zener Options)	25 mA	Tachometer LM2907-8, LM2917-8	±28V
Collector Voltage	28V	LM2907, LM2917	0.0V to +28V
Differential Input Voltage		Op Amp/Comparator	0.0V to +28V
Tachometer	28V	Power Dissipation	500 mW
Op Amp/Comparator	28V	Operating Temperature Range	−40°C to +85°C
		Storage Temperature Range	−65°C to +150°C
		Lead Temperature (Soldering, 10 seconds)	300°C

electrical characteristics V_{CC} = 12 V_{DC}, T_A = 25°C, see test circuit

PARAMETER	CONDITIONS	MIN	TYP	MAX	UNITS
TACHOMETER					
Input Thresholds	V_{IN} = 250 mVp-p @ 1 kHz (Note 2)	±10	±15	±40	mV
Hysteresis	V_{IN} = 250 mVp-p @ 1 kHz (Note 2)		30		mV
Offset Voltage	V_{IN} = 250 mVp-p @ 1 kHz (Note 2)				
LM2907/LM2917			3.5	10	mV
LM2907-8/LM2917-8			5	15	mV
Input Bias Current	V_{IN} = ±50 mV_{DC}		0.1	1	μA
V_{OH}	V_{IN} = +125 mV_{DC} (Note 3)		8.3		V
Pin 2					
V_{OL}	V_{IN} = −125 mV_{DC} (Note 3)		2.3		V
Output Current; I_2, I_3	V2 = V3 = 6.0V (Note 4)	140	180	240	μA
Leakage Current; I_3	I2 = 0, V3 = 0			0.1	μA
Gain Constant, K	(Note 3)	0.9	1.0	1.1	
Linearity	f_{IN} = 1 kHz, 5 kHz, 10 kHz, (Note 5)	−1.0	0.3	+1.0	%
OP/AMP COMPARATOR					
V_{OS}	V_{IN} = 6.0V		3	10	mV
I_{BIAS}	V_{IN} = 6.0V		50	500	nA
Input Common-Mode Voltage		0		V_{CC}−1.5V	V
Voltage Gain			200		V/mV
Output Sink Current	V_C = 1.0	40	50		mA
Output Source Current	V_E = V_{CC} − 2.0		10		mA
Saturation Voltage	I_{SINK} = 5 mA		0.1	0.5	V
	I_{SINK} = 20 mA			1.0	V
	I_{SINK} = 50 mA		1.0	1.5	V
ZENER REGULATOR					
Regulator Voltage	R_{DROP} = 470Ω		7.56		V
Series Resistance			10.5	15	Ω
Temperature Stability			+1		mV/°C
TOTAL SUPPLY CURRENT			3.8	6	mA

Note 1: For operating at elevated temperatures the device must be derated based on a +125°C maximum junction temperature and a thermal resistance of +187°C/W junction to ambient for both packages.

Note 2: Hysteresis is the sum +V_{TH}−(−V_{TH}), offset voltage is their difference. See test circuit.

Note 3: V_{OH} is equal to 3/4 x V_{CC} − 1 V_{BE}, V_{OL} is equal to 1/4 x V_{CC} − 1 V_{BE} therefore V_{OH} − V_{OL} = V_{CC}/2. The difference, V_{OH} − V_{OL}, and the mirror gain, I_2/I_3, are the two factors that cause the tachometer gain constant to vary from 1.0.

Note 4: Be sure when choosing the time constant R1 x C1 that R1 is such that the maximum anticipated output voltage at pin 3 can be reached with I_3 x R1. The maximum value for R1 is limited by the output resistance of pin 3 which is greater than 10 MΩ typically.

Note 5: Nonlinearity is defined as the deviation of V_{out} (@ pin 3) for f_{IN} = 5 kHz from a straight line defined by the V_{OUT} @ 1 kHz and V_{OUT} @ 10 kHz. C1 = 1000 pF, R1 = 68k and C2 = 0.22 mFd.

typical performance characteristics

Total Supply Current

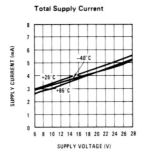

Zener Voltage Over Temperature

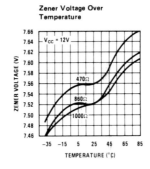

Tachometer Gain Constant Over Temperature

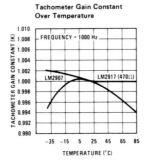

Tachometer Gain Constant Over Temperature

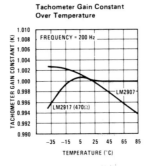

Tachometer Currents I_2 and I_3 Over Supply Voltage

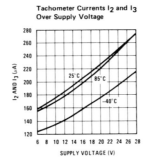

Tachometer Currents I_2 and I_3 Over Temperature

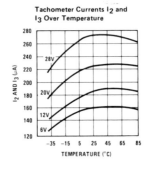

typical performance characteristics (con't)

**Tachometer Linearity
Over Temperature**

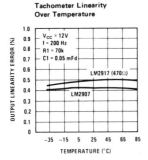

**Tachometer Linearity
Over Temperature**

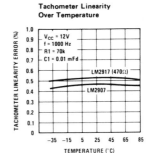

Tachometer Linearity vs R3

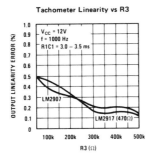

**Tachometer Input Hysteresis
vs Temperature**

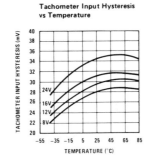

**Op Amp Output Transistor
Characteristics**

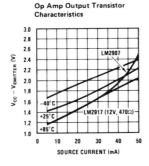

**Op Amp Output Transistor
Characteristics**

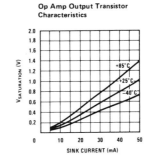

(D) SUFFIX
24-Lead Dual-In-Line
Side-Brazed Ceramic Package

CMOS Video Speed 8-Bit Flash Analog-to-Digital Converter

For Use in Low-Power Consumption,
High-Speed Digitization Applications

Features:
- *CMOS low power with SOS speed*
- *Parallel conversion technique*
- *15-MHz sampling rate (66-ns conversion time)*
- *8-bit latched 3-state output with overflow bit*
- *± ½ LSB accuracy (typ.)*
- *Single supply voltage (4 to 8 V)*
- *2 units in series allow 9-bit output*
- *2 units in parallel allow 30-MHz sampling rate*

The RCA CA3308* is a CMOS 200-mW parallel (FLASH) analog-to-digital converter designed for applications demanding both low-power consumption and high-speed digitization.

The CA3308 operates over a wide full-scale input-voltage range of 4 volts up to 8 volts with maximum power consumptions as low as 200 mW, depending upon the clock frequency selected. When operated from a 5-volt supply at a clock frequency of 15 MHz, the power consumption of the CA3308 is less than 150 mW.

The intrinsic high conversion rate makes the CA3308 ideally suited for digitizing high-speed signals. The overflow bit makes possible the connection of two or more CA3308s in series to increase the resolution of the conversion system. A series connection of two CA3308s may be used to produce a 9-bit high-speed converter. Operation of two CA3308s in parallel doubles the conversion speed (i.e., increases the sampling rate from 15 to 30 MHz). CA3308s may be combined with a high-speed 8-bit D/A converter, a binary adder, control logic, and an op amp to form a very high-speed 15-bit A/D converter.

256 paralleled auto-balanced voltage comparators measure the input voltage with respect to a known reference to produce the parallel-bit outputs in the CA3308.

255 comparators are required to quantize all input voltage levels in this 8-bit converter, and the additional comparator is required for the overflow bit.

The voltage supply for analog circuitry is termed V_{AA} and AGND. The voltage supply for digital circuitry is termed V_{DD} and V_{SS}.

The CA3308 type is available in a 24-lead dual-in-line ceramic package (D suffix).

---
* Formerly Developmental Type No. TA11279.

Applications:
- *The CA3308 is especially suited for high-speed conversion applications where low power is also important*
- *TV video digitizing (industrial/security/broadcast)*
- *High-speed A/D conversion*
- *Ultrasound signature analysis*
- *Transient signal analysis*
- *High-energy physics research*
- *High-speed oscilloscope storage/display*
- *General-purpose hybrid ADCs*
- *Optical character recognition*
- *Radar pulse analysis*
- *Motion signature analysis*
- *µP data acquisition systems*

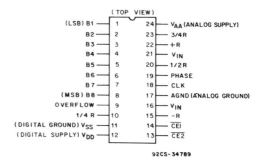

(TOP VIEW)

(LSB) B1	1	24	V_{AA} (ANALOG SUPPLY)
B2	2	23	3/4R
B3	3	22	+R
B4	4	21	V_{IN}
B5	5	20	1/2R
B6	6	19	PHASE
B7	7	18	CLK
(MSB) B8	8	17	AGND (ANALOG GROUND)
OVERFLOW	9	16	V_{IN}
1/4 R	10	15	–R
(DIGITAL GROUND) V_{SS}	11	14	$\overline{CE1}$
(DIGITAL SUPPLY) V_{DD}	12	13	$\overline{CE2}$

92CS-34789

TERMINAL ASSIGNMENT

MAXIMUM RATINGS, *Absolute-Maximum Values:*

DC SUPPLY VOLTAGE RANGE (V_{DD} AND V_{AA})

   (VOLTAGE REFERENCED TO V_{SS} TERMINAL) .... −0.5 to +8 V

INPUT VOLTAGE RANGE

   ALL INPUTS... −0.5 to V_{DD} +0.5 V

DC INPUT CURRENT

   CLK, PH, CE1, CE2, V_{IN} .... ±10 mA

POWER DISSIPATION PER PACKAGE (P_D)

FOR T_A=−40 to 55°C .... 315 mW

   FOR T_A=55°C to 85°C .... Derate linearly at 3.3 mW/°C

TEMPERATURE RANGE

   OPERATING .... −40 to +85°C

   STORAGE .... −65 to +150°C

LEAD TEMPERATURE (DURING SOLDERING)

   AT DISTANCE 1/16 ± 1/32 in. (1.59 ± 0.79 mm) FROM CASE FOR 10 s MAX. .... +265°C

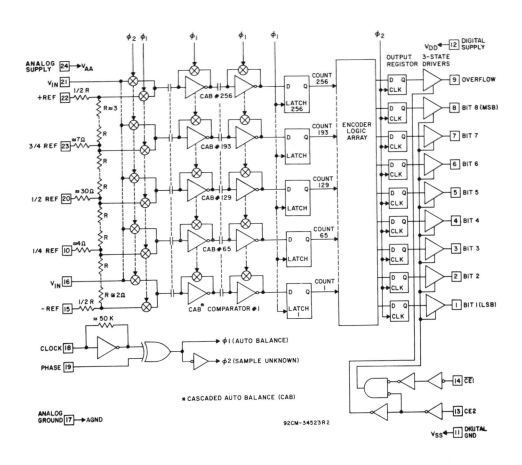

92CM-34523R2

393

ELECTRICAL CHARACTERISTICS

CHARACTERISTIC	TEST CONDITIONS $V_{AA} = V_{DD}$	LIMITS			UNITS
		MIN.	TYP.	MAX.	
Resolution		—	—	8	Bits
Linearity Error	V_{DD}=5 V, V_{REF}=6.4 V	—	—	±0.5	(CA3308AD)
	CLK=15 MHz, gain adjusted			±1	(CA3308D)
Differential Linearity Error	V_{DD}=5 V V_{REF}=6.4 V	—	—	±0.5	(CA3308AD)
	CLK=15 MHz			±1	(CA3308D)
Quantizing Error		$-\frac{1}{2}$	—	$\frac{1}{2}$	LSB
Analog Input:	V_{DD}=5 V				
Full Scale Range	CLK=15 MHz	4	—	8	V
Input Capacitance		—	50	—	pF
Input Current	V_{IN}= 6.4 V	—	1000	2000	μA
Maximum Conversion Speed	V_{DD}=5 V	15 M	17 M	—	SPS
Device Current (Excludes I_{REF})	V_{DD}=5 V (CLK=15 MHz)	—	50	—	mA
Ladder Impedance		300	600	900	Ω
Digital Inputs:					
Low Voltage		—	—	1.5	V
High Voltage	V_{DD}=5 V	3.5	—	—	V
Input Current (Except Pin 18)		—	±1	—	μA
Digital Outputs:					
Output Low (Sink) Current	V_{DD}=5 V, V_O=0.4 V	3.2	10	—	mA
Output High (Source) Current	V_{DD}=5 V, V_O=4.6 V	1.6	−6	—	
Digital Output Delay, t_d	V_{DD}=5 V	—	25	—	ns

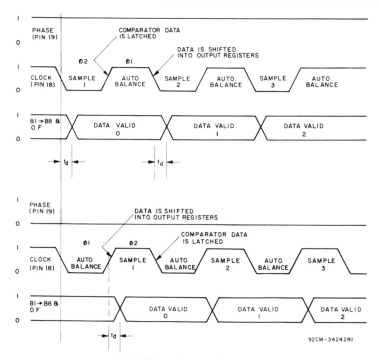

92CM-34242RI

Fig. 2-Timing diagram for the CA3308.

National Semiconductor

LM1524/LM2524/LM3524
Regulating Pulse Width Modulator

General Description

The LM1524 series of regulating pulse width modulators contains all of the control circuitry necessary to implement switching regulators of either polarity, transformer coupled DC to DC converters, transformerless polarity converters and voltage doublers, as well as other power control applications. This device includes a 5V voltage regulator capable of supplying up to 50 mA to external circuitry, a control amplifier, an oscillator, a pulse width modulator, a phase splitting flip-flop, dual alternating output switch transistors, and current limiting and shutdown circuitry. Both the regulator output transistor and each output switch are internally current limited and, to limit junction temperature, an internal thermal shutdown circuit is employed. The LM1524 is rated for operation from −55°C to +125°C and is packaged in a hermetic 16-lead DIP (J). The LM2524 and LM3524 are rated for operation from 0°C to +70°C and are packaged in either a hermetic 16-lead DIP (J) or a 16-lead molded DIP (N).

Features

- Complete PWM power control circuitry
- Frequency adjustable to greater than 100 kHz
- 2% frequency stability with temperature
- Total quiescent current less than 10 mA
- Dual alternating output switches for both push-pull or single-ended applications
- Current limit amplifier provides external component protection
- On-chip protection against excessive junction temperature and output current
- 5V, 50 mA linear regulator output available to user

Block and Connection Diagrams

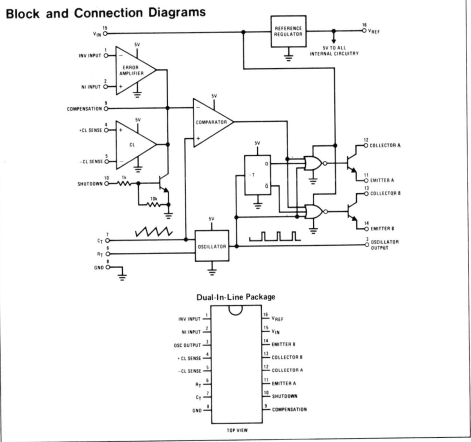

Dual-In-Line Package

TOP VIEW

Absolute Maximum Ratings

Input Voltage	40V	Maximum Junction Temperature	
Reference Voltage, Forced	6V	(J Package)	150°C
Reference Output Current	50 mA	(N Package)	125°C
Output Current (Each Output)	100 mA	Storage Temperature Range	−65°C to +150°C
Oscillator Charging Current (Pin 6 or 7)	5 mA	Lead Temperature (Soldering, 10 seconds)	300°C
Internal Power Dissipation (Note 1)	1W		
Operating Temperature Range			
LM1524	−55°C to +125°C		
LM2524/LM3524	0°C to +70°C		

Electrical Characteristics

Unless otherwise stated, these specifications apply for $T_A = -55°C$ to $+125°C$ for the LM1524 and $0°C$ to $+70°C$ for the LM2524 and LM3524, $V_{IN} = 20V$, and $f = 20$ kHz. Typical values other than temperature coefficients, are at $T_A = 25°C$.

PARAMETER	CONDITIONS	LM1524/LM2524 MIN	TYP	MAX	LM3524 MIN	TYP	MAX	UNITS
Reference Section								
Output Voltage		4.8	5.0	5.2	4.6	5.0	5.4	V
Line Regulation	$V_{IN} = 8-40V$		10	20		10	30	mV
Load Regulation	$I_L = 0-20$ mA		20	50		20	50	mV
Ripple Rejection	$f = 120$ Hz, $T_A = 25°C$		66			66		dB
Short-Circuit Output Current	$V_{REF} = 0$, $T_A = 25°C$		100			100		mA
Temperature Stability	Over Operating Temperature Range		0.3	1		0.3	1	%
Long Term Stability	$T_A = 25°C$		20			20		mV/khr
Oscillator Section								
Maximum Frequency	$C_T = 0.001$ μF, $R_T = 2$ kΩ		350			350		kHz
Initial Accuracy	R_T and C_T constant		5			5		%
Frequency Change with Voltage	$V_{IN} = 8-40V$, $T_A = 25°C$			1			1	%
Frequency Change with Temperature	Over Operating Temperature Range			2			2	%
Output Amplitude (Pin 3)	$T_A = 25°C$		3.5			3.5		V
Output Pulse Width (Pin 3)	$C_T = 0.01$ μF, $T_A = 25°C$		0.5			0.5		μs
Error Amplifier Section								
Input Offset Voltage	$V_{CM} = 2.5V$		0.5	5		2	10	mV
Input Bias Current	$V_{CM} = 2.5V$		2	10		2	10	μA
Open Loop Voltage Gain		72	80		60	80		dB
Common-Mode Input Voltage Range	$T_A = 25°C$	1.8		3.4	1.8		3.4	V
Common-Mode Rejection Ratio	$T_A = 25°C$		70			70		dB
Small Signal Bandwidth	$A_V = 0$ dB, $T_A = 25°C$		3			3		MHz
Output Voltage Swing	$T_A = 25°C$	0.5		3.8	0.5		3.8	V
Comparator Section								
Maximum Duty Cycle	% Each Output ON	45			45			%
Input Threshold (Pin 9)	Zero Duty Cycle		1			1		V
Input Threshold (Pin 9)	Maximum Duty Cycle		3.5			3.5		V
Input Bias Current			−1			−1		μA
Current Limiting Section								
Sense Voltage	$V_{(Pin 2)} - V_{(Pin 1)} \geq 50$ mV, Pin 9 = 2V, $T_A = 25°C$	190	200	210	180	200	220	mV
Sense Voltage T.C.			0.2			0.2		mV/°C
Common-Mode Voltage		−0.7		1	−0.7		1	V
Output Section (Each Output)								
Collector-Emitter Voltage		40			40			V
Collector Leakage Current	$V_{CE} = 40V$		0.1	50		0.1	50	μA
Saturation Voltage	$I_C = 50$ mA		1	2		1	2	V
Emitter Output Voltage	$V_{IN} = 20V$, $I_E = -250$ μA	17	18		17	18		V
Rise Time (10% to 90%)	$R_C = 2$ kΩ, $T_A = 25°C$		0.2			0.2		μs
Fall Time (90% to 10%)	$R_C = 2$ kΩ, $T_A = 25°C$		0.1			0.1		μs
Total Standby Current	$V_{IN} = 40V$, Pins 1, 4, 7, 8, 11 and 14 are grounded, Pin 2 = 2V, All Other Inputs and Outputs Open		5	10		5	10	mA

Note 1: For operation at elevated temperatures, devices in the J package must be derated based on a thermal resistance of 100°C/W, junction to ambient, and devices in the N package must be derated based on a thermal resistance of 150°C/W junction to ambient.

Typical Performance Characteristics

Maximum Average Power Dissipation (J Package)

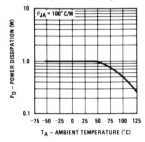

Maximum Average Power Dissipation (N Package)

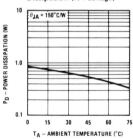

Maximum and Minimum Duty Cycle Threshold Voltage

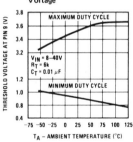

Output Transistor Saturation Voltage

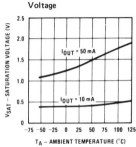

Output Transistor Emitter Voltage

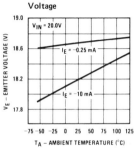

Reference and Switching Transistor Peak Output Current

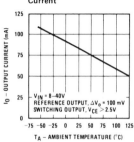

Standby Current

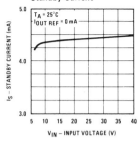

Standby Current

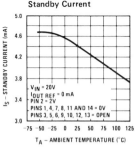

Current Limit Sense Voltage ($V_{Pin\ 4} - V_{Pin\ 5}$)

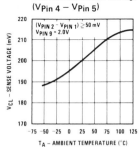

Test Circuit

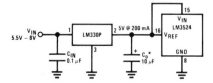

Functional Description

INTERNAL VOLTAGE REGULATOR

The LM3524 has on chip a 5V, 50 mA, short circuit protected voltage regulator. This voltage regulator provides a supply for all internal circuitry of the device and can be used as an external reference.

For input voltages of less than 8V the 5V output should be shorted to pin 15, V_{IN}, which disables the 5V regulator. With these pins shorted the input voltage must be limited to a maximum of 6V. If input voltages of 6–8V are to be used, a pre-regulator, as shown in *Figure 1*, must be added.

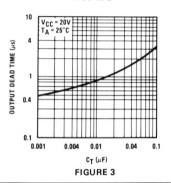

*Minimum C_O of 10 µF required for stability.

FIGURE 1

OSCILLATOR

The LM3524 provides a stable on-board oscillator. Its frequency is set by an external resistor, R_T and capacitor, C_T. A graph of R_T, C_T vs oscillator frequency is shown in *Figure 2*. The oscillator's output provides the signals for triggering an internal flip-flop, which directs the PWM information to the outputs, and a blanking pulse to turn off both outputs during transitions to ensure that cross conduction does not occur. The width of the blanking pulse, or dead time, is controlled by the value of C_T, as shown in *Figure 3*. The recommended

values of R_T are 1.8 kΩ to 100 kΩ, and for C_T, 0.001 µF to 0.1 µF.

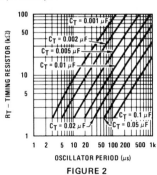

FIGURE 2

FIGURE 3

National Semiconductor

Industrial/Automotive/Functional Blocks/Telecommunications

LM122/LM222/LM322, LM2905/LM3905 Precision Timers

General Description

The LM122 series are precision timers that offer great versatility with high accuracy. They operate with unregulated supplies from 4.5V to 40V while maintaining constant timing periods from microseconds to hours. Internal logic and regulator circuits complement the basic timing function enabling the LM122 series to operate in many different applications with a minimum of external components.

The output of the timer is a floating transistor with built in current limiting. It can drive either ground referred or supply referred loads up to 40V and 50 mA. The floating nature of this output makes it ideal for interfacing, lamp or relay driving, and signal conditioning where an open collector or emitter is required. A "logic reverse" circuit can be programmed by the user to make the output transistor either "on" or "off" during the timing period.

The **trigger** input to the LM122 series has a threshold of 1.6V independent of supply voltage, but it is fully protected against inputs as high as ±40V — even when using a 5V supply. The circuitry reacts only to the rising edge of the trigger signal, and is immune to any trigger voltage during the timing periods.

An internal 3.15V regulator is included in the timer to reject supply voltage changes and to provide the user with a convenient reference for applications other than a basic timer. External loads up to 5 mA can be driven by the regulator. An internal 2V divider between the reference and ground sets the timing period to 1 RC. The timing period can be voltage controlled by driving this divider

with an external source through the V_{ADJ} pin. Timing ratios of 50:1 can be easily achieved.

The comparator used in the LM122 utilizes high gain PNP input transistors to achieve 300 pA typical input bias current over a common mode range of 0V to 3V. A **boost** terminal allows the user to increase comparator operating current for timing periods less than 1 ms. This lets the timer operate over a $3\mu s$ to multi-hour timing range with excellent repeatability.

The LM122 operates over a temperature range of $-55°C$ to $+125°C$. An electrically identical LM222 is specified from $-25°C$ to $+85°C$, and the LM322 is specified from $0°C$ to $+70°C$. The LM2905/LM3905 are identical to the LM122 series except that the **boost** and V_{ADJ} pin options are not available, limiting minimum timing period to 1 ms.

Features

- Immune to changes in trigger voltage during timing interval
- Timing periods from microseconds to hours
- Internal logic reversal
- Immune to power supply ripple during the timing interval
- Operates from 4.5V to 40V supplies
- Input protected to ±40V
- Floating transistor output with internal current limiting
- Internal regulated reference
- Timing period can be voltage controlled
- TTL compatible input and output

Typical Applications

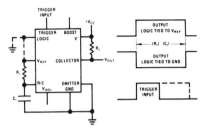

Basic Timer-Collector Output and Timing Chart

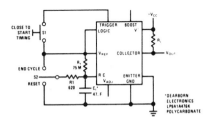

One Hour Timer with Reset and Manual Cycle End

Absolute Maximum Ratings

Power Dissipation	500 mW	Operating Temperature Range	
V^+ Voltage	40V	LM122	$-55°C \leq T_A \leq +125°C$
Collector Output Voltage	40V	LM222	$-25°C \leq T_A \leq +85°C$
V_{REF} Current	5 mA	LM322	$0°C \leq T_A \leq +70°C$
Trigger Voltage	±40V	LM2905	$-40°C \leq T_A \leq +85°C$
V_{ADJ} Voltage (Forced)	5V	LM3905	$0°C \leq T_A \leq +70°C$
Logic Reverse Voltage	5.5V		
Output Short Circuit Duration (Note 1)			
Lead Temperature (Soldering, 10 sec)	300°C		

Electrical Characteristics (Note 2)

PARAMETER	CONDITIONS	LM122/LM222			LM322			LM2905/LM3905			UNITS
		MIN	TYP	MAX	MIN	TYP	MAX	MIN	TYP	MAX	
Timing Ratio	$T_A = 25°C$, $4.5V \leq V^+ \leq 40V$	0.626	0.632	0.638	0.620	0.632	0.644	0.620	0.632	0.644	
	Boost Tied to V^+, (Note 3)	0.620	0.632	0.644	0.620	0.632	0.644				
Comparator Input Current	$T_A = 25°C$, $4.5V \leq V^+ \leq 40V$		0.3	1.0		0.3	1.5		0.5	1.5	nA
	Boost Tied to V^+		30	100		30	100				nA
Trigger Voltage	$T_A = 25°C$, $4.5V \leq V^+ \leq 40V$	1.2	1.6	2	1.2	1.6	2	1.2	1.6	2	V
Trigger Current	$T_A = 25°C$, $V_{TRIG} = 2V$		25			25			25		µA
Supply Current	$T_A \geq 25°C$, $4.5V \leq V^+ \leq 40V$		2.5	4		2.5	4.5		2.5	4.5	mA
Timing Ratio	$4.5V \leq V^+ \leq 40V$	0.62		0.644	0.61		0.654	0.61		0.654	
	Boost Tied to V^+	0.62		0.644	0.61		0.654				
Comparator Input Current	$4.5V \leq V^+ \leq 40V$	-5		5	-2		2	-2.5		2.5	nA
	Boost Tied to V^+, (Note 4)			100			150				nA
Trigger Voltage	$4.5V \leq V^+ \leq 40V$	0.8		2.5	0.8		2.5	0.8		2.5	V
Trigger Current	$V_{TRIG} = 2.5V$			200			200			200	µA
Output Leakage Current	$V_{CE} = 40V$			1			5			5	µA
Capacitor Saturation Voltage	$R_t \geq 1\,M\Omega$		2.5			2.5			2.5		mV
	$R_t = 10\,k\Omega$		25			25			25		mV
Reset Resistance			150			150			150		Ω
Reference Voltage	$T_A = 25°C$	3	3.15	3.3	3	3.15	3.3	3	3.15	3.3	V
Reference Regulation	$0 \leq I_{OUT} \leq 3\,mA$		20	50		20	50		20	50	mV
	$4.5V \leq V^+ \leq 40V$		6	25		6	25		6	25	mV
Collector Saturation Voltage	$I_L = 8\,mA$		0.25	0.4		0.25	0.4		0.25	0.4	V
	$I_L = 50\,mA$		0.7	1.4		0.7	1.4		0.7	1.4	V
Emitter Saturation Voltage	$T_A = 25°C$, $I_L = 3\,mA$		1.8	2.2		1.8	2.2		1.8	2.2	V
	$T_A = 25°C$, $I_L = 50\,mA$		2.1	3		2.1	3		2.1	3	V
Average Temperature Coefficient of Timing Ratio			0.003			0.003			0.003		%/°C
Minimum Trigger Width	$V_{TRIG} = 3V$		0.25			0.25			0.25		µs

Note 1: Continuous output shorts are not allowed. Short circuit duration at ambient temperatures up to 40°C may be calculated from $t = 120/V_{CE}$ seconds, where V_{CE} is the collector to emitter voltage across the output transistor during the short.

Note 2: These specifications apply for $T_{AMIN} \leq T_A \leq T_{AMAX}$ unless otherwise noted.

Note 3: Output pulse width can be calculated from the following equation: $t = (R_t)(C_t)(1 - 2(0.632 - r) - V_C/V_{REF})$ where r is timing ratio and V_C is capacitor saturation voltage. This reduces to $t = (R_t)(C_t)$ for all but the most critical applications.

Note 4: Sign reversal may occur at high temperatures ($> 100°C$) where comparator input current is predominately leakage. See typical curves.

Typical Performance Characteristics

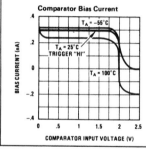

Comparator Bias Current

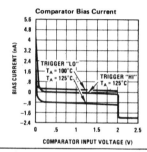

Comparator Bias Current

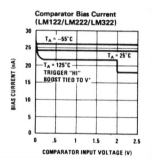

Comparator Bias Current (LM122/LM222/LM322)

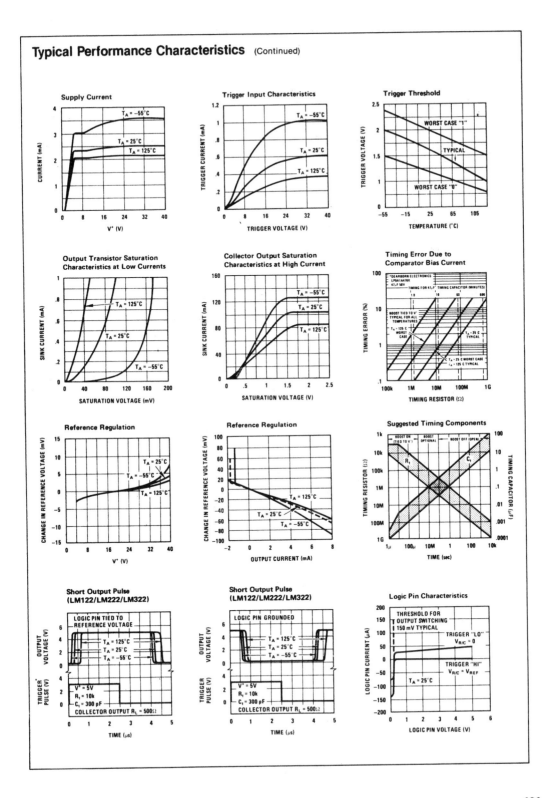

Schematic Diagram

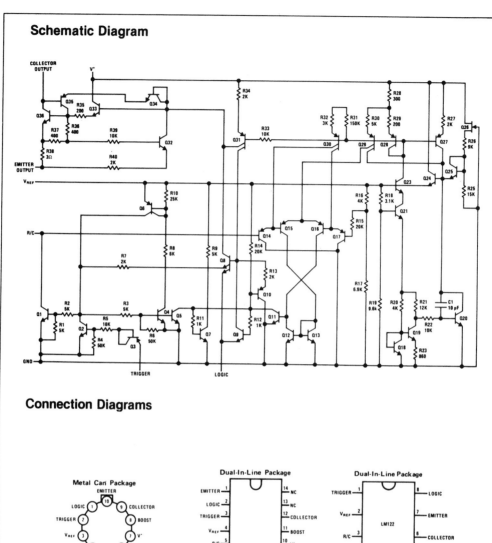

Connection Diagrams

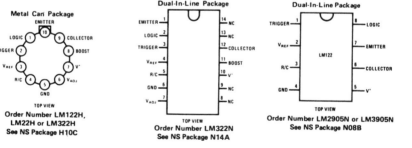

Metal Can Package

TOP VIEW

Order Number LM122H,
LM22H or LM322H
See NS Package H10C

Dual-In-Line Package

TOP VIEW

Order Number LM322N
See NS Package N14A

Dual-In-Line Package

TOP VIEW

Order Number LM2905N or LM3905N
See NS Package N08B

CD4051BM/CD4051BC single 8-channel analog multiplexer/demultiplexer
CD4052BM/CD4052BC dual 4-channel analog multiplexer/demultiplexer
CD4053BM/CD4053BC triple 2-channel analog multiplexer/demultiplexer

general description

These analog multiplexers/demultiplexers are digitally controlled analog switches having low "ON" impedance and very low "OFF" leakage currents. Control of analog signals up to 15 Vp-p can be achieved by digital signal amplitudes of 3 - 15 V. For example, if V_{DD} = 5 V, V_{SS} = 0 V and V_{EE} = –5 V, analog signals from –5 V to +5 V can be controlled by digital inputs of 0 - 5 V. The multiplexer circuits dissipate extremely low quiescent power over the full V_{DD} – V_{SS} and V_{DD} – V_{EE} supply voltage ranges, independent of the logic state of the control signals. When a logical "1" is present at the inhibit input terminal all channels are "OFF."

CD4051BM/CD4051BC is a single 8-channel multiplexer having three binary control inputs, A, B and C, and an inhibit input. The three binary signals select 1 of 8 channels to be turned "ON" and connect the input to the output.

CD4052BM/CD4052BC is a differential 4-channel multiplexer having two binary control inputs, A and B, and an inhibit input. The two binary input signals select 1 of 4 pairs of channels to be turned on and connect the differential analog inputs to the differential outputs.

CD4053BM/CD4053BC is a triple 2-channel multiplexer having three separate digital control inputs, A, B and C, and an inhibit input. Each control input selects one of a pair of channels which are connected in a single-pole double-throw configuration.

features

- Wide range of digital and analog signal levels: digital 3 - 15 V, analog to 15 Vp-p
- Low "ON" resistance: 80 Ω (typ) over entire 15 Vp-p signal-input range for V_{DD} – V_{EE} = 15 V
- High "OFF" resistance: channel leakage of ±10 pA (typ) at V_{DD} – V_{EE} = 10 V
- Logic level conversion for digital addressing signals of 3 - 15 V (V_{DD} – V_{SS} = 3 - 15 V) to switch analog signals to 15 Vp-p (V_{DD} – V_{EE} = 15 V)
- Matched switch characteristics: ΔR_{ON} = 5 Ω (typ) for V_{DD} – V_{EE} = 15 V
- Very low quiescent power dissipation under all digital-control input and supply conditions: 1 μW (typ) at V_{DD} – V_{SS} = V_{DD} – V_{EE} = 10 V
- Binary address decoding on chip

connection diagrams

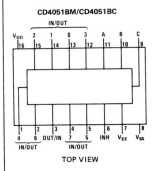

CD4051BM/CD4051BC

TOP VIEW

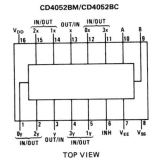

CD4052BM/CD4052BC

TOP VIEW

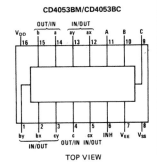

CD4053BM/CD4053BC

TOP VIEW

absolute maximum rating

V_{DD} DC Supply Voltage   –0.5 Vdc to +18 Vdc
V_{IN} Input Voltage   –0.5 Vdc to V_{DD} + 0.5 Vdc
T_S Storage Temperature Range   –65°C to +150°C
P_D Package Dissipation   500 mW
T_L Lead Temperature (soldering, 10 seconds)   300°C

recommended operating conditions

V_{DD} DC Supply Voltage   +5 Vdc to +15 Vdc
V_{IN} Input Voltage   0 V to V_{DD} Vdc
T_A Operating Temperature Range
     4051BM/4052BM/4053BM   –55°C to +125°C
     4051BC/4052BC/4053BC   –40°C to +85°C

dc electrical characteristics (Note 2)

PARAMETER		CONDITIONS	–55°C MIN	–55°C MAX	+25°C MIN	+25°C TYP	+25°C MAX	+125°C MIN	+125°C MAX	UNITS
I_{DD}	Quiescent Device Current	V_{DD} = 5 V		5			5		150	μA
		V_{DD} = 10 V		10			10		300	μA
		V_{DD} = 15 V		20			20		600	μA
Signal Inputs (V_{IS}) and Outputs (V_{OS})										
R_{ON}	"ON" Resistance (Peak for $V_{EE} \leqslant V_{IS} \leqslant V_{DD}$)	R_L = 10 kΩ (any channel selected) V_{DD} = 2.5 V, V_{EE} = –2.5 V or V_{DD} = 5 V, V_{EE} = 0 V		2000		270	2500		3500	Ω
		V_{DD} = 5 V, V_{EE} = –5 V or V_{DD} = 10 V, V_{EE} = 0 V		310		120	400		580	Ω
		V_{DD} = 7.5 V, V_{EE} = –7.5 V or V_{DD} = 15 V, V_{EE} = 0 V		220		80	280		400	Ω
ΔR_{ON}	Δ "ON" Resistance Between Any Two Channels	R_L = 10 kΩ (any channel selected) V_{DD} = 2.5 V, V_{EE} = –2.5 V or V_{DD} = 5 V, V_{EE} = 0 V				10				Ω
		V_{DD} = 5 V, V_{EE} = –5 V or V_{DD} = 10 V, V_{EE} = 0 V				10				Ω
		V_{DD} = 7.5 V, V_{EE} = –7.5 V or V_{DD} = 15 V, V_{EE} = 0 V				5				Ω
	"OFF" Channel Leakage Current, any channel "OFF"	V_{DD} = 7.5 V, V_{EE} = –7.5 V O/I = ±7.5 V, I/O = 0 V		±50		±0.01	±50		±500	nA
	"OFF" Channel Leakage Current, all channels "OFF" (Common OUT/IN)	Inhibit = 5 V CD4051		±200		±0.08	±200		±2000	nA
		V_{DD} = 7.5 V, CD4052		±200		±0.04	±200		±2000	nA
		V_{EE} = –7.5 V, O/I = 0 V, I/O = ±7.5 V CD4053		±200		±0.02	±200		±2000	nA
Control Inputs A, B, C and Inhibit										
V_{IL}	Low Level Input Voltage	V_{DD} = 5 V		1.5			1.5		1.5	V
		V_{DD} = 10 V		3			3		3	V
		V_{DD} = 15 V		4			4		4	V
V_{IH}	High Level Input Voltage	V_{DD} = 5 V	3.5		3.5			3.5		V
		V_{DD} = 10 V	7		7			7		V
		V_{DD} = 15 V	11		11			11		V
I_{IN}	Input Current	V_{DD} = 15 V, V_{EE} = 0 V V_{IN} = 0 V		–0.1		-10^{-5}	–0.1		–1.0	μA
		V_{DD} = 15 V, V_{EE} = 0 V, V_{IN} = 15 V		0.1		10^{-5}	0.1		1.0	μA

Note 1: "Absolute Maximum Ratings" are those values beyond which the safety of the device cannot be guaranteed. Except for "Operating Temperature Range" they are not meant to imply that the devices should be operated at these limits. The table of "Electrical Characteristics" provides conditions for actual device operation.

Note 2: All voltages measured with respect to V_{SS} unless otherwise specified.

dc electrical characteristics (con't) (Note 2)

PARAMETER		CONDITIONS		−40°C MIN	−40°C MAX	+25°C MIN	+25°C TYP	+25°C MAX	+85°C MIN	+85°C MAX	UNITS
I_{DD}	Quiescent Device Current	$V_{DD} = 5\,V$			20			20		150	μA
		$V_{DD} = 10\,V$			40			40		300	μA
		$V_{DD} = 15\,V$			80			80		600	μA
Signal Inputs (V_{IS}) and Outputs (V_{OS})											
R_{ON}	"ON" Resistance (Peak for $V_{EE} \leqslant V_{IS} \leqslant V_{DD}$)	$R_L = 10\,k\Omega$ (any channel selected)	$V_{DD} = 2.5\,V$, $V_{EE} = -2.5\,V$ or $V_{DD} = 5\,V$, $V_{EE} = 0\,V$		2100		270	2500		3200	Ω
			$V_{DD} = 5\,V$, $V_{EE} = -5\,V$ or $V_{DD} = 10\,V$, $V_{EE} = 0\,V$		330		120	400		520	Ω
			$V_{DD} = 7.5\,V$, $V_{EE} = -7.5\,V$ or $V_{DD} = 15\,V$, $V_{EE} = 0\,V$		230		80	280		360	Ω
ΔR_{ON}	Δ "ON" Resistance Between Any Two Channels	$R_L = 10\,k\Omega$ (any channel selected)	$V_{DD} = 2.5\,V$, $V_{EE} = -2.5\,V$ or $V_{DD} = 5\,V$, $V_{EE} = 0\,V$				10				Ω
			$V_{DD} = 5\,V$, $V_{EE} = -5\,V$ or $V_{DD} = 10\,V$, $V_{EE} = 0\,V$				10				Ω
			$V_{DD} = 7.5\,V$, $V_{EE} = -7.5\,V$ or $V_{DD} = 15\,V$, $V_{EE} = 0\,V$				5				Ω
	"OFF" Channel Leakage Current, any channel "OFF"	$V_{DD} = 7.5\,V$, $V_{EE} = -7.5\,V$ O/I = ±7.5 V, I/O = 0 V			±50		±0.01	±50		±500	nA
	"OFF" Channel Leakage Current, all channels "OFF" (Common OUT/IN)	Inhibit = 5 V $V_{DD} = 7.5\,V$, $V_{EE} = -7.5\,V$, O/I = 0 V, I/O = ±7.5 V	CD4051		±200		±0.08	±200		±2000	nA
			CD4052		±200		±0.04	±200		±2000	nA
			CD4053		±200		±0.02	±200		±2000	nA
Control Inputs A, B, C and Inhibit											
V_{IL}	Low Level Input Voltage	$V_{DD} = 5\,V$			1.5			1.5		1.5	V
		$V_{DD} = 10\,V$			3			3		3	V
		$V_{DD} = 15\,V$			4			4		4	V
V_{IH}	High Level Input Voltage	$V_{DD} = 5\,V$		3.5		3.5			3.5		V
		$V_{DD} = 10\,V$		7		7			7		V
		$V_{DD} = 15\,V$		11		11			11		V
I_{IN}	Input Current	$V_{DD} = 15\,V$, $V_{EE} = 0\,V$ $V_{IN} = 0\,V$			−0.1		-10^{-5}	−0.1		−1.0	μA
		$V_{DD} = 15\,V$, $V_{EE} = 0\,V$, $V_{IN} = 15\,V$			0.1		10^{-5}	0.1		1.0	μA

Note 1: "Absolute Maximum Ratings" are those values beyond which the safety of the device cannot be guaranteed. Except for "Operating Temperature Range" they are not meant to imply that the devices should be operated at these limits. The table of "Electrical Characteristics" provides conditions for actual device operation.

Note 2: All voltages measured with respect to V_{SS} unless otherwise specified.

ac electrical characteristics

$T_A = 25°C$, $t_r = t_f = 20\,ns$, unless otherwise specified.

PARAMETER		CONDITIONS	V_{DD}	MIN	TYP	MAX	UNITS
t_{PZH}, t_{PZL}	Propagation Delay Time from Inhibit to Signal Output (channel turning on)	$V_{EE} = V_{SS} = 0\,V$ $R_L = 1\,k\Omega$ $C_L = 50\,pF$	5 V 10 V 15 V		600 225 160	1200 450 320	ns ns ns
t_{PHZ}, t_{PLZ}	Propagation Delay Time from to Signal Output (channel turning off)	$V_{EE} = V_{SS} = 0\,V$ $R_L = 1\,k\Omega$ $C_L = 50\,pF$	5 V 10 V 15 V		210 100 75	420 200 150	ns ns ns
C_{IN}	Input Capacitance Control Input Signal Input (IN/OUT)				5 10	7.5 15	pF pF
C_{OUT}	Output Capacitance (common OUT/IN) CD4051 CD4052 CD4053	$V_{EE} = V_{SS} = 0\,V$	10 V 10 V 10 V		30 15 8		pF pF pF
C_{IOS}	Feedthrough Capacitance				0.2		pF
C_{PD}	Power Dissipation Capacitance CD4051 CD4052 CD4053				110 140 70		pF pF pF
Signal Inputs (V_{IS}) and Outputs (V_{OS})							
	Sine Wave Response (Distortion)	$R_L = 10\,k\Omega$ $f_{IS} = 1\,kHz$ $V_{IS} = 5\,Vp\text{-}p$ $V_{EE} = V_{SI} = 0\,V$	10 V		0.04		%
	Frequency Response, Channel "ON" (Sine Wave Input)	$R_L = 1\,k\Omega$, $V_{EE} = V_{SS} = 0\,V$, $V_{IS} = 5\,Vp\text{-}p$, $20 \log_{10} V_{OS}/V_{IS} = -3\,dB$	10 V		40		MHz
	Feedthrough, Channel "OFF"	$R_L = 1\,k\Omega$, $V_{EE} = V_{SS} = 0\,V$, $V_{IS} = 5\,Vp\text{-}p$, $20 \log_{10} V_{OS}/V_{IS} = -40\,dB$	10 V		10		MHz
	Crosstalk Between Any Two Channels (frequency at 40 dB)	$R_L = 1\,k\Omega$, $V_{EE} = V_{SS} = 0\,V$, $V_{IS}(A) = 5\,Vp\text{-}p$ $20 \log_{10} V_{OS}(B)/V_{IS}(A) = -40\,dB$ (Note 3)	10 V		3		MHz
t_{PHL}, t_{PLH}	Propagation Delay Signal Input to Serial Output	$V_{EE} = V_{SS} = 0\,V$ $C_L = 50\,pF$	5 V 10 V 15 V		25 15 10	55 35 25	ns ns ns
Control Inputs, A, B, C and Inhibit							
	Control Input to Signal Crosstalk	$V_{EE} = V_{SS} = 0\,V$, $R_L = 10\,k\Omega$ at both ends of channel. Input Square Wave Amplitude = 10 V	10 V		65		mV (peak)
t_{PHL}, t_{PLH}	Propagation Delay Time from Address to Signal Output (channels "ON" or "OFF")	$V_{EE} = V_{SS} = 0\,V$ $C_L = 50\,pF$	5 V 10 V 15 V		500 180 120	1000 360 240	ns ns ns

Note 3: A, B are two arbitrary channels with A turned "ON" and B "OFF."

INTERSIL

ICL7126
Single-Chip 3½-Digit
Low-Power A/D Converter

FEATURES

- Guaranteed zero reading for 0 Volts input on all scales
- True polarity at zero for precise null detection
- 1pA typical input current
- True differential input and reference
- Direct LCD display drive — no external components required
- Pin compatible with the ICL7106
- Low noise — less than 15μV p-p
- On-chip clock and reference
- Low power dissipation guaranteed less than 1mW
- No additional active circuits required
- Evaluation Kit available (ICL7126EV/KIT)
- 8,000 hours typical 9 Volt battery life

GENERAL DESCRIPTION

The Intersil ICL7126 is a high performance. very low power 3½ digit A/D converter. All the necessary active devices are contained on a single CMOS IC. including seven segment decoders, display drivers, reference, and clock. The 7126 is designed to interface with a liquid crystal display (LCD) and includes a backplane drive. The supply current is 100μA, ideally suited for 9V battery operation.

The 7126 brings together an unprecedented combination of high accuracy, versatility, and true economy. High accuracy, like auto-zero to less than 10μV, zero drift of less than 1μV/°C, input bias current of 10pA max., and rollover error of less than one count. The versatility of true differential input and reference is useful in all systems, but gives the designer an uncommon advantage when measuring load cells, strain gauges and other bridge-type transducers. And finally the true economy of single power supply operation allows a high performance panel meter to be built with the addition of only 7 passive components and a display.

The ICL7126 can be used as a plug-in replacement for the ICL7106 in a wide variety of applications, changing only the passive components.

PIN CONFIGURATION

```
                ICL7126
         V    |1      40| OSC 1
        D1    |2      39| OSC 2
        C1    |3      38| OSC 3
        B1    |4      37| TEST
(UNITS) A1    |5      36| REF HI
        F1    |6      35| REF LO
        G1    |7      34| C'REF
        E1    |8      33| C REF
        D2    |9      32| COMMON
        C2    |10     31| IN HI
(TENS)  B2    |11     30| IN LO
        A2    |12     29| A/Z
        F2    |13     28| BUFF
        E2    |14     27| INT
        D3    |15     26| V
(100's) B3    |16     25| G₂ (TENS)
        F3    |17     24| C₃
        E3    |18     23| A₃  (100's)
(1000)  AB4   |19     22| G₃
        POL   |20     21| BP
       (MINUS)
```

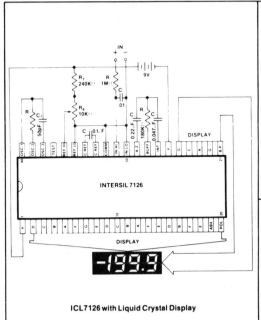

ICL7126 with Liquid Crystal Display

ORDERING INFORMATION

Part	Temp. Range	Package	Order Number
7126	0°C to +70°C	40-Pin Ceramic DIP	ICL7126CDL
7126	0°C to +70°C	40-Pin Plastic DIP	ICL7126CPL
7126	0°C to +70°C	40-Pin CERDIP	ICL7126CJL
7126 Kit		Evaluation Kits	ICL7126EV/KIT

407

ICL7126

ABSOLUTE MAXIMUM RATINGS

Supply Voltage (V⁺ to V⁻) 15V
Analog Input Voltage (either input) (Note 1) ..... V⁺ to V⁻
Reference Input Voltage (either input) .... V⁺ to V⁻
Clock Input .... TEST to V⁺

Power Dissipation (Note 2)
Ceramic Package 1000mW
Plastic Package 800mW
Operating Temperature .... 0°C to +70°C
Storage Temperature.................... −65°C to +160°C
Lead Temperature (Soldering, 60 sec).............. 300°C

Note 1: Input voltages may exceed the supply voltages provided the input current is limited to ±100μA.
Note 2: Dissipation rating assumes device is mounted with all leads soldered to printed circuit board.

*COMMENT: Stresses above those listed under "Absolute Maximum Ratings" may cause permanent damage to the devices. This is a stress rating only and functional operation of the devices at these or any other conditions above those indicated in the operational sections of the specifications is not implied. Exposure to absolute maximum rating conditions for extended periods may affect device reliability.

ELECTRICAL CHARACTERISTICS (Note 3)

CHARACTERISTICS	CONDITIONS	MIN	TYP	MAX	UNITS
Zero Input Reading	V_{IN} = 0.0V Full Scale = 200.0mV	−000.0	±000.0	+000.0	Digital Reading
Ratiometric Reading	V_{IN} = V_{REF} V_{REF} = 100mV	999	999/1000	1000	Digital Reading
Rollover Error (Difference in reading for equal positive and negative reading near Full Scale)	$-V_{IN}$ = $+V_{IN}$ ≃ 200.0mV	−1	±.2	+1	Counts
Linearity (Max. deviation from best straight line fit)	Full scale = 200mV or full scale = 2.000V	−1	±.2	+1	Counts
Common Mode Rejection Ratio (Note 4)	V_{CM} = ±1V, V_{IN} = 0V. Full Scale = 200.0mV		50		μV/V
Noise (Pk - Pk value not exceeded 95% of time)	V_{IN} = 0V Full Scale = 200.0mV		15		μV
Leakage Current @ Input	V_{IN} = 0V		1	10	pA
Zero Reading Drift	V_{IN} = 0 0° < T_A < 70°C		0.2	1	μV/°C
Scale Factor Temperature Coefficient	V_{IN} = 199.0mV 0 < T_A < 70°C (Ext. Ref. 0 ppm/°C)		1	5	ppm/°C
Supply Current (Does not include COMMON current)	V_{IN} = 0 (Note 6)		50	100	μA
Analog COMMON Voltage (With respect to pos. supply)	250KΩ between Common & pos. Supply	2.4	2.8	3.2	V
Temp. Coeff. of Analog COMMON (with respect to pos. Supply)	250KΩ between Common & pos. Supply		80		ppm/°C
Pk-Pk Segment Drive Voltage (Note 5)	V+ to V- = 9V	4	5	6	V
Pk-Pk Backplane Drive Voltage (Note 5)	V+ to V- = 9V	4	5	6	V
Power Dissipation Capacitance	vs. Clock Freq.		40		pF

Note 3: Unless otherwise noted, specifications apply at T_A = 25°C, f_{clock} = 16kHZ and are tested in the circuit of Figure 1.
Note 4: Refer to "Differential Input" discussion on page 4.
Note 5: Back plane drive is in phase with segment drive for 'off' segment, 180° out of phase for 'on' segment. Frequency is 20 times conversion rate. Average DC component is less then 50mV.
Note 6: During auto zero phase, current is 10-20μA higher. 48kHz oscillator, Figure 2, increases current by 8μA (typ).

LM79XX Series 3-Terminal Negative Regulators

General Description

The LM79XX series of 3-terminal regulators is available with fixed output voltages of −5V, −12V, and −15V. These devices need only one external component—a compensation capacitor at the output. The LM79XX series is packaged in the TO-220 power package and is capable of supplying 1.5A of output current.

These regulators employ internal current limiting safe area protection and thermal shutdown for protection against virtually all overload conditions.

Low ground pin current of the LM79XX series allows output voltage to be easily boosted above the preset value with a resistor divider. The low quiescent current drain of these devices with a specified maximum change with line and load ensures good regulation in the voltage boosted mode.

For applications requiring other voltages, see LM137 data sheet.

Features

- Thermal, short circuit and safe area protection
- High ripple rejection
- 1.5A output current
- 4% preset output voltage

Typical Applications

±15V, 1 Amp Tracking Regulators

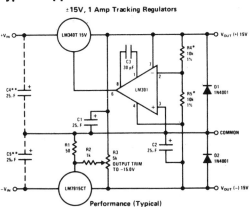

Performance (Typical)

	(−15)	(+15)
Load Regulation at ΔI_L = 1A	40 mV	2 mV
Output Ripple, C_{IN} = 3000μF, I_L = 1A	100μVrms	100μVrms
Temperature Stability	50 mV	50 mV
Output Noise 10 Hz $\leq$ f $\leq$ 10 kHz	150μVrms	150μVrms

*Resistor tolerance of R4 and R5 determine matching of (+) and (−) outputs
**Necessary only if raw supply filter capacitors are more than 3″ from regulators

Fixed Regulator

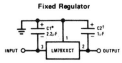

*Required if regulator is separated from filter capacitor by more than 3″. For value given, capacitor must be solid tantalum. 25μF aluminum electrolytic may be substituted.

†Required for stability. For value given, capacitor must be solid tantalum. 25μF aluminum electrolytic may be substituted. Values given may be increased without limit.

For output capacitance in excess of 100μF, a high current diode from input to output (1N4001, etc.) will protect the regulator from momentary input shorts.

Variable Output

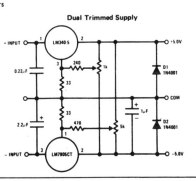

*Improves transient response and ripple rejection. Do not increase beyond 50μF.

$$V_{OUT} = V_{SET} \left(\frac{R1 + R2}{R2} \right)$$

Select R2 as follows

LM7905CT	300Ω
LM7912CT	750Ω
LM7915CT	1k

Dual Trimmed Supply

Absolute Maximum Ratings

Input Voltage
(V_O = 5V)　　　　　　　　　　　　　　　　　−35V
(V_O = 12V and 15V)　　　　　　　　　　　　−40V
Input-Output Differential
(V_O = 5V)　　　　　　　　　　　　　　　　　25V
(V_O = 12V and 15V)　　　　　　　　　　　　30V
Power Dissipation　　　　　　　　Internally Limited
Operating Junction Temperature Range　0°C to +125°C
Storage Temperature Range　　　　−65°C to +150°C
Lead Temperature (Soldering, 10 seconds)　　　230°C

Electrical Characteristics
Conditions unless otherwise noted: I_{OUT} = 500 mA, C_{IN} = 2.2μF, C_{OUT} = 1μF, 0°C ≤ T_J ≤ +125°C, Power Dissipation ≤ 15W.

PART NUMBER			LM7905C			UNITS
OUTPUT VOLTAGE			5V			
INPUT VOLTAGE (unless otherwise specified)			−10V			
PARAMETER		CONDITIONS	MIN	TYP	MAX	
V_O	Output Voltage	T_J = 25°C	−4.8	−5.0	−5.2	V
		5 mA ≤ I_{OUT} ≤ 1A,	−4.75		−5.25	V
		P ≤ 15W	(−20 ≤ V_{IN} ≤ −7)			V
ΔV_O	Line Regulation	T_J = 25°C, (Note 2)		8	50	mV
			(−25 ≤ V_{IN} ≤ −7)			V
				2	15	mV
			(−12 ≤ V_{IN} ≤ −8)			V
ΔV_O	Load Regulation	T_J = 25°C, (Note 2)				mV
		5 mA ≤ I_{OUT} ≤ 1.5A		15	100	mV
		250 mA ≤ I_{OUT} ≤ 750 mA		5	50	mV
I_Q	Quiescent Current	T_J = 25°C		1	2	mA
ΔI_Q	Quiescent Current Change	With Line			0.5	mA
			(−25 ≤ V_{IN} ≤ −7)			V
		With Load, 5 mA ≤ I_{OUT} ≤ 1A			0.5	mA
V_n	Output Noise Voltage	T_A = 25°C, 10 Hz ≤ f ≤ 100 Hz		125		μV
	Ripple Rejection	f = 120 Hz	54	66		dB
			(−18 ≤ V_{IN} ≤ −8)			V
	Dropout Voltage	T_J = 25°C, I_{OUT} = 1A		1.1		V
I_{OMAX}	Peak Output Current	T_J = 25°C		2.2		A
	Average Temperature Coefficient of Output Voltage	I_{OUT} = 5 mA, 0°C ≤ T_J ≤ 100°C		0.4		mV/°C

Electrical Characteristics (Continued)

Conditions unless otherwise noted: I_{OUT} = 500 mA, C_{IN} = 2.2μF, C_{OUT} = 1μF, $0°C \leq T_J \leq +125°C$, Power Dissipation = 1.5W.

PART NUMBER		LM7912C			LM7915C			UNITS
OUTPUT VOLTAGE		12V			15V			
INPUT VOLTAGE (unless otherwise specified)		−19V			−23V			
PARAMETER	**CONDITIONS**	MIN	TYP	MAX	MIN	TYP	MAX	
V_O Output Voltage	T_J = 25°C	−11.5	−12.0	−12.5	−14.4	−15.0	−15.6	V
	5 mA $\leq I_{OUT} \leq$ 1A,	−11.4		−12.6	−14.25		−15.75	V
	P $\leq$ 15W	($-27 \leq V_{IN} \leq -14.5$)			($-30 \leq V_{IN} \leq -17.5$)			V
ΔV_O Line Regulation	T_J = 25°C, (Note 2)		5	80		5	100	mV
		($-30 \leq V_{IN} \leq -14.5$)			($-30 \leq V_{IN} \leq -17.5$)			V
			3	30		3	50	mV
		($-22 \leq V_{IN} \leq -16$)			($-26 \leq V_{IN} \leq -20$)			V
ΔV_O Load Regulation	T_J = 25°C, (Note 2)		15	200		15	200	mV
	5 mA $\leq I_{OUT} \leq$ 1.5A		15	200		15	200	mV
	250 mA $\leq I_{OUT} \leq$ 750 mA		5	75		5	75	mV
I_Q Quiescent Current	T_J = 25°C		1.5	3		1.5	3	mA
ΔI_Q Quiescent Current Change	With Line			0.5			0.5	mA
		($-30 \leq V_{IN} \leq -14.5$)			($-30 \leq V_{IN} \leq -17.5$)			V
	With Load, 5 mA $\leq I_{OUT} \leq$ 1A			0.5			0.5	mA
V_n Output Noise Voltage	T_A = 25°C, 10 Hz $\leq f \leq$ 100 Hz		300			375		μV
Ripple Rejection	f = 120 Hz	54	70		54	70		dB
		($-25 \leq V_{IN} \leq -15$)			($-30 \leq V_{IN} \leq -17.5$)			V
Dropout Voltage	T_J = 25°C, I_{OUT} = 1A		1.1			1.1		V
I_{OMAX} Peak Output Current	T_J = 25°C		2.2			2.2		A
Average Temperature Coefficient of Output Voltage	I_{OUT} = 5 mA, $0°C \leq T_J \leq$ 100°C		−0.8			−1.0		mV/°C

Note 1: For calculations of junction temperature rise due to power dissipation, thermal resistance junction to ambient (θ_{JA}) is 50°C/W (no heat sink) and 5°C/W (infinite heat sink).

Note 2: Regulation is measured at a constant junction temperature by pulse testing with a low duty cycle. Changes in output voltage due to heating effects must be taken into account.

National Semiconductor

LF13006, LF13007 Digital Gain Set

General Description

The LF13006, LF13007 are precision digital gain sets used for accurately setting non-inverting op amp gains. Gains are set with a 3-bit digital word which can be latched in with $\overline{WR}$ and $\overline{CS}$ pins. All digital inputs are TTL and CMOS compatible.

The LF13006 shown below will set binary scaled gains of 1, 2, 4, 8, 16, 32, 64, and 128. The LF13007 will set gains of 1, 2, 5, 10, 20, 50, and 100 (a common attenuator sequence). In addition, both versions have several taps and two uncommitted matching resistors which allow customization of the gain.

The gains are set with precision thin film resistors. The low temperature coefficient of the thin film resistors and their excellent tracking result in gain ratios which are virtually independent of temperature.

The LF13006, LF13007 used in conjunction with an amplifier not only satisfies the need for a digitally programmable amplifier in microprocessor based systems, but is also useful for discrete applications, eliminating the need to find 0.5% resistors in the ratio of 100 to 1 which track each other over temperature.

Features

- TTL and CMOS compatible logic levels
- Microprocessor compatible
- Gain error 0.5% max
- Binary or scope knob gains
- Wide supply range +5V to ±18V
- Packaged in 16-pin DIP

Block Diagram and Typical Application (LF13006)

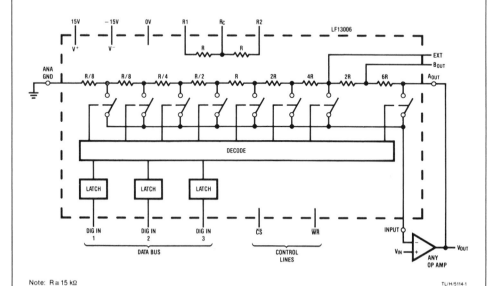

Note: R ≅ 15 kΩ

TL/H/5114-1

Absolute Maximum Ratings

Supply Voltage, V^+ to V^- 36V
Supply Voltage, V^+ to GND 25V
Voltage at Any Digital Input .... V^+ to GND
Analog Voltage .... V^+ to $V^- + 2V$
Operating Temperature Range .... $-40°C$ to $+85°C$

Electrical Characteristics (Note 1)

Parameter	Conditions	Typ	Tested Limit (Note 2)	Design Limit (Note 3)	Units (Limit)
Gain Error	$A_{OUT} = \pm 10V$ ANA GND = 0V $I_{INPUT} < 10$ nA	0.3	0.5	**0.5**	% (max)
Gain Temperature Coefficient	$A_{OUT} = \pm 10V$ ANA GND = 0V	0.001			%/°C
Digital Input Voltage Low High		1.4 1.6	0.8 2.0	**0.8** **2.0**	V (max) V (min)
Digital Input Current Low High	$V_{IL} = 0V$ $V_{IH} = 5V$	-35 0.0001	-100 1	**-100** **1**	μA (max) μA (max)
Positive Power Supply Current	All Logic Inputs Low	3	5	**5**	mA (max)
Negative Power Supply Current	All Logic Inputs Low	-2	-5	**-5**	mA (max)
$\overline{\text{Write}}$ Pulse Width, t_W	$V_{IL} = 0V$, $V_{IH} = 5V$	40		**100**	ns (min)
$\overline{\text{Chip Select}}$ Set-Up Time, t_{CS}	$V_{IL} = 0V$, $V_{IH} = 5V$	60		**120**	ns (min)
$\overline{\text{Chip Select}}$ Hold Time, t_{CH}	$V_{IL} = 0V$, $V_{IH} = 5V$	0		**0**	ns (min)
DIG IN Set-Up Time, t_{DS}	$V_{IL} = 0V$, $V_{IH} = 5V$	80		**150**	ns (min)
DIG IN Hold Time, t_{DH}	$V_{IL} = 0V$, $V_{IH} = 5V$	0		**0**	ns (min)
Switching Time for Gain Change	(Note 4)	200			ns (max)
Switch On Resistance		3			kΩ
Unit Resistance, R		15	12–18		kΩ
R1 and R2 Mismatch		0.3	0.5	**0.5**	% (max)
R1/R2 Temperature Coefficient		0.001			%/°C

Note 1: Parameters are specified at $V^+ = 15V$ and $V^- = -15V$. Min V^+ to ground voltage is 5V. Min V^+ to V^- voltage is 5V. **Boldface numbers apply at temperature extremes.** All other numbers apply at $T_A = T_j = 25°C$.

Note 2: Guaranteed and 100% production tested.

Note 3: Guaranteed (but not 100% production tested) over the operating temperature. These limits are not used to calculate outgoing quality levels.

Note 4: Settling time for gain change is the switching time for gain change plus settling time (see section on Settling Time).

Note 5: $\overline{\text{WR}}$ minimum high threshold voltage increases to 2.4V under the extreme conditions when all three digital inputs are simultaneously taken from 0V to 5V at a slew rate of greater than 500V/μs.

GAIN TABLE

Digital Input	Gain			
	LF13006		LF13007	
	A_{OUT}	B_{OUT}	A_{OUT}	B_{OUT}
000	1	1	1	1
001	2	1.25	1.25	1
010	4	2.5	2	1.6
011	8	5	5	4
100	16	10	10	8
101	32	20	20	16
110	64	40	50	40
111	128	80	100	80

Connection Diagram

Dual-In-Line Package

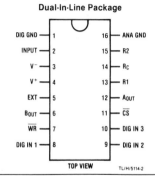

TOP VIEW

TL/H/5114-2

413

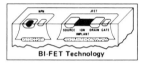
BI-FET Technology

LF11508/LF12508/LF13508
8-channel analog multiplexer
LF11509/LF12509/LF13509
4-channel differential analog multiplexer

general description

The LF11508/LF12508/LF13508 is an 8-channel analog multiplexer which connects the output to 1 of the 8 analog inputs depending on the state of a 3-bit binary address. An enable control allows disconnecting the output, thereby providing a package select function.

This device is fabricated with National's BI-FET technology which provides ion-implanted JFETs for the analog switch on the same chip as the bipolar decode and switch drive circuitry. This technology makes possible low constant "ON" resistance with analog input voltage variations. This device does not suffer from latch-up problems or static charge blow-out problems associated with similar CMOS parts. The digital inputs are designed to operate from both TTL and CMOS levels while always providing a definite break-before-make action.

The LF11509/LF12509/LF13509 is a 4-channel differential analog multiplexer. A 2-bit binary address will connect a pair of independent analog inputs to one of any 4 pairs of independent analog outputs. The device has all the features of the LF11508 series and should be used whenever differential analog inputs are required.

features

- JFET switches rather than CMOS
- No static discharge blow-out problem
- No SCR latch-up problems
- Analog signal range 11V, −15V
- Constant "ON" resistance for analog signals between −11V and 11V
- "ON" resistance 380 Ω typ
- Digital inputs compatible with TTL and CMOS
- Output enable control
- Break-before-make action: t_{OFF} = 0.2 μs; t_{ON} = 2 μs typ

functional diagrams and truth tables

LF11508/LF12508/LF13508

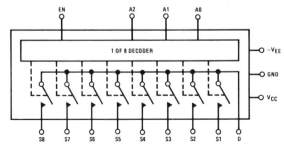

EN	A2	A1	A0	SWITCH ON
H	L	L	L	S1
H	L	L	H	S2
H	L	H	L	S3
H	L	H	H	S4
H	H	L	L	S5
H	H	L	H	S6
H	H	H	L	S7
H	H	H	H	S8
L	X	X	X	NONE

LF11509/LF12509/LF13509

EN	A1	A0	SWITCH PAIR ON
L	X	X	None
H	L	L	S1
H	L	H	S2
H	H	L	S3
H	H	H	S4

© 1977 National Semiconductor Corp.

absolute maximum ratings

	LF11508, LF11509	LF12508, LF12509	LF13508, LF13509
Positive Supply − Negative Supply ($V_{CC} - V_{EE}$)	36V	36V	36V
Positive Analog Input Voltage (Note 1)	V_{CC}	V_{CC}	V_{CC}
Negative Analog Input Voltage (Note 1)	$-V_{EE}$	$-V_{EE}$	$-V_{EE}$
Positive Digital Input Voltage	V_{CC}	V_{CC}	V_{CC}
Negative Digital Input Voltage	$-5V$	$-5V$	$-5V$
Analog Switch Current	$|I_S| < 10$ mA	$|I_S| < 10$ mA	$|I_S| < 10$ mA
Power Dissipation (P_D at 25°C) and Thermal Resistance (θ_{jA}), (Note 2)			
Molded DIP (N) P_D	−	−	500 mW
θ_{jA}	−	−	150°C/W
Cavity DIP (D) P_D	900 mW	900 mW	900 mW
θ_{jA}	100°C/W	100°C/W	100°C/W
Maximum Junction Temperature (T_{jMAX})	150°C	110°C	100°C
Operating Temperature Range	$-55°C \leq T_A \leq +125°C$	$-25°C \leq T_A \leq +85°C$	$0°C \leq T_A \leq +70°C$
Storage Temperature Range	$-65°C$ to $+150°C$	$-65°C$ to $+150°C$	$-65°C$ to $+150°C$
Lead Temperature (Soldering, 60 seconds)	300°C	300°C	300°C

electrical characteristics (Note 3)

SYMBOL	PARAMETER	CONDITIONS		LF11508, LF11509 MIN	LF11508, LF11509 TYP	LF11508, LF11509 MAX	LF12508, LF12509, LF13508, LF13509 MIN	LF12508, LF12509, LF13508, LF13509 TYP	LF12508, LF12509, LF13508, LF13509 MAX	UNITS
R_{ON}	"ON" Resistance	$V_{OUT} = 0V$, $I_S = 100\,\mu A$	$T_A = 25°C$		380	500		380	650	Ω
					600	750		500	850	Ω
ΔR_{ON}	ΔR_{ON} with Analog Voltage Swing	$-10V \leq V_{OUT} \leq +10V$, $I_S = 100\,\mu A$	$T_A = 25°C$		0.01	1		0.01	1	%
R_{ON} Match	R_{ON} Match Between Switches	$V_{OUT} = 0V$, $I_S = 100\,\mu A$	$T_A = 25°C$		20	100		20	150	Ω
$I_{S(OFF)}$	Source Current in "OFF" Condition	Switch "OFF", $V_S = 11$, $V_D = -11$, (Note 4)	$T_A = 25°C$			1			5	nA
					10	50		0.09	50	nA
$I_{D(OFF)}$	Drain Current in "OFF" Condition	Switch "OFF", $V_S = 11$, $V_D = -11$, (Note 4)	$T_A = 25°C$			10			20	nA
					25	500		0.6	500	nA
$I_{D(ON)}$	Leakage Current in "ON" Condition	Switch "ON" $V_D = 11V$, (Note 4)	$T_A = 25°C$			10			20	nA
					35	500		1	500	nA
V_{INH}	Digital "1" Input Voltage			2.0			2.0			V
V_{INL}	Digital "0" Input Voltage					0.7			0.7	V
I_{INL}	Digital "0" Input Current	$V_{IN} = 0.7V$	$T_A = 25°C$		1.5	20		1.5	30	μA
						40			40	μA
$I_{INL(EN)}$	Digital "0" Enable Current	$V_{EN} = 0.7V$	$T_A = 25°C$		1.2	20		1.2	30	μA
						40			40	μA
t_{TRAN}	Switching Time of Multiplexer	(Figure 1), (Note 5)	$T_A = 25°C$		2.0	3		1.8		μs
t_{OPEN}	Break-Before-Make	(Figure 3)	$T_A = 25°C$		1.6			1.6		μs
$t_{ON(EN)}$	Enable Delay "ON"	(Figure 2)	$T_A = 25°C$		1.6			1.6		μs
$t_{OFF(EN)}$	Enable Delay "OFF"	(Figure 2)	$T_A = 25°C$		0.2			0.2		μs
$I_{SO(OFF)}$	"OFF" Isolation	(Note 6)	$T_A = 25°C$		-66			-66		dB
CT	Crosstalk	LF11509 Series, (Note 6)	$T_A = 25°C$		-66			-66		dB
$C_{S(OFF)}$	Source Capacitance ("OFF")	Switch "OFF", $V_{OUT} = 0V$, $V_S = 0V$	$T_A = 25°C$		2.2			2.2		pF
$C_{D(OFF)}$	Drain Capacitance ("OFF")	Switch "OFF", $V_{OUT} = 0V$, $V_S = 0V$	$T_A = 25°C$		11.4			11.4		pF
I_{CC}	Positive Supply Current	All Digital Inputs Grounded	$T_A = 25°C$		7.4	10		7.4	12	mA
					9.2	13		7.9	15	mA
I_{EE}	Negative Supply Current	All Digital Inputs Grounded	$T_A = 25°C$		2.7	4.5		2.7	5	mA
					2.9	5.5		2.8	6	mA

notes

Note 1: If the analog input voltage exceeds this limit, the input current should be limited to less than 10 mA.

Note 2: The maximum power dissipation for these devices must be derated at elevated temperatures and is dictated by T_{jMAX}, θ_{jA}, and the ambient temperature, T_A. The maximum available power dissipation at any temperature is $P_D = (T_{jMAX} - T_A)/\theta_{jA}$ or the 25°C P_{DMAX}, whichever is less.

Note 3: These specifications apply for $V_S = \pm15V$ and over the absolute maximum operating temperature range ($T_L \leq T_A \leq T_H$) unless otherwise noted.

Note 4: Conditions applied to leakage tests insure worse case leakages. Exceeding 11V on the analog input may cause an "OFF" channel to turn "ON".

Note 5: Lots are sample tested to this parameter. The measurement conditions of *Figure 1* insure worse case transition time.

Note 6: "OFF" isolation is measured with all switches "OFF" and driving a source. Crosstalk is measured with a pair of switches "ON", driving channel A and measuring channel B. R_L = 200, C_L = 7 pF, V_S = 3 Vrms, f = 500 kHz.

connection diagrams

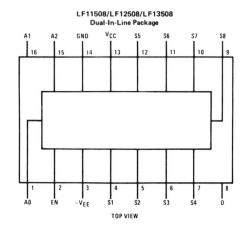

LF11508/LF12508/LF13508
Dual-In-Line Package

TOP VIEW

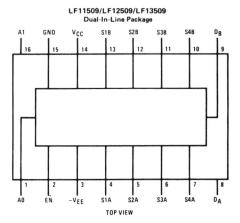

LF11509/LF12509/LF13509
Dual-In-Line Package

TOP VIEW

ac test circuits and switching time waveforms

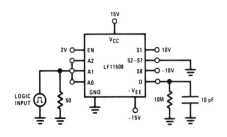

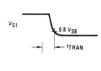

FIGURE 1. Transition Time

typical performance characteristics

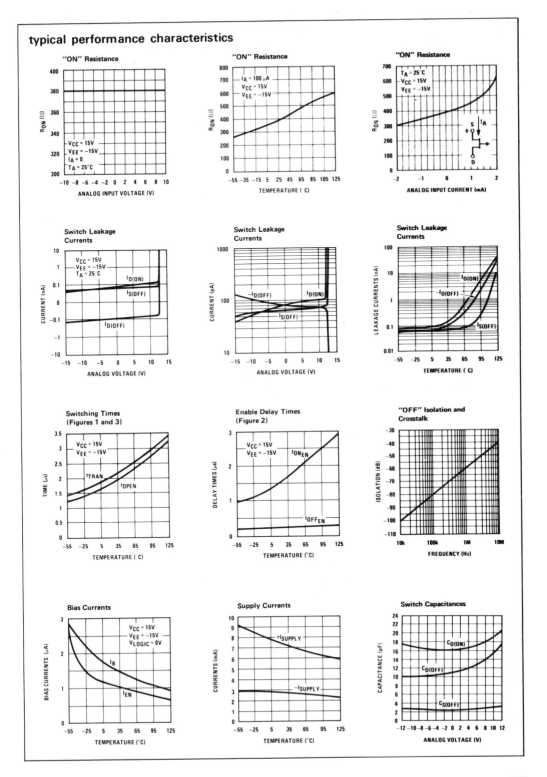

National Semiconductor

MM54104 DIGITALKER™ Speech Synthesis System

General Description

The DIGITALKER is a speech synthesis system consisting of multiple N-channel MOS integrated circuits. It contains an MM54104 speech processor chip (SPC) and speech ROM and when used with external filter, amplifier, and speaker, produces a system which generates high quality speech including the natural inflection and emphasis of the original speech. Male, female, and children's voices can be synthesized.

The SPC communicates with the speech ROM, which contains the compressed speech data as well as the frequency and amplitude data required for speech output. Up to 128k bits of speech data can be directly accessed. This can be expanded with minimal external logic.

With the addition of an external resistor, on-chip debounce is provided for use with a switch interface.

An interrupt is generated at the end of each speech sequence so that several sequences or words can be cascaded to form different speech expressions.

Encoding (digitizing) of custom word or phrase lists must be done by National Semiconductor. Customers submit to the factory high quality recorded magnetic reel to reel tapes containing the words or phrases to be encoded. National Semiconductor will sell kits consisting of the SPC and ROM(s) containing the digitized word or phrases.

Features

- Designed to be easily interfaced to most popular microprocessors
- 256 possible addressable expressions
- Male, female, and children's voices
- Any language
- Natural inflection and emphasis of original speech
- Addresses 128k of ROM directly
- TTL compatible
- MICROBUS™ and COPS™ compatible
- On-chip switch debounce for interfacing to manual switches independent of a microprocessor
- Easily expandable to greater than 128k ROM
- Interrupt capability for cascading words or phrases
- Crystal controlled or externally driven oscillator
- Ability to store silence durations for timing sequences

Applications

- Telecommunications
- Appliance
- Automotive
- Teaching aids
- Consumer products
- Clocks
- Language translation
- Annunciators

Typical Applications

Minimum Configuration Using Switch Interface

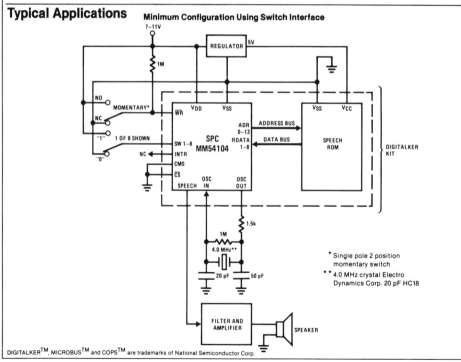

* Single pole 2 position momentary switch

** 4.0 MHz crystal Electro Dynamics Corp. 20 pF HC18

DIGITALKER™, MICROBUS™ and COPS™ are trademarks of National Semiconductor Corp.

© 1980 National Semiconductor Corp.

Absolute Maximum Ratings

Storage Temperature Range	− 65°C to + 150°C	Voltage at Any Pin	12V
Operating Temperature Range	− 40°C to 85°C	Operating Voltage Range, V_{DD}-V_{SS}	7V to 11V
V_{DD}-V_{SS}	12V	Lead Temperature (Soldering, 10 seconds)	300°C

DC Electrical Characteristics $T_A = 0°C$ to $70°C$, $V_{DD} = 7V–11V$, $V_{SS} = 0V$, unless otherwise specified.

Symbol	Parameter	Conditions	Min	Typ	Max	Units
V_{IL}	Input Low Voltage		− 0.3		0.8	V
V_{IL}	Input Low Voltage	$T_A = − 40°C$ to $85°C$	− 0.3		0.6	V
V_{IH}	Input High Voltage		2.0		V_{DD}	V
V_{IH}	Input High Voltage	$T_A = − 40°C$ to $85°C$	2.2		V_{DD}	V
V_{OL}	Output Low Voltage	$I_{OL} = 1.6$ mA			0.4	V
V_{OH}	Output High Voltage	$I_{OH} = − 100 \mu A$	2.4		5.0	V
V_{ILX}	Clock Input Low Voltage		− 0.3		1.2	V
V_{IHX}	Clock Input High Voltage		5.5		V_{DD}	V
V_{OLX}	Clock Output Low Voltage	Typical Crystal Configuration and 10M Load on Pin 2			1.2	V
V_{OHX}	Clock Output High Voltage	Typical Crystal Configuration and 10M Load on Pin 2	5.5		V_{DD}	V
I_{DD}	Power Supply Current				45	mA
I_{DD}	Power Supply Current	$T_A = − 40°C$ to $85°C$			50	mA
I_{IL}	Input Leakage				± 10	μA
I_{ILX}	Clock Input Leakage				± 10	μA
V_S	Silence Voltage			$0.45 V_{DD}$		V
V_{OUT}	Peak to Peak Speech Output	$V_{DD} = 11V$		2.0		V
R_{EXT}	External Load on Speech Output	R_{EXT} Connected Between Speech Output and V_{SS}	50			kΩ

AC Electrical Characteristics $T_A = 0°C$ to $70°C$, $V_{DD} = 7V–11V$, $V_{SS} = 0V$, unless otherwise specified.

Symbol	Parameter	Min	Max	Units
t_{aw}	CMS Valid to Write Strobe	350		ns
t_{csw}	Chip Select ON to Write Strobe	310		ns
t_{dw}	Data Bus Valid to Write Strobe	50		ns
t_{wa}	CMS Hold Time after Write Strobe	50		ns
t_{wd}	Data Bus Hold Time after Write Strobe	100		ns
t_{ww}	Write Strobe Width (50% Point)	430		ns
t_{red}	$\overline{ROMEN}$ ON to Valid ROM Data		2	μs
t_{wss}	Write Strobe to Speech Output Delay		410	μs
f_t	External Clock Frequency	3.92	4.08	MHz

Note: Rise and fall times (10% to 90%) of MICROBUS signals should be 50 ns maximum.

Answers to Odd Problems

CHAPTER 1: **1.** 1.4 V. **3.** 0.3 V peak—inverted. **5.** 1.5 μV. **7.** 22.5mΩ.
9. +0.6 V. **11.** 60 dB. **13.** 23.76. **15.** 1.26 V/μs, rise time limited. **17.** 1.66 μV.
19. LM363—has least combined output offset of 10.3 mV. **21.** 3.6mΩ. **23.** 1.59
μF, use 2 μF.

CHAPTER 2: **1.** +3 V. **3.** $V_s = 200$ mV, $e_n = 79.8$ μV. **5.** −7 mV. **7.** R_2 and
$R_3 = 3.18$ kΩ, $R_1 = 6.2$ kΩ and $R_F = 124$ kΩ. Op-amp should have an open loop
gain of at least 2000 at 5 kHz. **9.** $C = 0.47$ μF, $r_2 = 560$ Ω. Let $R = 10$. kΩ.

CHAPTER 3: **1.** 14.6 W. **3.** Use 12 Vrms 100 mA transformer, 1 A 100 V_{PRV}
bridge rectifier, $R_{TOTAL} = 0.58$ Ω, $C = 100$ μF with 25 V rating, $R_1 + R_2 = 10$
kΩpot. and $R_{CL} = 30$ Ω (reference Fig. 3-4.). **5.** Use 24 Vrms 100 mA transformer,
1 A 200 V_{PRV} bridge rectifier, $R_{TOTAL} = 1.23$ Ω, $C = 30$ μF with 50 V rating,
$R_1 + R_2 = 10$ kΩ pot. and $R_{CL} = 60$ Ω (reference Fig. 3-4.) **7.** K factor = 1.5,
$R_2 = 160$ kΩ and $R_{CL} = 20$ Ω. **9.** Use 18 Vrms 1-A transformer, 1-A 100 V_{PRV}
bridge rectifiers, $R_{TOTAL} = 0.9$ Ω, $C = 2500$ μF with 35 V rating, $R_2 = 2$ kΩ pot.
11. Use LM340k-5 regulator, 4 A 100 V_{PRV} bridge rectifier (100A surge),
$R_{TOTAL} = 0.14$ Ω, $C = 3000$ μF with 25 V rating.

CHAPTER 4: **1.** 20 kHz. **3.** 20V. **5.** 8 kΩ. **7.** 4 A. **9.** $R_1 = 15$ kΩ, with
$C_T = 0.01$ μF and $R_T = 10$ kΩ. **11.** Use a LM3524 regulator with external
transistors. $R_1 = 12$ kΩ, with $C_T = 0.01$ μF, $R_T = 5$ kΩ, $L_1 = 0.27$ mH, $C_4 = 3300$
μF. $R_{CL} = 0.143$ Ω.

CHAPTER 5: **1.** 19.2 mV. **3.** See text. **5.** 1.88 mV. **7.** Use a LM136-type
reference rather than a LM385 in the circuit of Fig. 5-3. Minimum output voltage is
6.25 V. **9.** Decrease by 1.408 nA. **11.** None, since the required reference voltage is
set with the 10 kΩ pot.

CHAPTER 6: **1.** See text. **3.** See text. **5.** Connect top end of a 15-kΩ load resistor
to +15-V supply and bottom end to pin 7. Also connect pin 1 to the −15-V supply.
7. Reference Fig. 6-4. $R_L = 900$ Ω, $R_1 = 179$ kΩ, $R_2 = 1$ kΩ. **9.** 2200 μF.

CHAPTER 7: **1.** Trigger pulse amplitude at least −10 V. Coupling capacitor should
be 0.1 μF. **3.** 0 V to +12.75 V. **5.** See Fig. 7-9. Change R_T to 180 kΩ. **7.** Reference
Fig. 7-11. $R_T = 6.2$ MΩ. Ground emitter of Q_3 and connect L.E.D. in collector
circuit in series with a 500 Ω resistor. Ground logic pin for one minute "ON" time.

For 30 second "OFF" time, tie logic to V_{REF} and change R_T to 3 MΩ. **9.** See Fig. A-1. **11.** 61 kΩ.

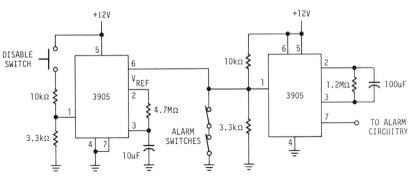

Figure A-1.

CHAPTER 8: **1.** Frequency would lower. **3.** Use circuit of Fig. 8-3 and change R_T to 5 kΩ for a 5-kHz center frequency. Use a pushbutton switch to connect a second 5 kΩ in parallel with R_T to achieve a 10 kHz center frequency. **5.** 30 mV Peak. **7.** 7.145 kHz and 8.85 kHz.

CHAPTER 9:**1.** See text. **3.** 0°. **5.** $R_1 = 2.5$ kΩ, $f_L = 6.5$ kHz. **7.** $C_2 = 0.047$ μF, rejection 26 dB. **9.** (See Fig. A-2) Demodulated output = 201 mV. **11.** See Fig. 9-7. Use 500 kHz for first crystal (f_1) and 4.5 MHz for f_2.

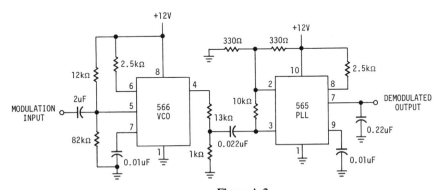

Figure A-2.

CHAPTER 10: **1.** Use 0.68 μF for C_2. **3.** Use 0.068 μF for C_1. **5.** 1.8 mV per μS. **7.** See Text. **9.** Reference Fig. 10-3. Use five tone decoders and six 3-input NOR gates. **11.** 15 kHz.

CHAPTER 11: **1.** 10-bit ($2^{10} = 1024$). **3.** 2.18 V (range 2.17 V to 2.19 V).
5.a. Successive approximation (ADC0807), maximum 257 μS. **b.** Flash Converter
(CA3308) 3.33 μS. **c.** Dual Slope (ICL7126) 53 mS. **d.** Tracking (8-bit) 500 μS.
7. 100 nS. **9.** Reference Fig. 11-14, $f_{OSC} = 32$ kHz, use 1 V reference, $C_{INT} = 0.1$
μF.

CHAPTER 12: **1.** Current. **3.** See text. **5.** 8.3 V. **7.** See Fig A-3. Total output
excursion $\pm$ 11.95 V. **9.** 6.6%.

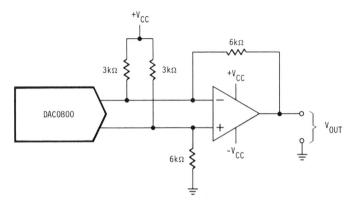

Figure A-3.

CHAPTER 13: **1.** 200 μS. **3.** See Fig. 13-3. Input pulse must be greater than 1.4 V
and wider than 20 μS. **5.** ADC0804 is an eight-bit converter so LF398 should have
gain error of less than 0.05% (11-bits). **7.** See Fig. A-4. **9.** 50 mV. **11.** A small
amount (225 pA) of the multiplexer leakage will flow into the LF356.

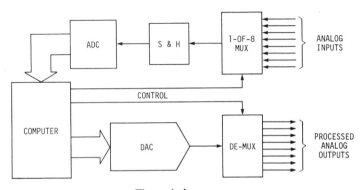

Figure A-4.

CHAPTER 14: **1.** See text. **3.** $f_c = 6$ kHz, $R_2 = 5$ kΩ and $R_3 = 50$ kΩ. **5.** $f_p = 556$ Hz, $R_2 = 40$ kΩ, $R_3 = 54$ kΩ and $R_4 = 98.8$ kΩ, $H_{op} = 1.155$. **7.** The higher frequency components of speech, although of lower amplitude, contain the intelligibility, so they are "emphasized" prior to transmission for signal to noise advantage. **9.** Reference Fig. 14-13. Connect input signal to analog ground (pin 16) and ground the non-inverting side of the op-amp. Set code to 100. **11.** Use the DGS as the timing resistor by connecting A_{OUT} to the supply voltage and the input to the timing capacitor.

CHAPTER 15: **1.** See text. **3.** They don't indicate tolerances. **5.** TO-3 package with 20 W rating.

Index